The Kantian Legacy
in Nineteenth-Century Science

DIBNER
INSTITUTE
FOR THE HISTORY
OF SCIENCE AND
TECHNOLOGY

George Smith, general editor

Jed Z. Buchwald and I. Bernard Cohen, editors, *Isaac Newton's Natural Philosophy*

Jed Z. Buchwald and Andrew Warwick, editors, *Histories of the Electron: The Birth of Microphysics*

Geoffrey Cantor and Sally Shuttleworth, editors, *Science Serialized: Representations of the Sciences in Nineteenth-Century Periodicals*

Michael Friedman and Alfred Nordmann, editors, *The Kantian Legacy in Nineteenth-Century Science*

Anthony Grafton and Nancy Siraisi, editors, *Natural Particulars: Nature and the Disciplines in Renaissance Europe*

J. P. Hogendijk and A. I. Sabra, editors, *The Enterprise of Science in Islam: New Perspectives*

Frederic L. Holmes and Trevor H. Levere, editors, *Instruments and Experimentation in the History of Chemistry*

Agatha C. Hughes and Thomas P. Hughes, editors, *Systems, Experts, and Computers: The Systems Approach in Management and Engineering, World War II and After*

N. L. Swerdlow, editor, *Ancient Astronomy and Celestial Divination*

The Kantian Legacy in Nineteenth-Century Science

—

Michael Friedman and Alfred Nordmann, editors

The MIT Press
Cambridge, Massachusetts
London, England

MIT Press books may be purchased at special quantity discounts for business or sales promotional use. For information, please email special_sales@mitpress.mit.edu or write to Special Sales Department, The MIT Press, 55 Hayward Street, Cambridge, MA 02142.

This book was set in Bembo by SPI Publisher Services, and was printed and bound in the United States of America.

Library of Congress Cataloging-in-Publication Data

The Kantian legacy in nineteenth-century science / Michael Friedman and Alfred Nordmann, editors.
   p.   cm.—(Dibner Institute studies in the history of science and technology)
Includes bibliographical references and indexes.
ISBN 0-262-06254-2 (hc: alk. paper)
1. Science—Philosophy—History—19th century. 2. Philosophy and science—History—19th century. 3. Kant, Immanuel, 1724–1804. I. Friedman, Michael, 1947–. II. Nordmann, Alfred, 1956–. III. Series.

Q174.8.K36   2006
501—dc22

                                                        2005058035

10 9 8 7 6 5 4 3 2 1

Contents

I

Editors' Introduction

The papers in this volume originated at an international conference devoted to Kant's influence on nineteenth-century science and scientific philosophy held at the Dibner Institute for the History of Science and Technology at MIT in November of 2000. Many of the papers represent substantially revised versions of what was presented there; some of them are entirely new. The conference was originally conceived as organized around five main focal points of nineteenth-century scientific thought: *Naturphilosophie*, Fries, Helmholtz, neo-Kantianism, and Poincaré, and the papers published here broadly reflect this organization. Moreover, both the conference and the volume reflect the conviction that the time is ripe for a fruitful collaboration among historians of philosophy, historians of science, and historians of mathematics in developing a richer and more nuanced picture of the Kantian legacy in the nineteenth century and beyond. It is now a commonplace, for example, that the development of modern mathematics, mathematical logic, and the foundations of mathematics can be profitably seen as an evolution "from Kant to Hilbert" (see, e.g., Ewald 1996). It is our conviction, in addition, that the development of modern scientific thought more generally—including the physical sciences, the life sciences, and the relationships between both of these and the mathematical sciences—can also be greatly illuminated when viewed as an evolution from Kant, through Poincaré, to Einstein and the logical empiricists and beyond. Concentrating on the nineteenth century in particular allows us to put aside the seductive, but in the end much too crude, question of whether Kantian philosophical ideas still have relevance in the context of the great conceptual revolutions wrought by twentieth-century science. Our project, rather, is the more complex and subtle tracing of the multiple intellectual transformations that have actually led, step by step, from Kant's original scientific situation to the new scientific problems of the twentieth century.

Recent work by historians of philosophy has highlighted the centrality and importance of Kant's scientific preoccupations for understanding both Kant's own philosophy and the post-Kantian philosophies that reacted, in various ways, to Kant's original doctrines and concerns. Recent Kant scholarship, for example, beginning with Buchdahl (1969), has paid serious attention to Kant's *Metaphysical Foundations of Natural Science* (1786), which was published between the first (1781) and second (1787) editions of the *Critique of Pure Reason*, as well as Kant's further thoughts on the philosophy of nature, beginning with the *Critique of Judgment* (1790) and extending to the late unpublished reflections contained in the *Opus postumum* (1796–1803).[1] Similarly, recent scholarship in post-Kantian German idealism has rediscovered the central importance of the philosophy of nature in the "absolute idealism" of Schelling and Hegel—which begins, in particular, with Schelling's attempt to extend and transform the dynamical theory of matter Kant had presented in the *Metaphysical Foundations* to embrace newly discovered phenomena in chemistry and electricity and magnetism.[2]

This work by historians of philosophy complements recent work by historians of science on the role of *Naturphilosophie* in such early nineteenth-century scientific developments as Oersted's discovery of electromagnetism and the impact of German romanticism—especially Goethe—on biology.[3] Jakob Friedrich Fries was an important figure in the development of biology in this period, and he also played a central role in rethinking and recasting Kantian philosophy of mathematics and mathematical physics in light of more recent developments in these fields as well.[4] Although by no means a *Naturphilosoph* himself, Fries thus had one foot in the intellectual context within which German romanticism and *Naturphilosophie* flourished, and another in the intellectual problems later explored by the neo-Kantian reaction against these movements. We hope that closer attention to the work of Fries, against the background of a more sophisticated understanding of both Kant's natural philosophy and its further articulation and transformation in German romanticism and *Naturphilosophie*, will help to overcome the simple-minded dichotomy and sense of opposition between these tendencies of thought and the later neo-Kantian movement. It will help us, in particular, to appreciate some of the important intellectual continuities linking the preoccupations of *Naturphilosophie* to later thinkers like Helmholtz, who explicitly rejected it.

Nevertheless, there is no doubt that the neo-Kantian or "back to Kant!" movement, as represented by natural scientific thinkers like Helmholtz and academic philosophers like Cohen or Riehl, understood itself as standing in stark opposition to German romanticism and *Naturphilosophie*—as a reaction against extravagant "metaphysical" speculation and a return to the more sober "methodological" concerns of Kant himself.[5] Helmholtz developed his neo-Kantian epistemological stance in connection with his scientific research in the psychophysiology of sense perception, which was intimately connected, in turn, with his work on the conservation of energy and the foundations of geometry.[6] Helmholtz's own antimetaphysical reaction against the perceived speculative excesses of the *Naturphilosophie* of Schelling and Hegel was therefore intensely fruitful from a scientific point of view; and it led in particular to a profound interaction between the latest results in the physical and life sciences and some of the deepest results in contemporary mathematics (namely, the Riemannian theory of manifolds). In the academic neo-Kantianism of Cohen or Riehl, by contrast, the primary focus was on the further systematic development of the Kantian or "critical" philosophy itself, so as to adapt it, more specifically, to the new scientific situation of the mid- to late nineteenth century. They understood their project to be one of beginning with "the fact of science" as it existed in the nineteenth century and then of uncovering the methodological preconditions or presuppositions of this fact—just as Kant's achievement, as they understood it, was to do precisely this for the basically Newtonian science of the eighteenth century. Nevertheless, because late nineteenth-century science was much less unified and coherent than eighteenth-century Newtonianism, fundamental difficulties inevitably arose in actually carrying out their project.

In the case of Poincaré, finally, we are again confronted with a practicing scientist and mathematician, but one who, like Helmholtz, was driven by the character of his own scientific work to engage in further reflection on scientific epistemology. Whereas Poincaré, unlike Helmholtz, did not officially identify himself with the "back to Kant!" movement, he nevertheless defended broadly Kantian and neo-Kantian philosophical positions. Most famously, of course, he defended the idea that the fundamental principles of mathematical reasoning (especially mathematical induction) are synthetic a priori rather than purely logical against the contrary contemporary position represented by the new mathematical logic of Frege and Russell. And, although his "conventionalist" philosophy of geometry did not embrace the original Kantian idea that Euclidean geometry, too, is synthetic a priori, Poincaré still adopted a more intuitive conception of geometry than the "formalistic" conception now associated with the work of Hilbert. Poincaré's commitment to preserving Euclidean geometry and the fundamental principles of Newtonian physics in the face of the radically new physical developments associated with Einstein's theory of relativity, moreover, represented (at least in comparison with Einstein himself) a conservative philosophical impulse that also further aligned him with contemporary neo-Kantianism.[7]

It is ironic, then, that the logical empiricist movement of the early twentieth century conceived itself as a new scientific philosophy appropriate to both the modern mathematical logic of Frege and Russell and the modern mathematical physics of Einstein—and at the same time took Poincaré (along with Helmholtz) as one of its main sources of philosophical inspiration. It is ironic, in particular, that the logical empiricists understood modern mathematical logic and mathematical physics to have decisively undercut, once and for all, all traces of the Kantian synthetic a priori, while they simultaneously appropriated the work of a previous thinker who had no such revolutionary ambitions.[8] This phenomenon provides an excellent illustration, in fact, of the point already emphasized above: namely, that Kant's scientific legacy in the twentieth century is much more complex and subtle than it first appears. Indeed, recent research into the historical background and evolution of twentieth-century logical empiricism has not only uncovered important remaining Kantian and neo-Kantian ideas inherited from Helmholtz and Poincaré, but also significant direct influences from academic neo-Kantianism as well—especially as represented by the Marburg School of neo-Kantianism founded by Cohen.[9]

The subject of the present volume, however, is Kant's scientific legacy in the nineteenth century, and the evolution of logical empiricism in particular is only touched on in passing. Nevertheless, it is useful to keep the early twentieth-century context also in mind, in order to get at least a preliminary sense of the way in which the nineteenth-century developments on which we focus here have actually led from Kant's scientific situation in the eighteenth century to the new scientific problems of the twentieth. We hope, more generally, to give at least a preliminary sense of the incredible richness of mutual interaction between philosophical ideas and discoveries in the natural and mathematical sciences throughout this period. Without in any way pretending that the specific events and figures on which we have chosen to focus represent the only viable route through this maze (or even the only important such route), the line of thought leading from Kant, through romanticism and *Naturphilosophie*, through Fries, to Helmholtz and academic neo-Kantianism, and finally to Poincaré does represent, in our view, a particularly important such route. It represents one important point of convergence, in any case, between contemporary research by historians of philosophy, historians of science, and historians of mathematics. We hope that other scholars will follow our example and do better.

We want to thank first of all Jed Buchwald who brought us together at the Dibner Institute and contributed to our discussions. We also thank

George Smith, who very carefully read (and edited) the manuscript for MIT Press. Also at MIT Press, we are indebted to Judy Feldmann and copy editor Suzanne Wolk. In Darmstadt, Toni Petrino helped with the bibliography, and at Stanford, Alexei Angelides and Samuel Kahn prepared the index.

NOTES

1. For the current state of this scholarship, see Friedman 1992, Förster 2000, and Watkins 2001, together with the references contained therein. Pollok 2001 is a detailed and up-to-date "critical commentary" on the *Metaphysical Foundations* and provides very useful information, in particular, about the contemporaneous scientific and philosophical context of this work.

2. Beiser 2002 emphasizes the central importance of the philosophy of nature in Schelling and Hegel, as a reaction against and complement to the more properly "idealistic" post-Kantian system of Fichte. The translation of Schelling's *Ideas for a Philosophy of Nature* (Schelling 1988), together with Stern's introduction, gives a good sense of the context and development of Schelling's transformation of Kant's original dynamical theory.

3. Discussions of the impact of *Naturphilosophie* on the development of physics begins with Stauffer 1957 and extends through Kuhn 1959, Williams 1965 and 1966, Gower 1973, Williams 1973, and Caneva 1997. See also the recent translation of Oersted's selected scientific works 1998, including Wilson's introduction. For German romanticism and the development of biology, see Lenoir 1981, 1989, and 1992, Gregory 1984, 1989, and 1990, Engelhardt 1998, as well as Richards 1992, 2002, and forthcoming.

4. See Gregory 1997 and Pulte 1999a and 1999b.

5. The papers collected in Cahan 1993 present a good picture of Helmholtz's scientific and philosophical work. For the development of academic neo-Kantianism, see Köhnke 1991.

6. For psychophysiology and sense perception, see Sherman 1981, Hatfield 1990, Lenoir 1993, Turner 1994. For the foundations of geometry, see Richards 1977, DiSalle 1993. For the role of both of these, together with the conservation of energy, in the development of Helmholtz's epistemology, see Friedman 1997.

7. Poincaré's scientific and philosophical work is the subject of Greffe et al. 1996. For Poincaré's mathematical work, see Torretti 1978, Gray 1999. For philosophy of mathematics, see Folina 1992. For Poincaré's conception of geometry against the background of Kant and Helmholtz, see Friedman 2000b. For Poincaré and Einstein, see Miller 1981.

8. For Poincaré and the logical empiricists, see Friedman 1996. For an analogous irony arising in the logical empiricist appropriation of Helmholtz, see Friedman 1997.

9. Giere and Richardson 1996 presents a good picture of recent research into the background and evolution of logical empiricism. For the influence of Marburg on neo-Kantianism, in particular, see Friedman 1996b, Richardson 1997a, Friedman 1999, together with the further references cited therein.

II

Kant and *Naturphilosophie*

Frederick Beiser

A Relapse into Dogmatism?

Probably no other aspect of romantic *Naturphilosophie* has more aroused the wrath of its neo-Kantian critics than the organic concept of nature. These critics have dismissed this concept as a relapse into the worst kind of dogmatic metaphysics. They charge that this idea violates the regulative constraints that Kant so wisely placed upon teleology. Supposedly, the romantic *Naturphilosophers*—thinkers like Schelling, Hegel, Schlegel, and Novalis—naively and dogmatically gave the idea of an organism a constitutive status. True to their penchant for grand speculation and a priori reasoning, the romantics recklessly assumed that nature really is an organism, thus failing to observe Kant's critical teaching that it must be investigated only as if it were one.

Were the romantics really so naive and so careless? Or did they have some rationale for transgressing the Kantian limits? And, if they did, how plausible was it? These are the main questions I wish to discuss here. Part of my task is simply exegetical and historical: to reconstruct Kant's arguments against the constitutive status of teleology and the romantic replies to them. In doing so, I hope to show not only that the romantics were aware of the need to justify their new metaphysics, but also that they developed a rather sophisticated defense of it.[1] Another part of my task is more philosophical: to assess the romantic case, to determine whether they had an adequate response to Kant. While I contend that the romantics have a more plausible case than their neo-Kantian detractors think, I also hope to show that their response to Kant is ultimately inadequate, incapable of matching the deeper skepticism underlying Kant's regulative doctrine. Still, it will become clear that this gives the neo-Kantians no grounds for complacency and "I-told-you-sos." The problem is that *Naturphilosophie* grew out of a deep *aporeia* in the Kantian system—namely, its failure to explain the interaction between the intellectual and sensible, the noumenal and phenomenal. Indeed, the romantics' most interesting and plausible argument for their organic concept of nature exploits a very common Kantian strategy: It attempts to provide something like a transcendental deduction of the idea of an organism. In other words, it attempts to show that the constitutive status of the idea of an organism is the necessary condition of the possibility of experience. Whatever the ultimate merits of such a daring and difficult argument, neo-Kantian detractors of *Naturphilosophie* have failed to recognize it, let alone assess it.

If there is any general moral involved in my reexamination of these old disputes, it is that we have to break with the two dominant models of post-Kantian idealism and Romanticism, that these movements mark either a progression or regression from Kant. The neo-Kantian model of an irresponsible relapse into metaphysical dogmatism, and the neo-Hegelian model of the inevitable march toward absolute wisdom, are both worthless in assessing the philosophical complexities of these disputes. Alas, at the end of the day, philosophical commitment is as difficult as any decision in life: We have to compare incommensurables, play off one *aporeia* against another, and then take a leap. In the case of Kant versus the *Naturphilosophers,* we have to trade off the difficulties of the Kantian dualisms against the dangers of romantic speculation. Which is better and which is worse? I, for one, cannot find a clear answer.

THE NEO-KANTIAN STEREOTYPE

Before I begin to consider the Kantian arguments against teleology and the romantic response to them, let me first set the record straight about one basic issue, namely, Kant's relation to *Naturphilosophie.* To this day the name of Kant is still invoked as a talisman to frighten off the specter of *Naturphilosophie.* Among some prominent historians of science, Kant is still seen as the friend of natural science and as the foe of metaphysical speculation; his regulative doctrine is indeed held up as the very touchstone of scientific propriety.[2] To be sure, there is some basis for a positivistic reading of Kant. Kant did condemn the metaphysics of vital materialism—a central doctrine for most *Naturphilosophers*—and he did stress that philosophy must remain within the boundaries of possible experience. Nowhere is this protopositivistic side of Kant more in evidence than in his caustic reviews of Herder's *Ideas for a Philosophy of the History of Humankind* (*Ideen zu einer Philosophie der Geschichte der Menschheit*) (VIII, 45–60).

Still, this interpretation is a simplistic stereotype. It stresses one aspect of a much more complex picture, whose other aspects bring Kant much closer to the *Naturphilosophers*. There are at least three problems with the neo-Kantian interpretation.

First, in fundamental respects, Kant was the father of *Naturphilosophie*. Kant's dynamic theory of matter in the *Metaphysical Foundations of Natural Science* was a formative influence upon the first generation of *Naturphilosophers*, more specifically upon Friedrich Schelling, Karl Adolf Eschenmayer, Heinrich Link, and Alexander Scherer. These thinkers took Kant's dynamic theory a step further by applying it to the new chemistry and all the new discoveries in electricity and magnetism (Durner 1994). Furthermore, Kant's methodological views—especially his demand for systematic unity and his insistence upon synthetic a priori principles—were also very important for some *Naturphilosophers*. It is indeed somewhat ironic to find the neo-Kantians criticizing the *Naturphilosophers* for a priori speculation and system building when so much of their inspiration for these activities came from Kant himself! Even the method of analogy, for which *Naturphilosophie* had been so severely criticized, has its Kantian roots. It was Herder who, in the first instance, had set the example for the use of this method; but Herder was simply following in the footsteps of his teacher, Kant himself, who had used it in his *Universal Natural History and Theory of the Heavens* (*Allgemeine Naturgeschichte und Theorie des Himmels*) (Adler 1968), a work Herder greatly admired.

Second, Kant's regulative doctrine was *not* the foundation of physiological and biological research in the late eighteenth and early nineteenth centuries, as has sometimes been argued (Lenoir 1989, 6). Rather, the very opposite was the case. It is striking that virtually all the notable German physiologists of the late eighteenth century—Albrecht von Haller, Johann Friedrich Blumenbach, Christian Friedrich Kielmeyer, Kaspar Friedrich Wolff, Alexander von Humboldt—conceived of vital powers as causal agents rather than regulative principles (Larson 1979). Their aim was to do for the organic world what Newton had done for the inorganic: to determine its fundamental laws of motion. Although they forswore knowledge of the causes of these laws, much as Newton had declined to speculate about the cause of gravity, they still saw these causes as vital agents behind organic growth.

Third, Kant himself was deeply ambivalent about his regulative doctrines.[3] Nowhere are his vacillations more apparent than in the Appendix to the Transcendental Dialectic of the first *Critique*. Here Kant explicitly rejects the merely hypothetical and heuristic status of the principles of the systematicity of nature, and he expressly affirms that we must assume there *is* some systematic order in nature, so that the concept of the unity of nature is "inherent in the objects" (*den Objekten selbst anhängend*) (B 678). Proceeding simply according to an as-if assumption, Kant argues, is not sufficient to justify or motivate enquiry (B 679, 681–682, 685, 688). Kant then blurs his distinction between the regulative and the constitutive, and indeed between reason and the understanding, by suggesting that the assumption of systematicity is necessary for the application of the categories themselves. Without the idea of systematic unity, he says, there would not be "coherent employment of the understanding," and not even a "sufficient criterion of empirical truth" (B 679). To some extent the same equivocation extends into the *Critique of Judgment* itself, where Kant sometimes states that we could not have a coherent experience at all without the application of the maxims of reflective judgment (§V, V, 185).

For all these reasons, it seems to me inadvisable to make a sharp and fast distinction between *Naturphilosophie* and the tradition of German physiology and biology, as if *Naturphilosophie* were a corrupt metaphysics flouting Kant's regulative guidelines, and as if physiology and biology were hard empirical science heeding them. Ultimately, there is only a distinction in degree, and not in kind, between Schelling, Hegel, and Novalis on the one hand and Blumenbach, Kielmeyer, and Humboldt on the other hand. Any qualitative distinction underestimates not only Kant's profound influence upon *Naturphilosophie* but also the deep tension between Kant's regulative constraints and late eighteenth-century physiology. Even worse, it exaggerates the speculative and a priori dimension of *Naturphilosophie,* as if it had no concern with observation and experiment, while it downplays the metaphysical interests of those engaged in observation and experiment.

It is one of the more unfortunate aspects of the neo-Kantian legacy that it has succeeded for generations in portraying *Naturphilosophie* as an aberration from true science, which follows the path of experiment and observation. Fortunately, in recent decades it has become clear that this picture of *Naturphilosophie* is profoundly anachronistic. It cannot come to terms with some very basic facts: that there was no clear distinction between philosophy and science in this period, and that there was no such thing as a pure empirical science limited to only observation and experiment. In the late eighteenth and early nineteenth centuries, *Naturphilosophie* was not a metaphysical perversion of, or deviation from, normal empirical science. Rather, it was "normal science" itself. From our contemporary perspective it is hard to imagine a scientist who is also a poet and philosopher. But this is just what is so fascinating and challenging about *Naturphilosophie,* which has to be understood in the context of its own time as the science of its day.

KANT'S ARGUMENTS FOR REGULATIVE CONSTRAINT

Whatever Kant's doubts and hesitations about the regulative doctrine, there can be little doubt that, at least sometimes, he affirmed it. For Kant repeatedly insists in the third *Critique* that the idea of the purposiveness of nature has only a regulative validity. It was on this very important point, of course, that he came into conflict with the *Naturphilosophers*. Kant and the *Naturphilosophers* share a very similar concept of the purposiveness of nature; yet Kant denies, while the *Naturphilosophers* affirm, its constitutive status. The question then arises: Why did the romantics make this assumption in the face of the Kantian critique of knowledge?

To assess the legitimacy of their case, it is first necessary to consider the challenge facing them: Kant's powerful arguments for the regulative status of teleology. There are at least three such arguments, which in some respects dovetail with one another. It is noteworthy that two of them appear in their most explicit form not in the *Critique of Judgment* but in two more obscure works of the 1780s: the 1786 *Metaphysical Foundations of Natural Science* and the 1788 essay "On the Use of Teleological Principles in Philosophy."[4] It is also interesting to note that some of these arguments were, at least by implication, directed against *Naturphilosophie,* given that their target was Herder's vital materialism in his *Ideas for a Philosophy of the History of Humankind,* a seminal work for the romantic generation.[5]

It is important to see that Kant's arguments against teleology concern one very specific concept: what Kant calls a "natural purpose" (*Naturzweck*). This concept is explicitly defined in §65 of the *Critique of Judgment,* a seminal text for the romantics' own conception of teleology. Something is a natural purpose, Kant explains, only if it satisfies two essential conditions. First, it must have organic unity, where each part is inseparable from the whole, and where the very idea of the whole determines the place of each part within it. Second, it must be self-generating and self-organizing, so that all its parts are reciprocally the cause and effect of one another, and so that it has no external cause. Kant argues that this second condition is also necessary, since the first condition alone (organic unity) is not sufficient to regard something as a natural purpose, given that it can also be satisfied by a machine. Only things that also meet the second condition are *natural* purposes because they produce themselves and do not have some external cause or designer.

Kant's analysis of the concept of a natural purpose was decisive for the romantics, who accept both its main points. Their organic concept of nature

begins from Kant's concept of natural purpose, and then generalizes it for
nature as a whole. All of nature then becomes a natural purpose, a giant natural purpose which consists in myriad smaller natural purposes. According
to this concept, there is no fundamental difference in kind between the ideal
and real, the mental and physical, since they are only different degrees of
organization and development of living force. Mind is very organized and
developed matter, and matter is less organized and developed mind. It is
important to see that such an organic concept does not abrogate the
mechanical, whose laws remain in force as much as ever; but it does see
the mechanical as a limiting case of the organic. While the organic explains
the parts of nature with respect to the whole, the mechanical simply treats
these parts in relation to one another, as if they were somehow self-
sufficient. The mechanical explains one event by prior events acting on it,
and so on ad infinitum; the organic explains why these parts act on one
another in the first place.

Of course, it is just this flight of speculative fancy whose wings Kant
was so intent to clip. The conflict between Kant and the *Naturphilosophers*
could not be more clear and precise: Kant denies and they affirm the constitutive validity of the concept of a natural purpose. Kant's denial works
against applying the concept on both a macro- and a microcosmic scale.

Kant's first argument against giving constitutive status to the idea of a
natural purpose, which appears in its most detailed form in the 1788 essay
on teleological principles, is essentially skeptical. It states that we have no
means of knowing whether objects in nature, such as vegetables and animals, are really purposive; in other words, we have no way to prove that such
objects are really organisms rather than just very complicated machines.
According to Kant, we understand the power to act from purposes only
from our own human experience, and more specifically when we create
something according to our will, which consists in "the power to produce
something according to an idea" (VIII, 181). If, therefore, something cannot
act according to ideas, we have no right to assume that it has the power to
act for ends. Hence the concept of a natural purpose, of a being that acts
purposively yet does not have a will, is "completely fictitious and empty"
(*völlig erdichtet und leer*) (VIII, 181).

In drawing such a conclusion Kant is not saying that the concept is
completely meaningless—in that case it could hardly have even a regulative
status—but simply that it has no reference. His point is simply that we *know*
of purposiveness only in the cases of beings that act with will and understanding, and that we therefore cannot make verifiable claims about the
purposiveness of beings that do not have will and understanding. In a

nutshell, Kant's argument is that intentionality—in the sense of conscious end or goal-directed action—is the criterion of purposiveness, and that such a criterion cannot be satisfied owing to the intrinsic limits of human knowledge.

Kant's second argument, which appears in §68 of the third *Critique,* might be called his anti-Frankenstein stratagem. This argument consists in a simple application of the central principle of the critical philosophy, or what Kant calls the principle behind its "new method of thought" (B xviii, also A xx). According to this principle, which Kant explicitly restates at §68, "we have complete insight only into that which we can make ourselves and according to our own concepts" (V, 384). This principle means that organisms are incomprehensible to us, Kant argues, because we do not have it within our means to create or produce them. We can indeed create some *material* thing, just as nature can produce one. But we have no power to produce the infinitely complex structure of an organism. Hence, if we know only what we can produce, and if we cannot produce organisms, it follows that we cannot know organisms.

Kant's third argument is directed against hylozoism or vital materialism, the doctrine that matter consists in *vis viva* or living force. To ascribe natural purposes to living things, it is not necessary to be a hylozoist, because it is possible to hold that such purposes are characteristic only of living or organic matter. Hylozoism is the stronger thesis that living force is essential to *matter as such;* it therefore implies that there is no difference between the organic and inorganic. Still, hylozoism is sufficient, even if not necessary, to justify the ascription of purposes to things, for it maintains that living forces are purposive.

Kant's argument against hylozoism proceeds from his analysis of matter in the *Metaphysical Foundations.* According to Kant's second law of mechanics, the law of inertia, every change in matter must have an external cause, that is, it persists in rest or motion, in the same direction and with the same speed, unless there is some external cause to make it change its direction and speed (IV, 543). This law implies, therefore, that changes in matter cannot be internal, or that matter has no intrinsic grounds of determination. This means for Kant that matter is essentially lifeless; for he defines life as the faculty of a substance to act from an internal principle, its power to change itself. Kant vehemently insists that the very possibility of natural science rests upon fully recognizing these implications of the law of inertia; in his view, hylozoism is at best speculative and at worst anthropomorphic; hence he condemns it as nothing less than "the death of all Naturphilosophie" (*der Tod aller Naturphilosophie*).

Kant's polemic against hylozoism appears to be ambiguous, vacillating unstably between two very different contentions: (1) that ascribing life to material things is *meaningless,* because it is contrary to the very idea of matter, and (2) that attributing life to material things is *problematic* because we cannot ever know whether they are really purposive. The first contention would mean that teleology is a *fiction,* while the second would give it a *hypothetical* status. Although Kant indeed makes both these contentions, the tension is only apparent, because they are directed against two very different versions of hylozoism, which he himself distinguishes (§65, V, 374–375; §73, V, 394–395). The first contention is directed against the doctrine that matter as such or by its very nature is living. Kant maintains that this doctrine is flatly contrary to the essence of matter, which is inertia. The second contention is targeted against the doctrine that, though matter itself is not living, there is still some living force or substance within it that somehow directs and organizes its activity. Against this doctrine Kant makes two points: First, we have no empirical evidence that there is such a principle in matter, because experience only validates the law of inertia; and, second, the concept of a living force inherent in matter is essentially circular, because we explain its appearances by appeal to living force and then explain living force by its appearances.

On the basis of these arguments, Kant concludes that the concept of an organism or a natural purpose has only a regulative status. To avoid some common misunderstandings, it is important to see precisely what such a conclusion means. Except for the most radical version of vital materialism, Kant is not saying that this concept is only a fiction, as if it were false that there are organisms in nature. Rather, he is saying that this concept has only a problematic status; in other words, we have no evidence or reason to assume the existence or nonexistence of natural purposes. While it is indeed possible that there are such purposes, it is also possible that there are none at all because they might be, for all we know, really only very complicated machines. True to his vocation as a critical philosopher, whose only goal is to determine the limits of our cognitive powers, Kant neither affirms nor denies the *sui generis* status of organisms; alternatively, he neither affirms nor denies the possibility of mechanism. Thus he states explicitly at §71 of the third *Critique:* "We are quite unable to prove that organized natural products cannot be produced through the mechanism of nature" (V, 388). When Kant denies the possibility of a complete mechanical explanation of organisms, when he famously denies that there will never be a Newton to explain the growth of a single blade of grass, he does so not because he thinks that organisms *are* extramechanical—for that too would be a dogmatic claim to knowledge—but only because he thinks that it is a necessary limitation of the human understanding that we cannot fully understand an organism mechanically and that we must resort to teleology to make them comprehensible to ourselves.

THE FIRST LINE OF DEFENSE

How did the *Naturphilosophers* defend themselves against this array of arguments?[6]

Their first strategy would be to diminish the size of their target, purging the concept of a natural purpose of all its traditional theological associations. Schelling, Hegel, Schlegel, and Novalis did not wish to retain or revive the old metaphysical notion of providence, according to which everything in nature follows a divine plan. Rather, they believed that their teleology was completely intrinsic, limited to the ends of nature itself. According to their view, nature is an end in itself, and the purpose of all of nature is not to realize any end beyond itself.

While this strategy scores an important point—that teleology need not carry the traditional baggage of physico-theology—it still does not blunt Kant's main arguments. Although Kant sometimes wrote as if the concept of the objective purposiveness of nature inevitably led to a physico-theology (§75, V, 398–399), the thrust of his arguments is directed against the concept of a natural purpose (*Naturzweck*), and therefore against the idea that nature alone is *self*-generating and *self*-organizing. Hence his target was indeed the central doctrine of the *Naturphilosophers*: an immanent teleology.

Limiting the question to the realm of nature itself, it seems that the *Naturphilosophers* could still avoid Kant's arguments. All they need to do is to point out that the concept of a natural purpose need not involve any of the shaky assumptions Kant attributed to it. More specifically, they could make two replies to Kant. First, they could maintain that the idea of a natural purpose does not necessarily imply intentionality, that is, the attribution of a *will* to a living thing. To state that an object is a natural purpose is not to assume that there is some intention behind its creation, still less that there is some kind of will within the object itself. Rather, all that it means is that the object is an organic unity, where the idea of the whole precedes its parts, and that the parts are mutually interacting, the cause and effect of one another. These are indeed the necessary and sufficient characteristics of a natural purpose on Kant's own account of that concept. So, by his own reckoning, there should be no need to demonstrate the existence of intentionality.[7] Second, the *Naturphilosophers* could deny that the idea of living matter entails that there is some kind of *soul* or *spirit* within matter itself that somehow directs and organizes its growth. It is important to see that, like Kant, the *Naturphilosophers* were also opposed to any form of animism and vitalism that postulated some *supernatural force* or immaterial substance behind organic growth. Like almost all physiologists of the late eighteenth century, they too wanted to avoid the dilemma of materialism versus vitalism.[8] While they held that materialism is too reductivist because it cannot explain the *sui generis* structure of organisms, they also rejected vitalism because it is too obscurantist, involving an appeal to some occult force or supernatural agency.[9]

It is important to see that the teleology of the *Naturphilosophers* is first and foremost a form of methodological holism. It involves a very different paradigm or concept of explanation from that of mechanism. It claims that to explain an object by its natural purpose is to explain it holistically, where the purpose is the idea of the whole. This whole is an organic unity, irreducible to its parts, each of which is intelligible only from its place within the whole. Such holistic explanation is the opposite of mechanical explanation, which involves an antithetical concept of the whole. According to this concept, the whole is the mere sum total of its parts, each of which is self-sufficient apart from the whole. The difference between these forms of explanation then amounts to two different whole-part conceptions. Either the whole precedes its parts or the parts precede the whole. This is the difference between what Kant calls a *totum* or *compositum*,[10] or, in the language of the third *Critique,* a synthetic or analytic universal.

All this makes it seem as if there is really no dispute between Kant and the *Naturphilosophers* after all. Kant is denying the attribution of purposiveness to objects in nature only in a very strong sense, one that implies the existence of intentionality or spiritual powers in nature, whereas the *Naturphilosophers* are affirming it in a weaker sense, one that has no such implications. Furthermore, Kant agrees with the *Naturphilosophers* that teleological explanations are irreducible to mechanical ones (§75, V, 398; §82, V, 429).

But the appearances of sweet harmony here are very deceptive. To conclude that there are no differences between Kant and the *Naturphilosophers* would be all too hasty. It would fail to appreciate the full force of Kant's arguments, and indeed the main point at issue between him and the *Naturphilosophers*. To see why, let us take a closer look at the *status controversiae*.

The Limits of Experience

Although the *Naturphilosophers* deny that the concept of a natural purpose refers to some kind of occult substance or supernatural power, they still give it some ontological status or objective reference. It is not that teleology is only a distinctive form of explanation, one that is logically irreducible to mechanism. On the contrary, teleology has a constitutive status, an objective reference, in two fundamental respects. First, it refers to a distinctive structure, function, or form of the organic; second, it also denotes a force behind this structure, function, or form. To be sure, this force is not supernatural, and still less is it a kind of entity; but it is a form of causal agency, a force whose manifestations are organic structures, functions, or forms.

Of course, it is precisely these ontological assumptions that Kant contests. He doubts not only that there is a specific kind of causal agency behind organic growth, but also that there is a distinctive structure, form, or function of an organism. The whole point behind his regulative doctrine is precisely to bracket both of these assumptions. Hence even if we drop the ontology of vital spirits and supernatural forces, Kant is still at odds with the *Naturphilosophers*.

To be more precise, Kant could concede that the idea of a natural purpose does not involve any assumptions about an intention, soul, or spirit. Furthermore, he could acknowledge—as he indeed insists—that the idea of a whole in a natural purpose is irreducible to its parts. Still, even if he makes these concessions, Kant still disputes that the concept of a natural purpose has objective validity. For even if teleological explanations are logically irreducible to mechanical ones, we still have to ask: What right do we have to assume that these explanations refer to some unique form of structure or causality in the natural world? After all, Kant insists that teleology is a necessary method of explaining nature *for us,* given the limitations of our human understanding; in other words, we cannot know that there really are unique structures, forms, or forces in nature that are irreducible to mechanical causes. For all we know, organisms might be simply very complicated mechanisms. Again, Kant was quite explicit and emphatic about this point: "We are quite unable to prove that organized natural products cannot be produced through the mechanism of nature" (§71, V, 388–389; compare §75, V, 400).

The first naive and natural response to Kant's challenge is to claim that observation and experiment do confirm the existence of unique living structures or forms. At the end of the eighteenth century, this line of reasoning was indeed very prevalent among some of the leading physiologists in Germany, thinkers like Blumenbach, Wolff, and Kielmeyer, who had a profound influence upon the development of *Naturphilosophie*.[11] Blumenbach, Wolff, and Kielmeyer maintained that it is possible to provide convincing empirical evidence that some living things actually generate and organize themselves. Their confidence is easily comprehensible once it is placed in historical context.

During the first half of the eighteenth century there were two fundamental theories about the origin and development of life: preformation and epigenesis.[12] According to preformation, organisms are already preformed in the embryo, and their development involves little more than an increase in size. According to epigenesis, however, organisms originally exist only as inchoate "germs" or "seeds," and their development consists in the actual generation of an organism's characteristic structure and organization. Toward the close of the eighteenth century, the theory of preformation had been severely discredited, chiefly because it could not account for the many empirical facts amassing against it—for example, to take the most spectacular case, the regeneration of freshwater polyps. Hence, in his influential tract *On the Formative Drive* (*Über den Bildungstrieb*), Blumenbach confessed that he had to abandon his previous allegiance to preformation because of the sheer weight of experimental evidence against it, which he then proceeded to describe in great detail (1791, 44–77). And, in his famous dispute with Albrecht Haller, Wolff contended that his theory of epigenesis did not rest upon the inference that what could not be observed (a preformed embryo) did not exist, as Haller contended, but upon the simple observation of what did exist. After months of painstaking observations, Wolff concluded that he saw under his microscope nothing less than the generation of intestinal tissue in chick embryos; there was no preformed structure observable anywhere, and all that could be seen was the formation of an inchoate mass into a differentiated structure. So, in Wolff's view, those who denied epigenesis were in the same embarrassing predicament as Galileo's critics: They simply refused to look through the microscope (Roe 1981, 80–83, 86).

These developments are striking and dramatic, and they are crucial in explaining the rise of the organic concept of nature among the *Naturphilosophers*. Still, they need not be intimidating for a Kantian. Unperturbed, he will maintain that, even if observation and experiment show that something is self-organizing without a preformed structure, this still does not prove that it is a natural purpose. The problem is that, for all we know, the thing might still be acting entirely because of mechanical causes. The attribution of purposes to nature implies that there is some other form of causality not strictly reducible to mechanism; but no amount of experience can ever be sufficient to exclude entirely the operation of hidden mechanical causes.

Of course, one of the reasons the physiologists and *Naturphilosophers* inferred the existence of natural purposiveness from their observations is that they believed they had already refuted mechanism. Because they had independent arguments against mechanism, they believed they could safely exclude the possibility that self-generation and self-organization arose from hidden mechanical causes. Usually they made two kinds of argument against mechanism. First, they claimed that the structure of an organism is too complex to have arisen from mechanical causes alone. Second, they contended that a natural purpose is very different from a mechanism, because it implies that an object is the cause of itself, whereas mechanism implies that all causes are external to a thing.

Neither of these arguments is conclusive, however. Both make invalid inferences—and, indeed, the very kind of dogmatic inferences that Kant exposes in the third *Critique*. The first assumes that because we cannot conceive the structure arising from mechanical causes it could not have done so, which is to make a dogmatic inference from the limits of our cognitive powers to what must exist. The second assumes that because the concept of a purpose involves a different form of explanation from that of mechanism it therefore must also refer to a special kind of structure or cause; but this is just the point that Kant questions when he maintains that the concept of a purpose has a strictly problematic status. Although Kant himself argues that we cannot use mechanism to explain organism, he does so for a very different reason from the *Naturphilosopher:* not because of any presumed insight into the objective nature of organisms but because of the limitation of our powers of knowledge. The *Naturphilosophers* were dogmatic in their critique of mechanism, however, because they assumed that the impossibility of explaining an organism according to mechanical causes came from the objective nature of an organism itself. They therefore excluded the very possibility that Kant wanted to keep open: that organisms, for all we know, could still be produced by mechanism.

THE TRANSCENDENTAL DEDUCTION OF THE ORGANIC

As I have explained the controversy so far, it seems as if Kant has trumped the *Naturphilosophers,* who were guilty of dogmatism after all. Although they were perhaps not as naive as the neo-Kantians portrayed them as being, their defense of *Naturphilosophie* still cannot justify giving constitutive status to the idea of an organism. The Kantians, it seems, can just keep on wearing their perukes, powdering and curling them with tender care.

But this is not the end of the story, the most interesting and important chapter of which remains to be told. It is now necessary to consider the romantics' fundamental rationale for *Naturphilosophie,* an argument implicit in all their writings on nature, and one that would have been explicitly given by all of them if we were to only ask. To understand this argument it is necessary to go back in history and to reconstruct the context behind the development of the organic view of nature in the late 1790s. Much of that context was set by the early criticism of Kant's philosophy, and especially by the reaction against his dualisms.

Many of Kant's early critics charged that his dualisms—whatever the initial rationale for them—make it impossible for him to solve his own problems.[13] According to Salomon Maimon (1790, 62–65, 182–183, 362–364), whose reaction was typical and influential, Kant's dualisms are so severe that they undermine any attempt to answer the central question of transcendental philosophy: "How is synthetic a priori knowledge possible?" If the understanding and sensibility are such heterogeneous faculties—if the understanding is an active, purely intellectual faculty, which is beyond space and time, and if sensibility is a passive, purely empirical faculty, which is within space and time—then how do they interact with each other to produce knowledge? Kant had stressed that there must be the most intimate interchange between these faculties if knowledge is to be possible—"Concepts without intuitions are empty, intuitions without concepts are blind," as he put it in a famous slogan—yet he had so radically divided them that any interchange between them seemed impossible. The problem here, Maimon maintained, was analogous to, and indeed just as severe as, Descartes' classic difficulty regarding his mind-body dualism.

It was in this context that the *Naturphilosophers* first developed their own organic concept of nature. One of the main motivations behind this concept was to surmount Kant's problematic dualisms, and so to resolve the outstanding problem of transcendental philosophy. The young romantics held that it is possible to bridge these dualisms only by giving *constitutive status* to the concept of an organism. Of course, Kant himself had already set the stage for such an argument in the third *Critique* by proposing that the concept of an organism mediate between the noumenal and the phenomenal. The only sticking point between him and the *Naturphilosophers* then concerned the regulative versus the constitutive status of this concept. But here the *Naturphilosophers* would insist that transcendental philosophy itself demanded giving this concept constitutive status; for only under the assumption that there is an organism is it possible to explain the actual inter-action between the subjective and objective, the ideal and the real, the noumenal and phenomenal. To leave the concept with a purely regulative status simply left the mystery of their actual interaction. Hence, for these reasons, the *Naturphilosophers* believed that the concept of an organism had its own transcendental deduction: it was nothing less than a necessary condition of possible experience. Here we witness once again a phenomenon often seen in the history of post-Kantian thought: that it was necessary to transcend Kant's limits to solve his own problems.[14]

It is important to add that this transcendental argument in behalf of an organism was not simply a possible strategy; it is not merely a historical reconstruction of an implicit line of reasoning. Rather, it can be found more or less explicitly in the early writings of Schelling and Hegel. It was Schelling who first suggested this argument in the introduction to his 1797 *Ideas for a Philosophy of Nature*. Hegel later developed it, in his characteristically dense and obscure prose, in his work on *The Difference between the Fichtean and Schellingian Systems of Philosophy*. Since Schelling's arguments are clearer, and the prototype for Hegel's, I will focus upon them here.

It is striking that, in the introduction to his *Ideas,* Schelling raises the question, "What problems must a philosophy of nature resolve?" and answers it by referring to the basic problem of transcendental philosophy: "How a world outside us, how a nature and with it experience, is possible?" (1797, 665). Schelling makes it perfectly explicit, therefore, that *Naturphilosophie* has a transcendental task: Its basic objective is to solve the problem of knowledge. The solution to this problem is especially difficult, Schelling explains, because all knowledge requires some form of correspondence or connection between the subjective and objective, the ideal and the real, or the transcendental and empirical. Such a connection or correspondence seems impossible, however, because these realms appear to be completely heterogeneous from one another. To explain the possibility of knowledge, then, it is necessary to unite these realms, to forge a bridge between them.

Schelling then argues at length that this problem cannot be resolved from conventional Kantian premises (1797, 666, 675–676). He contends that the orthodox Kantian distinction between the form and matter of experience simply reinstates the dualism that gave rise to the problem of knowledge in the first place. The Kantians cannot bridge the gulf between these realms, he explains, because they so sharply distinguish between the form and matter of experience that they cannot explain how the intellectual, ideal, and subjective forms interact with the empirical, real, and objective matter. They simply state that the forms are imposed upon this matter, though they offer no explanation of how that is possible.

In the *Ideas* Schelling only offers some suggestions about how the Kantian dualisms could be overcome; and while he is critical of the Kantian regulative constraints, he also does not dare to abolish them (1797, 54). However, in some later works, especially his 1799 *Outline of a System of Natural Philosophy* (*Entwurf eines Systems der Naturphilosophie*) and his 1800 *System of Transcendental Idealism* (*System des transcendentalen Idealismus*), he puts forward a solution to the problem of the Kantian dualisms that clearly goes beyond Kant's regulative limits. Schelling's solution is nothing less than his organic concept of nature. If nature is an organism, then it follows that there is no distinction in kind but only one of degree between the mental and the physical, the subjective and objective, the ideal and the real. They are then simply different degrees of organization and development of a single living force, which is found everywhere within nature. These apparent opposites can then be viewed as interdependent. The mental is simply the highest degree of organization and development of the living powers of the body; and the body is only the lowest degree of organization and development of the living powers of the mind.

Doubtless Schelling's organic concept of nature is very bold and speculative. Schelling himself admitted its metaphysical status, knowing perfectly well that it could not be derived from experience. Nevertheless, he insisted that the concept is metaphysical in the sense that it is transcendental, a necessary condition of possible experience.

Whatever the merits or flaws of Schelling's organic concept, it should be clear by now that it simply begs the question against him to dismiss the concept as transcendent metaphysics. This famous complaint of the neo-Kantians ignores the transcendental strategy of *Naturphilosophie;* even worse, it also begs important questions about how to solve the problems posed by the Kantian dualisms. Alas, those who demand that we go back to Kant often seem to forget why philosophers were compelled to go beyond him in the first place.

A FINAL SETTLING OF ACCOUNTS

Now that we have seen the dialectical struggle between Kant and the *Naturphilosophers,* how should we assess its outcome? What was at stake?

For all the difficulties of Kant's dualisms, it would be wrong to think that Kant had been superseded by his romantic successors. Perhaps the constitutive status of the organic is the only means to overcome the difficulties of his dualisms, just as Schelling and Hegel argue. Yet Kant himself was perfectly aware of these difficulties; and he believed that it was necessary to live with them all the same. Unlike the romantics, he was content to leave the connection between understanding and sensibility, the intellectual and empirical, a mystery.[15] To be sure, he sometimes wrote about a single source of these faculties; but he regarded any theory about this source as very speculative, and in any case as not strictly necessary to explain the possibility of empirical knowledge. If a transcendental deduction presupposes the fact of some interaction between these heterogeneous faculties, it does not follow that it has to explain that fact. After all, it is not as if the skeptic can dispute the fact of interconnection; he gains a foothold only in disputing the extravagant theories about it.

It is precisely here that we seem to come to the last crossroads, the final parting of the ways. The diverging paths are Kantian modesty versus post-Kantian curiosity, Kantian skepticism versus post-Kantian speculation. But, once again, this proves to be too simplistic. For if in one respect Schelling's organic concept goes beyond Kant, flying in the face of his regulative limits, in another respect it is entirely warranted by him, entirely in keeping with his spirit, if not his letter. This concept was the inevitable result of joining together two very Kantian lines of thought. First, the dynamic theory of matter, which claims that matter is not inert extension but active force. The fundamental premise behind this theory is that mechanism is insufficient to explain matter, which consists in forces of attraction and repulsion. Second, the idea that nature is a unity, a systematic whole, where the idea of the whole precedes all its parts. On the basis of these Kantian themes Schelling already had sufficient rationale for his organic concept of nature. For if mechanism cannot explain matter itself, let alone life and mind, then it fails as the paradigm to explain all of nature. The only other plausible candidate is organicism. The great advantage of organicism is that it does justice to the unity and systematicity of nature. It explains matter and mind according to a single principle, seeing both as different degrees of organization and development of living force. There is no need to distinguish between the realms of the mechanical-material and the organic-immaterial since the mechanical is only a limiting case of the organic. We can now see clearly why Schelling wanted to go that extra step beyond Kant in demanding not only a *dynamic* but also a *vital* conception of matter. For if we insist on the principle of the unity of nature, only a vital concept unites the organic and inorganic, the mental and the physical, into one natural world.

Of course, Kant himself would never take this step, and indeed he fought against it with all the passion and energy at his command. The reason is not hard to fathom. For him, organicism came only at an enormous price: the loss of moral freedom. If we accept organicism, then we must abandon that dualism between the noumenal and phenomenal that Kant saw as the precondition of moral action and responsibility. Indeed, the organic concept does mean extending the realm of natural explanation not only to the domain of life and mind but also to the realm of the noumenal or rational. From its antidualistic perspective, there can be only an artificial and arbitrary border line between reason and the mental, the noumenal and the living. There is a continuum throughout all of nature extending from the most primitive matter to the most subtle and sophisticated forms of consciousness; the rational in all its forms is nothing less than the highest organization and development of the living forces inherent in all of nature. Although it gives pride of place to human rationality in the hierarchy of nature, the organic concept still sees rationality as one more manifestation of the forces within nature. Inevitably, the dominion of natural necessity then intrudes into the realm of the moral.

Of course, Schelling and the romantics gave a different assessment of the whole problematic. For them, dualism was not the solution but the problem. It was dualism that came with such a heavy price. Dualism meant the end of the unity of nature, the sancrosanct *lex continui;* it made a mystery out of moral decision and action; and it left unintelligible the interaction between the intellectual and sensible involved in all knowledge. But abandoning dualism made it necessary for the romantics to explain the very issue for which Kant had defended it in the first place: the possibility of freedom.

What the romantics have to say about freedom is another issue, which it would be far beyond our scope to investigate here. Suffice it to say for now that the issues involved in legitimating their organic concept of nature are much more complicated than they first appear. They raise all kinds of questions about the limits of knowledge, the meaning of the organic, the relationship between the mental and physical, and even the possibility of freedom itself. If I have shown that the organic concept is more than naive speculation—and if I have also shown that the Kantian critique is more than positivist dogmatism—I will have achieved my purposes here.

NOTES

1. The crucial texts for the romantic defense of the organic view of nature are the introduction to Schelling's *Ideas for a Philosophy of Nature* (1797, 661–723) and his preface to *On the World Soul* (1798, 415–419); Schlegel's *Lectures on Transcendental Philosophy* (1800, 3–43, 91–105); Novalis's "Allgemeines Brouillon" (1798/1799, 69, 338, 820, 460, 477, 820), and *Studies* (*Vorarbeiten*) (1798, 118, 125); finally the sections on Fichte's and Schelling's systems in Hegel's *The Difference between the Fichtean and Schellingian Systems of Philosophy* (1801), as well as the section on "Kantian Philosophy" in Hegel's *Faith and Knowledge* (1802, 301–333). Though Hegel himself later broke with the romantics, his Jena writings prior to 1804 are some of the most important defenses of the romantic position.

2. For example, this assumption is operative in the work of Timothy Lenoir (1980, 1981, and 1989). Lenoir's strategy is to free late eighteenth- and early nineteenth-century German physiology from the associations of *Naturphilosophie* by showing the impact of Kant upon such physiologists as Blumenbach, Kielmeyer, Treviranus, and Humboldt, who he thinks were responsive to Kant's criticisms of teleology and vitalism. For an assessment of Lenoir's views, see Caneva 1990 and Richards forthcoming.

3. Concerning some of Kant's equivocations, see Guyer 1990a and 1990b. On Kant's hidden proximity to the position of Hegel, see Tuschling 1991.

4. See "On the Use of Teleological Principles in Philosophy" (*Über den Gebrauch teleologischer Prinzipien in der Philosophie*) (VIII, 181–182); and *Metaphysical Foundations of Natural Science* (IV, 543–545). The argument of the first text recurs in the *Critique of Judgment* at §65 (V, 375); the argument of the second reappears at §65 (V, 374–375) and §73 (V, 394–395).

5. These works were written after Kant's review of Herder's *Ideas* (VIII, 45–66). The polemics in the review anticipate the later arguments.

6. Here one historical caveat is necessary. The romantics did not explicitly, self-consciously, and methodically reply point for point to Kant's arguments. It is therefore necessary for the historian to reconstruct their response to Kant, which means drawing out some of the implications of their general position. This means that it is necessary to consider what they would or could have said in response to Kant. My reconstruction is based upon the texts cited in note 1 above and note 9 below.

7. This point is involved in Schelling's and Hegel's claim that the rationality of nature is the result of its intelligible structure rather than its self-consciousness alone. This point became essential to their defense of an objective idealism against the subjective idealism of Kant and Fichte. The argument is especially apparent in Hegel's work on the *Difference between the Fichtean and Schellingian Systems* and their joint article "On the True Concept of Philosophy of Nature and the Correct Manner of Solving Its Problems" (*Über den wahren Begriff der Naturphilosophie und die richtige Art, ihre Probleme aufzulösen*) (Schelling 1801, 713–737).

8. Here I take issue with Lenoir (1980, 108), who accepts the standard caricature of *Naturphilosophie* as a species of vitalism, and distinguishes the tradition of vital materialism from *Naturphilosophie*.

9. On Schelling's attempt to steer a middle path between these extremes, see his *On the World Soul* (1798, 413–421), and his *First Outline of a System of the Philosophy of Nature* (*Erster Entwurf eines Systems der Naturphilosophie*) (1799, 74–78).

10. A 438 and *Inaugural Dissertation* §15, Corollarium (II, 405). In *Reflexion* 3789 Kant formulates the distinction as one between a *totum analyticum* and a *totum syntheticum* (XVII, 293). In *Reflexion* 6,178 he formulates it as a distinction between intuitive and discursive universality (XVIII, 481). See also §76 of the *Critique of Judgment* (V, 401–404).

11. On the influence of Blumenbach, see Lenoir 1981, 128–154, and on the influence of Kielmeyer, see Durner 1991, 95–99.

12. For a brief and useful survey, see Roe 1981, 1–20; Hankins 1985, 113–157; and Richards 1992, 5–16.

13. This point has been underestimated, I believe, by Kant's defenders. The monistic aspirations of post-Kantian philosophy arose from an *internal* critique of Kant and not from any prior metaphysical commitments. Cf. Ameriks 2000, 118–119, and Guyer 2000.

14. For this reason, as well as those cited above, Lenoir's distinction between transcendental and metaphysical *Naturphilosophie* becomes very shaky (1981, 146, 149). Such a distinction ignores the extent to which those who stressed the constitutive status of an organism used transcendental methods to justify it.

15. For a more detailed account of this issue, see Henrich 1994.

III

Nature Is the Poetry of Mind, or How Schelling Solved Goethe's Kantian Problems

Robert J. Richards

In 1853, two decades after Goethe's death, Hermann von Helmholtz, who had just become professor of anatomy at Königsberg, delivered an evaluation of the poet's contributions to science. The young Helmholtz lamented Goethe's stubborn rejection of Newton's prism experiments. Goethe's theory of light and color simply broke on the rocks of his poetic genius. The tragedy, though, was not repeated in biological science. In Helmholtz's estimation, Goethe had advanced in this area two singular and "uncommonly fruitful" ideas (1853, 34–38). The poet recognized, first, that the anatomical structures of various kinds of animals revealed a unity of type underlying the superficial differences arising from variability of food, habit, and locality. His second lasting achievement was the related theory of the metamorphosis of organisms: the thesis that the various articulations within an organism developed out of a more basic kind of structure—that, for instance, the different parts of plants were metamorphosed leaves or that the various bones of the animal skull were but transformed vertebrae. These two general morphological conceptions, according to Helmholtz, grounded the biology flourishing at midcentury. Goethe came to these ideas, Helmholtz shrewdly maintained, as the result of a poetically intuitive conception (*anschauliche Begriffe*) (1853, 43). He described, for instance, Goethe's immediate recognition, while playfully tossing around a sheep's skull on the Lido in Venice, that the fused bones of the battered cranium consisted of transmuted vertebrae. This experience resulted in the poet's vertebral theory of the skull, which became a standard conception in later morphology.[1] Poetic intuition thus liberated an idea initially embedded in matter and made it available to the analytic understanding of the scientist.

Forty years later, in 1892, at the meeting of the Goethe Society in Weimar, Helmholtz returned to reexamine the poet's scientific accomplishments, and, it would seem, implicitly his own; for by the end of his career, Helmholtz himself had achieved a position in German culture only a few steps below that of Goethe. His evaluation of Goethe's achievements in physical science was now more complex than his earlier assessment had been. While allowing that Goethe too rapidly dismissed Newton's analyses, Helmholtz admitted the considerable difficulty in experimentally finding one's way to an adequate theory of light and color (1892, 30–32). And, remarkably, in this second essay, he conceded that Goethe was intuitively right to have rejected Newton's particulate theory of light (1892, 32–33). Had Goethe only known of Christian Huygens's wave theory, Helmholtz suggested, he might well have moved toward a more satisfactory conception. Helmholtz reinforced his earlier judgment about the significance of Goethe's morphological ideas, and maintained that the poet's acute proposals led to an accelerated acceptance of Darwin's theory of evolution, particularly in Germany.

After fifty years of a career that ranged from physics to physiology, from optics to theories of artistic representation, Helmholtz more sensitively assessed Goethe's aesthetic approach to nature. In this second essay, he emphasized a principle operative in Goethe's work that I believe served as a fundamental organizing conception in the philosophy of the early romantics. This was the aesthetic-epistemic principle of the complementarity of the poetic and scientific conceptions of nature. Helmholtz came to agree with Goethe that "artistic representation" provided another way into the complexities of the physical world (1892, 7). Both aesthetic intuition and scientific comprehension drove down to the type, to the underlying force that gave form to the surface of things. Exercising aesthetic intuition within the realm of science, therefore, would not introduce anything foreign, but only aid the scientist in comprehending the fundamental structures and powers of nature.

Helmholtz was unaware of the metaphysical and epistemological barriers Goethe had to overcome in order to establish the principle of complementarity. Once established, the principle became instrumental, not only in smoothing the way for Darwin but in allowing Goethe himself to move along the path on which the Englishman would later travel. The barriers that Goethe initially encountered derived from Kant. As the result of the urging of his friend, the poet Friedrich Schiller, Goethe became grudgingly convinced of the Kantian epistemology, which seemed to block access to the real world. The escape route, however, came through the intellectual aid provided by another friend, the young idealist philosopher Friedrich Schelling. Goethe and Schelling became quite close, and each had a marked impact on the thought of the other. Schelling led his older mentor beyond Kant and ultimately to the kind of evolutionary theory that Kant had rejected; Goethe, in his turn, helped anchor Schelling's drifting idealism. Let me initially make clear the dimensions of the Kantian obstacles before I undertake to examine how Goethe's young protégé showed the way back into the heart of nature.

GOETHE'S KANTIAN PROBLEMS

By reason of inveterate attitude and poetic disposition, Goethe strongly inclined toward realism. His poetry expressed the immediate experience of nature and attempted to re-create that experience for the reader. During the 1780s, while a civil administrator for Carl August, duke of Saxe-Weimar-Eisenach, the young poet became devoted to Spinoza. Under the tutelage of his friend Johann Gottlieb Herder, Goethe, in the company of the enticing Charlotte von Stein, undertook a systematic study of the philosopher, who inspired him to explore empirical phenomena in order to discover those adequate ideas that determined the essential structures of natural objects. Archetypes of plants and animals, Goethe became convinced, animated nature; and he believed scientific experiment and systematic observation might bring those structures to intuitive recognition. This kind of rationalistic realism, though, met a formidable challenge.

In 1789 Goethe undertook a study of Kant's first *Critique* with the help of Karl Leonhard Reinhold, the main supporter of Kant in the philosophy faculty at Jena.[2] The poet, though, stumbled over the book's principal epistemological position, namely, that an impenetrable barrier stood between the mind and the world beyond. Initially, as he recalled, "sometimes my poetical abilities hindered me, sometimes my mundane understanding, and I felt I had not gotten very far" (1820, 95). Goethe's marginalia and notes indicate clearly enough that he understood Kant's claim; he did not appreciate, though, the rationale for the claim.[3] A few years later, his friend Friedrich Schiller would finally convince him of the validity of the Kantian epistemology. But this only exacerbated his frustration, since he remained constitutionally disposed to realism. He felt nature directly and immediately and expressed that intimate experience in the flood of poetry pouring from his pen. Thus the question became poignant: How could one have an immediate and aesthetically responsive contact with nature existing beyond mind if the Kantian position held firm? For Goethe, this unpleasantness quickly spread beyond the aesthetic to the scientific.

While traveling in Italy during the two-year period 1786–1788, Goethe became convinced that he had solved a deep problem in biology, and when he returned to Weimar he began working on a tract to explain his discovery. His *Metamorphosis of Plants* (1790) describes the development of plants in terms of an ideal structure that expresses the essence of all plants. Yet this archetype, as Goethe construed it, served not only as an ideal type but also as a force actually productive of natural organisms (see Richards 2002). Over this discovery the Kantian pall likewise fell: How could one be sure the archetypal idea corresponded with anything real, with a force actually resident in nature?

Goethe's hesitating difficulties with Kant, however, were initially blanketed in a cloud of enthusiasm when he took up, in 1790, the newly published *Critique of Judgment*. He thought that this new critique specified in a most perspicuous way the connection between aesthetic judgment and biological judgment. It showed explicitly what Goethe knew implicitly to be right, namely, the existence of an intimate relationship between the realm of art and the realm of science. The third *Critique* also offered to Goethe the confirmation that the work of art and the product of nature both existed in their own right and for themselves. Organisms might be shaped by the external environment, but their internal structures were neither explained nor justified by that environment nor by any other external cause, human or divine. The *Critique of Judgment*'s analysis also freed art from the oppression of final causes: art objects had aesthetic value independently of their moral worth, theological subject, or decorative character.[4] Yet in the midst of philosophical plenty, Goethe again collided with the Kantian barrier, now blocking two avenues. Kant would allow archetypal ideas—such as that of the ideal plant or the vertebrate structure—and even grant that the naturalist could assume these archetypes had creative efficacy; but they could only function, according to Kant, *as if* (*als ob*), that is, they could only serve as regulative heuristics. We might assume an archetypal intellect created natural objects, but this assumption could have no purchase on nature or valid science. Goethe yet believed these ideals operated as real causes. Further, the Königsberg sage had refused to recognize a natural process of which the poet had become convinced—the evolutionary transition of species.

These, then, are some of the difficulties that Goethe faced. The individual who did the most to convince him of the power of the Kantian view was his close friend, the poet Friedrich Schiller.

SCHILLER'S KANTIAN ADVOCACY

Goethe and Schiller differed in temperament and intellectual attitudes, approaching common issues from quite distant poles. As the older poet later recalled, "Schiller preached the gospel of freedom; I wanted to preserve the rights of nature" (1820, 97). This simple, but trenchant, characterization crystallized several facets of their intellectual differences: Schiller displayed a kind of religious fervor, Goethe a cooler, almost legal demeanor; Schiller emphasized the creative freedom of the artist, Goethe the constraints imposed by nature; Schiller looked inward, Goethe, outward; Schiller was a

Kantian idealist, Goethe—initially at least—a Spinozistic realist. But as their friendship matured, their ideas and attitudes began to migrate toward more common ground.

After their first serious intellectual encounter, in July 1794, Schiller sent Goethe a remarkable letter diagnosing their intellectual and artistic differences. The analysis flattered Goethe for his genius, yet did not unduly dim Schiller's estimate of his own virtues. The letter suggested how intellects of such diverse carriage might yet travel the same path. He wrote:

> What is difficult for you to realize (since genius is always a great mystery to itself) is the wonderful agreement of your philosophical instincts with the pure results of speculating reason. Certainly at first, it seems that there could not be a greater opposition than that between the speculative mind, which begins with unity, and the intuitive, which starts from the manifold [of sense]. If the first seeks experience with a chaste and true sense, and the second seeks the law with a self-active and free power of thought [*Denkkraft*], then they cannot fail to meet each other half way. To be sure, the intuitive mind is only concerned with the individual, the speculative only with the kind [*Gattung*]. But if the intuitive has genius and seeks in the empirical realm the character of the necessary, it will always produce the individual, but with the character of the kind; and if the speculative mind has genius and does not lose sight of experience—which that sort of mind rises above—then it will always produce the kind but animated with the possibility of life and with a fundamental relationship to real objects. (Schiller to Goethe, 23 August 1794, in Goethe 1990, 15)

Schiller further suggested that Goethe had a southern, virtually Greek, temperament, which could only have realized its potential after coming into contact with original, ancient sources. The lived reality allowed his imagination, "in a rational fashion, to give birth internally to a Grecian land" (Goethe 1990, 14). Yet after this rational re-creation occurred, according to Schiller, it had to be turned back into intuitions and feelings, which would then guide artistic production (Goethe 1990, 14–15).[5]

Shortly after he penned his letter, Schiller elaborated these categories of the intuitive mind and the speculative mind in his great treatise *Naive and Sentimental Poetry* of 1795 (*Naive und sentimentalische Dichtung*). He pictured Goethe as the naive poet, who intuitively responded to nature, and himself as the sentimental, who had to struggle reflectively with ideas in the execution of his art. Schiller derived inspiration for his diagnosis from Kant's third *Critique,* a book that both he and Goethe (quite independently) had been reading since its appearance four years earlier. The analysis depended on Kant's notion of genius, which the *Critique* describes in this way: "Genius is the talent (natural gift) that gives the rule to art [*Kunst*]. Now since talent is an inborn, productive ability of the artist, it belongs to nature. So we might also express it this way: Genius is the inborn mental trait (*ingenium*) through which nature gives the rule to art" (§46, V, 307). The definition suggested to Schiller that ineffable rules for the creation of beauty arose from the artist's nature, which was of a piece with nature writ large. In producing a painting or sculpture, for instance, the artist of genius plays with different forms in imagination. In this free play, according to Kant, certain expressions will seem aesthetically right; and the artist will experience aesthetic pleasure as he or she renders the artistic object. The harmony of forms and the pleasure they induce would be, in Kant's estimation, the outward signs of nonspecifiable rules of beauty. The naive artist, according to Schiller, follows these rules of beauty immediately and unreflectively, his pen or brush being guided by the sheer sense of aesthetic rightness. The difference between the naive poet and the sentimental poet, as Schiller reconstructs their activities, lies not, therefore, in the use of ideas—both employ ideas that at a deep level join their natures with external nature; it is, rather, that the naive poet does not reflectively struggle with the ideas in the manner of the sentimental poet.

Before his contact with Schiller, Goethe had inchoately assumed that the beauty of nature simply rushed into his eyes and gushed out of his pen. Schiller began teaching him that constructive concepts intervened, that his aesthetic appreciation of nature required the creative potency of ideas, of rules of beauty, even if those rules, as implied by Kant's definition of genius, lay below the limen of consciousness, buried deeply within the nature of the artist. While the metaphysics of this implication could have little justification within the confines of Kant's own epistemology, both Schiller and Goethe resonated to it. That metaphysical conviction lay behind Goethe's aphorism that "an unknown, law-like something in the object corresponds to an unknown, law-like something in the subject" (1991, 942, no. 1344). This kind of metaphysics enticed Goethe the way several of his women friends did at this time: alluringly and seductively, with the poet giving way even while recognizing the impropriety of his indulgence.

GOETHE'S MORPHOLOGY

In 1794, just after he had established his friendship with Schiller, Goethe composed a short essay on morphology that showed the clear impress of Kant's third *Critique* and the discussions he had with his new friend. In his *Essay on a General Theory of Comparison* (*Versuch einer allgemeinen Vergleichungslehre*), Goethe highlighted a particular aspect of Kant's proposal concerning teleological judgment, namely, that organisms, while they displayed an internal teleology, should not be regarded as elements of an external teleology—final causes in a more cosmological sense. In this Kantian light, Goethe urged in his essay that the anatomist reject the notion that plant and animal structures had been designed for divine purposes, rather that the researcher should understand those structures as having their raison d'être in the functional organization of the entire creature. To adapt Voltaire's piquant example, we should not marvel at a superior wisdom that supposedly designed the bark of the cork tree so that fine wines could be preserved; rather we should try to understand how that part functioned within the organization of the tree itself and how it was affected by its geographical circumstances. The relationship of organisms to their environment, according to Goethe, had to be regarded as nonintentional: the environment had an impact on creatures, changing their outward shape to conform to particular requirements, giving an organism "its purposiveness in respect to that external environment" (*seine Zweckmässigkeit nach aussen*). But that was only part of the story; for the internal structures of plants and animals showed another force at work. There was also an "inner kernel" (*innere Kern*) that provided a general corporeal pattern for an organism, which extrinsic forces might particularize in different ways: the seal, for instance, had a body formed by its aquatic environment, but its skeleton displayed the same general configuration as that of land mammals. Goethe thus concluded that the "ultimate form [of a plant or animal] is constructed likewise from an inner kernel, which is given its particularity through the determination of external elements. In this way, an animal obtains its purposiveness in respect to the outer environment, since it is formed from the external as well as from the internal" (Goethe 1790, 4.2, 182).

Living organisms thus derived their structures from two forces, an intrinsic one, which provided a general pattern (*Muster*), and an extrinsic one, which shaped an organism to its particular circumstances. This latter, environmental force, Goethe conceived much as Lamarck and Darwin would, namely, as a direct effect on the organism that adapted it to particular circumstances ("purposiveness in respect to that external environment"). Goethe had replaced divine teleology with natural causality, though a causality that yet retained a telic feature. Some time later, Georges Cuvier and Richard Owen would reach comparable conclusions, though they still detected the ultimate intentions of the Creator expressed in such proximate causes.

From 1794 through the turn of the century, Goethe composed several other essays in morphology, most reflecting his engagement with Kant. For instance, in 1796, he worked on a series of lectures—undelivered—that would sketch out his full conception of morphology. These *Lectures on a General Introduction to Comparative Anatomy* (*Vorträge, über eine allgemeine Einleitung in die vergleichende Anatomie*) argued that the theory of the archetype did not rest on mere hypothesis, since it followed from "the concept of a living, determined, independent, and spontaneously effective natural being" (Goethe 1796, 203). Such a being would have its parts mutually dependent upon one another and comprehensible only in relation to the whole. Now, Kant, of course, held much the same view, though he specified that such a teleological concept, while characteristic of the human mode of thought, was only regulative, not determinative of external nature. For Kant the concept of a living being simply could not function—as Goethe thought it must—in authentic science, which could refer only to mechanical causes in the explanation of natural phenomena. Goethe used the Kantian framework, nonetheless, to picture another conception, which he put in this way:

> We are thus assured [by reason of our concept] of the unity, variety, purposiveness, and lawfulness of our object. If we are thoughtful and forceful enough to approach our object and to consider and treat it with a simple, though comprehensive mode of representation [*Vorstellungsart*], one that is lawfully free [*gesetzmäßig-frei*], lively, yet regular—if we are in a position, employing the mental powers that one usually calls genius (which often produces rather dubious effects), to penetrate to the certain and unambiguous genius of productive nature—then, we should be able to apply this meaning of unity in multiplicity to this tremendous object. If we do so, then something must arise with which we as men ought to be delighted. (Goethe 1796, 203)[6]

In this turbulently flowing passage, several eddies of Kantian meaning form around the bedrock of Goethe's instinctive realism. First, he suggests here that the necessary concept we have of living beings assures us of their fundamental unity of type, a unity that nonetheless permits great variety in realization. Second, he thinks of this concept as being lawfully free (*gesetzmäßig-frei*), a phrase and conception that seem to come straight from Kant's description of aesthetic judgment as stemming from "the free lawfulness of the understanding" (*die freie Gesetzmäßigkeit des Verstandes*) (§22, V, 240–241).[7] Finally, when Goethe refers to the "productive genius of nature," he obviously plays off of Kant's own definition of genius as "nature giving the rules to art." And he seems to suggest two aspects of this notion: that archetypes in nature constitute a productive force bringing particular organisms into existence; and that these archetypes are comprehended by the artist in the creation of beautiful objects (a matter considered below).

Kant, of course, considers aesthetic judgments, as well as judgments of intrinsic teleology, not as determinative but as reflective; and such reflective judgments in the case of organisms (i.e., teleological judgments) are regulative, that is, virtually hypothetical. Yet the archetype for Goethe, as he here thinks of it, is not merely a regulative consideration, since it is the same one that the productive genius of nature employs—to use Kantian terms, it would be determinative. Goethe's residual Spinozism, according to which productive ideas reside in nature, imparts a decided turn to his newly adopted Kantianism, a twist that most Kantians would find simply confusing if not confused. After all, why should we assume that our mode of conceiving nature has purchase on nature herself?

The answer to this question would be worked out by a young philosopher whom Goethe adopted virtually as a son, Friedrich Wilhelm Joseph Schelling. Shortly after his arrival in Jena late in the spring of 1797, Schelling became entangled with a group of poets, historians, and critics who were forming what became known, according to their own designation, as the romantic circle. While Goethe initially stood at the periphery of this circle, he was drawn rapidly into its embrace.

THE GATHERING OF THE ROMANTICS AT JENA

Goethe realized that the university at Jena suffered a tremendous loss in 1794 when Karl Leonhard Reinhold accepted a position at Kiel. The privy counselor found, however, opportunity rapidly to make good the departure with the man whom Kant himself initially admired, Johann Gottlieb Fichte. The philosopher's first weeks at the university produced tremendous excitement. He and Schiller hit it off immediately, and Goethe took to reading the new arrival's works. Goethe's grasp of the philosophy impressed Fichte: "He [Goethe] has represented my system so succinctly and clearly that I myself could not have done it more clearly."[8] During his five years at Jena, Fichte raised the philosophical stakes of discussion; and he made Goethe aware—as much through force of personality as through his difficult ideas—of the liabilities of crucial aspects of the Kantian philosophy, especially the thing-in-itself. By 1799 Fichte's relationships with other faculty members at the university, many jealous of his standing and irritated by his imperious attitudes, became so frayed as to leave him dangling dangerously before less sophisticated detractors. A devoted cadre of students stayed loyal, but the fraternities continued to give him a hard time for his grousing about their rowdiness. Fichte finally stumbled in his relationship to his employer, Duke Carl August, when a charge of atheism was brought against him. On this occasion, even Goethe (who had some sympathy for the philosopher's religious position) could not save Fichte from himself. Fichte was dismissed from the university in June 1799.

Two years after Fichte's arrival in Jena, August Wilhelm Schlegel moved to the city, carrying with him many literary ambitions, as well as his new wife, Caroline Michaelis Böhmer Schlegel, and her daughter Auguste from a previous marriage. He had been recruited to join Schiller and Goethe as a co-worker on the new journal *Die Horen*. In 1798, after a falling out with Schiller, Schlegel became professor at the university. Goethe immediately discovered in this literary historian and critic one whose aesthetic judgment coincided with his own, even if he was initially wary of his new friend's "democratic tendencies" (Goethe 1988, 2:221–222). They conferred about Schlegel's translations of Shakespeare, and Goethe found in this genial man a literary confidant second only to Schiller. Schlegel, for his part, came, in the words of his brother, to "worship" Goethe (Schlegel 1800, 410). About a month after August Wilhelm and Caroline Schlegel had settled in Jena, Friedrich Schlegel—more volatile, brash, and passion-ridden than his older brother—set up house not far from the family. Goethe had read with great interest the younger Schlegel's collection of essays on Greek and Roman poetry, which had a preface that used Schiller's *Naive and Sentimental Poetry* to formulate the basic meaning for the romantic mode of thought.[9] Goethe soon invited this exuberant historian of ancient literature to accompany him on afternoon walks. Though Friedrich Schlegel, too, became an admirer of Goethe, his relationship to Schiller went quickly sour when he published some political tracts in the republican journal *Deutschland,* and then added personal insult to these political injuries with essays critical of Schiller's poetry and judgment. The suspicion that Caroline Schlegel gave succor to these attitudes was hardly amiss; and Schiller, in his turn, cultivated a hearty disdain for "Madam Lucifer." In June 1797, as his social circumstances became quite uncomfortable in Jena, Friedrich Schlegel left for Berlin, where he met and embraced the friendship of Friedrich Daniel Schleiermacher, who had begun on that philosophical-religious trajectory that would make a lasting impact on Protestant thought in Germany. Schlegel returned to Jena in 1799, shortly followed by Dorothea Mendelssohn Veit, the woman he would live with in imitation of Goethe's arrangement with his mistress Christiane Vulpius. The complexities of these social configurations did not prevent the renewal of friendship between the older man and his younger admirer. In January 1800 Schlegel began serializing his *Dialogue on Poetry* (*Gespräch über die Poesie*) in the brothers' journal *Athenaeum*. In that monograph he maintained that only Goethe could stand with such "romantic" poets as Cervantes and Shakespeare. Indeed *Wilhelm Meister* had achieved that ideal of beauty, a fusion of classical and romantic styles, so as to make it, in Schlegel's estimation, a tendency of the age, along with the French Revolution and Fichte's *Science of Knowledge* (*Wissenschaftslehre*) (Schlegel 1967, 198). Shortly after Goethe's death, Heinrich Heine judged, with some hyperbole, that the great poet "had owed the largest part of his fame to the Schlegels" (1836, 388). And so, by virtue of being the type-specimen, Goethe's work became, nolens-volens, definitional of the romantic ideal.

The romantic circle achieved philosophical completion when the twenty-three-year old Friedrich Schelling arrived in Jena to take up a newly created post that the combined forces of Fichte, Schiller, and Niethammer strove to make possible. Goethe was at first hesitant because he suspected a thinker too strongly enticed by the idealistic unreality conjured by Fichte. But when he met Schelling at a party thrown by Schiller, he came away singularly impressed with this very young philosopher, who showed a knowledge of real natural science, particularly Goethe's own works, and who seemed untainted by the kind of Jacobin inclinations displayed by Fichte. On 21 June 1797, Goethe wrote privy counselor Voigt to extol this new star on the philosophical horizon:

> Schelling's short visit was a real joy for me. For both him and us, it would be a wish realized were he to be brought here. For him, so that he might enter an active and energetic company—since he has had a rather isolated life in Leipzig—a company in which he might be guided in experience and experiments, and prosecute an enthusiastic study of nature, so that his beautiful mental talents might be applied with appropriate purpose. For us, his presence would likewise be an advantage—the activity of the Jena circle would, by the presence of so worthy a member, be greatly improved; and my own work would be considerably advanced. (1988, 2:615)

Goethe and Schelling grew quite close during the six years of the philosopher's stay in Jena. Caroline Schlegel regarded the poet's solicitude as quite paternal. So when Schelling declined into a depressive melancholy after the death of Caroline's daughter Auguste, the grieving mother felt she could ask the great man to care for her lover during the Christmastide of 1800–1801. Goethe also facilitated the divorce between Caroline and August Wilhelm Schlegel, so that she and Schelling could marry. The deep personal relationship between Goethe and Schelling inevitably affected their intellectual lives.

Goethe and Schelling: Nature as the Poetry of the Mind

Goethe is usually portrayed as utterly rejecting the scientific and meta-physical aspirations of the romantics.[10] Historians who make this judgment do so by considering under the rubric of "romantic writer" such diverse individuals as Schelling, Henrik Steffens, Gotthilf Heinrich Schubert, and Lorenz Oken. Goethe certainly thought the latter three often indulged in *Schwämerei* and obscurity.[11] But those individuals who might carry (or have pinned on them) the banner of romanticism expressed distinctive philo-sophical views and dispositions. Goethe reacted to each differently, and his feelings for each altered over time. When the Schlegels adopted more orthodox religious sentiments after the turn of the century, Goethe became disappointed, suspicious, and irritated with them. For Schelling, though, he harbored quite warm affection, which hardly abated over the years. And his enthusiasm for the young philosopher's ideas continued to grow from their first meeting until the time when Schelling left Jena in 1803. Thereafter he kept up with Schelling's changing philosophical interests, always indicating positive regard if not complete acceptance.[12] In the near term, though, the young philosopher secured the lines of Goethe's drifting metaphysical views, providing many of his instinctive attitudes hard rational demonstra-tions, while shifting others into more dangerous currents. And the recipro-cal pull on Schelling's own philosophy was hardly less dramatic.

During Schelling's years in Jena, he and Goethe met frequently to dis-cuss philosophical, scientific, and artistic matters. Goethe—the poet, scien-tist, and Weimar genius—like a whirlpool of creative energy carried the young philosopher into the center of his interests and flooded him with reorienting conceptions. His diverting power had its effect almost immedi-ately. In the winter term of 1798–1799, Schelling began lecturing at Jena on *Naturphilosophie,* lectures that would yield during Eastertide his *First Sketch of a System of Philosophy of Nature* (*Erster Entwurf zu einem System der Natur-philosophie*). In November he and Goethe met to discuss the character of *Naturphilosophie* and particularly the problems of organic metamorphosis.[13] After the publication of his lectures, Schelling, under the influence of Goethe, felt the need to clarify and develop an aspect of *Naturphilosophie* that he had neglected, namely, the role of experiment and observation. Dur-ing a particularly intense period, from the middle of September to the mid-dle of October 1799, the two met frequently to discuss this problem, and together they spent almost a week going over Schelling's *Introduction to the Sketch of a System of Philosophy of Nature* (*Einleitung zu dem Entwurf eines Sys-tems der Naturphilosophie*).[14] Schelling declared that the conversations had produced a great "fluorescence of ideas" for him.[15] The *Introduction* stated unequivocally the necessity of experiment in discovering the laws of nature. And indeed, Schelling—the knight errant of idealism—proclaimed that "all of our knowledge stems from experience" (1799, 278). It is hard to doubt that Goethe did anything but stimulate, promote, and encourage this appeal to experience as the true Excalibur of natural science. The *Introduction* clearly marks the deviant path of Schelling's idealism, which led him, within two years, to develop the kind of Spinozistic objectivism that Fichte scorned. Though many diverse pressures operated on Schelling, giving his thought direction, there can be little doubt that powerful Goethean forces pulled him sharply toward the ideal-realism that he would finally espouse.

Goethe, as well, shifted orientation; he began to rethink the relationship between art and science, especially their underlying connection in an identity between mind and nature. He had begun moving in that direction, as we have seen, owing to his reading of Kant and the insistent ideas of his dear friend Schiller. Already in 1796 he had pricked the ears of Friedrich Jacobi with the remark that his friend would "not find me any more the rigid realist" (1988, 2:240). Schelling, however, accelerated Goethe's move toward idealism. It began with their meeting in late May 1798 to conduct optical experiments together.[16] Then in early June, Goethe took up Schelling's book *On the World Soul* (1798). Like Jacob with the angel, he had to struggle mightily with the tome (1888, 2:109). He later recalled this time when "I found much to think about, to examine, and to do in natural science. Schelling's *World Soul* required my utmost mental attention. I saw it everywhere incorporated into the eternal metamorphosis of the external world" (1798, 58). This angel, having been brought to submission, inspired in Goethe a poem, itself entitled "World Soul" (*Weltseele*). It sang of that dream of the gods, the word-soul, which

> Grasps quickly after the unformed earth
> And with creative youth does not cease
> To animate and to bring to birth
> Ever more life in measured increase,

so that finally "each mote of dust lives" (1986, 53–54).[17] These few lines capture in poetic form the kind of "dynamic evolution" that Schelling himself portrayed in his tract and that Goethe would endorse (see Richards 2002, chaps. 3 and 11).

Goethe's poem not only indicates a locus of interest that Schelling's work had for him; it also signals a transformation in his attitude about the relationship between art and science. In an essay composed in the early

1790s—his "The Experiment as Mediator between Object and Subject" (*Der Versuch als Vermittler von Objekt und Subjekt*)—Goethe had drawn sharp methodological distinctions between art and science, even if he transgressed these boundaries in practice (1792, 321–332).[18] At that time he regarded the two enterprises as quite conceptually distinct. Schelling, by contrast, began developing his philosophy precisely along the lines prescribed in the romantic mandate of Friedrich Schlegel: "All art should become science and all science art; poetry and philosophy should be made one" (1958, 1:249). Despite his initial attitude, Goethe became, owing to his interactions with Schelling, more self-conscious of the way in which poetry and science could and must come together. Schelling's *System of Transcendental Idealism* demonstrated this for Goethe in a most compelling way.

The *System* began as a series of lectures Schelling gave in the winter term of 1799–1800 and published at Easter. He sent Goethe a copy, and the poet immediately responded that as far as he had quickly read, he thought he understood his young friend's argument. He was sure that "in this kind of presentation, there would be great advantage for anyone who was inclined to practice art and observe nature" (Goethe to Schelling, in Goethe 1988, 2:405). Goethe could rarely be moved to dispense patronizing flattery, and this certainly was not a case. He expressed his admiration for Schelling's ideas quite openly, as Friedrich Schlegel mentioned, with some pique, to his brother: Goethe, he wrote to Wilhelm, "talks of Schelling's *Naturphilosophie* constantly with particular fondness" (Schlegel 1800, 431). But what exactly in the young philosopher's *System* made such an impression on Goethe?

Schelling held that the ultimate aim of transcendental philosophy was to bring to intuitive identity the conscious and unconscious activity that constituted the unity of the self. With Fichte, he rejected the Kantian notion of a thing-in-itself as inconsistent and unjustifiable. He argued, rather, that the unconscious activity of an absolute self created both an empirical self and nature as the self's reciprocal correlate. Transcendental philosophy had the task of making this creative activity reflectively certain, on the one hand, and, on the other, to unite in intuitive synthesis nature, the unconscious product of self, and the conscious self, which stood over against nature—the objective realm with the subjective, necessity with freedom. The philosopher had the task, therefore, of bringing the intellectual intuition—that activity productive of self and nonself—up from the darkness of unconscious operation into the light of reflective awareness. But this intellectual intuition, according to Schelling, could be most perspicuously modeled on the aesthetic act, which united the unconscious laws of beauty and the conscious intentions of the artist. Close examination of the creation of the aesthetic object by the artist of genius, then, might illuminate the creation of the natural world and the empirical self by the genius of the absolute self.

Schelling's own theory of genius depended upon ideas drawn from Kant, Friedrich Schiller, and Friedrich Schlegel. He interpreted Kant's definition of genius (i.e., "the inborn mental trait [*ingenium*] through which nature gives the rule to art") to mean that the artist's unconscious nature determines those principles of beauty that express themselves in nonconceptual feelings, which in turn drive conscious actions.[19] The artist applies paint to canvas not in light of a conscious set of rules but by relying on aesthetic feeling—the palpable surface of underlying unconscious laws—to guide the brush. In Schelling's view, the inarticulable laws governing an aesthetic production surge forth with irresistible determinacy from the unconscious nature of the genius. These laws, or rules, must be followed. Yet every necessary blow of the sculptor's chisel, every perfect metaphor of the poet, nonetheless flow from the free will of the genius. Insistent forces thus well up from the unconscious nature of the artist and rush in turbulent cascades through the narrows of consciousness. This creates, according to Schelling, violent eddies of contradiction that "set in motion the artistic urge" (1800, 616–617). Such contradictions can be calmed only in the execution of the work of art. As the artist comes to rest in the finished, objective product, he or she will sense the union of nature and self, of necessity and freedom, of—finally—the unconscious and the conscious self. Thus will the goal of transcendental philosophy be reached: what is originally an identical self fragmented, as it were, through a kind of dialectical development in which self-reflection issues in the subjective structures of intelligence and the objective structures of nature—will have returned to its original identity. The intelligence "will feel surprised and very happy by this union, that is, it will see this union as a generous gift of a higher nature, which through this connection has made the impossible possible" (1800, 615).

In his analysis of the nature of artistic genius, Schelling attempted to portray, in another key, the creative essence of the self, as it constructs both itself and nature. He summarized this analysis by contending that "the aesthetic intuition is simply the intellectual intuition become objective" (1800, 625). Art, for Schelling, thus became the model for nature. And so he could introduce the romantic conceit that "nature is a poem that lies enclosed in a secret, wonderful script" (1800, 628; see Richards 2002, chap. 3). This was a philosophical position that could only attract Goethe's admiration.

Two weeks after sending a copy of his *System* to Goethe, Schelling left Jena with Caroline Schlegel and Auguste for Bamberg, where, in midsummer, tragedy befell them—Auguste, Caroline's daughter and a young woman of infinite promise, died of typhus. When Goethe learned of Auguste's death and perceived the vindictive atmosphere now permeating Jena because of Schelling's romantic involvement with Caroline, he feared that the philosopher might never come back. He decided nevertheless to continue study of this new brand of idealism, different as it was from Fichte's. He wrote Schelling to encourage his return and to suggest they would come to complete philosophical harmony:

> Since I have shaken off the usual sort of natural research and have withdrawn into myself like a monad and must hover over the mental regions of science, I have only occasionally felt a tug this way or that; but I am decisively inclined toward your doctrine. I wish for complete harmony, which I hope to have effected sooner or later through the study of your writings, or preferably from your personal presence; and I hope, as well, through the formation [*Ausbildung*] of myself in respect of the universal to have an impact sooner or later; this formation must become accordingly more pure—indeed, the more slowly I absorb this, the truer I remain to my own mode of thinking. (1988, 2:408)

Goethe mentioned in his letter that he had been taking instruction in the new idealism with Niethammer (who was on the philosophy faculty at Jena). The two met almost daily for a month, from early September to early October 1800.[20] After Schelling returned in the fall to Weimar, his philosophical élan slowly died away and was replaced by a growing depression over the death of Auguste. By Christmas he was in such a state that Caroline believed he might commit suicide, and she arranged for Goethe to take in her despairing lover. Schelling spent the Christmas holiday with Goethe, who apparently wrought the right kind of psychological cure. Schelling recovered, and his gratitude for Goethe's personal solicitude mixed sweetly with admiration for the older man's genius.[21] When Schelling lectured on the philosophy of art shortly thereafter, he did not hesitate to proclaim the poet's *Faust* "nothing else than the most intrinsic, purest essence of our age" (1802, 446). While Schelling remained at Weimar, he and Goethe continued to meet; they would discuss the philosopher's new projects, such as the *Bruno,* which gave forceful and fairly accessible expression to the Spinozistic identity theory that Schelling was developing—a theory certainly encouraged by Goethe.

There can, I believe, be little doubt of Goethe's admiration for Schelling or his enthusiasm for the new philosophy. Goethe explained to Schiller why he was so engaged with Schelling's ideas: "since one cannot escape considerations of nature and art, it is of the greatest urgency that I come to know this dominating and powerful mode of thought" (1990, 814). This "dominating and powerful mode of thought" solved for Goethe several deep problems concerning nature and art about which he constantly worried. Let me indicate specifically just how Schelling's philosophy accomplished this.

SCHELLING'S RESOLUTIONS OF GOETHE'S KANTIAN PROBLEMS

First, and most important, Schelling's philosophical view, especially as developed in his *System of Transcendental Idealism,* theoretically demonstrated that scientific understanding and artistic intuition did not play out in opposition to one another, as Goethe once thought, but reflected complementary modes of penetrating to nature's underlying laws. For Goethe this liberated his sense of the intimate connection between the scientific and the artistic approaches to nature, which he consequently expressed, as was his wont, in a poem—"Nature and Art"—that he composed at this time:

> Nature and art seem each other to flee,
> Yet, each finds the other before one can tell;
> The antagonism has departed me as well,
> And now they both attract me equally.
> (Goethe 1800) [22]

If, as Schelling maintained, "the world is the original, yet unconscious poetry of the mind [*Geist*]" (Schelling 1800, 349), then the poet might construct beautiful representations employing those same principles that went into the original creation of the natural world. So, again, there would be philosophical justification for the assumption of complementarity of scientific and aesthetic judgment: the poet, through creative genius, could compose those works that would have the authority of nature herself—an authority that Goethe always deeply felt but could not justify. And, reciprocally, the aesthetic might lead us to nature's concealed laws, to those archetypes according to which nature creatively expressed herself. As Goethe epigrammatically formulated it: "The beautiful is a manifestation of secret laws of nature, which without its appearance would have remained forever hidden" (1991, 749). This Goethean conception ran counter to the deep separation that Kant constructed between determinate judgments of nature and regulative judgments of art.[23] It was Schelling, though, who demonstrated how aesthetic judgment opened the heart of nature for scientific examination.

Schelling had also argued in the *System* that Kant should not have restricted genius to the aesthetic realm. He showed that genius could also be found in science. As Goethe met resistance from professionals to his work in optics and morphology, he would undoubtedly rest more comfortably in the knowledge, thanks to his young friend's analysis, that his genius in science need not adhere to conventional scientific wisdom. His aesthetic intuitions might probe more deeply, might lead more surely to new discoveries in science than could the plodding, tradition-bound studies of his critics. Moreover, Schelling had argued that the laws of nature, which the poet-scientist might comprehend, would be also laws of free creativity. Though Goethe probably did not inquire too deeply after Schelling's argument that the free creativity of mind conformed exactly to the fixed laws of nature, the argument nonetheless gave solace to his settled belief, which extended from ethics and politics to aesthetics, that true freedom, at least of the human variety, could only be realized in limitation. As he expressed it in the concluding stanza of *Natur und Kunst:*

> He who would be great must act with fine aplomb;
> In constraint he first shows himself the master,
> And only the law can give us full freedom. (Goethe 1800, 780)[24]

I have several times suggested that Schelling's philosophical principles would resolve for Goethe the conundrum that plagued all who became persuaded, as he did, of the Kantian epistemology—namely, how might we have authentic understanding of external nature, if we were shielded by our own representations from reality? If our mental constructions erected only a faux nature? The resolution, from Schelling's perspective, was simply that mind may indeed construct nature, but that there was no thing-in-itself standing behind the construction. Nature really was as she appeared to be. So the bright colors and forms that dazzled the eye were not meretricious and superficial traits—they inhered in nature. In Schelling's view the true idealism was the most authentic realism. Moreover, as Schelling drove his philosophy to an absolute ideal-realism, his position merged with that of Spinoza: the ideas that constituted nature's creations were not captives of individual minds but stood beyond empirical self and nature, though they were realized in both. Hence the solution to the puzzle of Goethe's epigram that "an unknown, law-like something in the object corresponds to an unknown, law-like something in the subject" (1991, 942). The connection between object and subject occurred through the organic activity of absolute mind and its ideas, which latter functioned as those archetypal concepts at the foundations of morphology.

Schelling's impact on Goethe reverberated through the years, and again became particularly manifest during the time (1817–1824) he composed *On Morphology*. The 1820 essay "Intuitive Judgment" (*Anschauende Urteilskraft*) provides a good example of this lasting influence. The essay returns to Kant's third *Critique*, as Goethe himself did at this time, to consider the philosopher's distinction between reflective and determinative judgment. It will be recalled that Kant classified judgments of beauty and judgments appropriate to biology (ends-means assessments) as reflective. Such judgments arose in attempting to understand the relationship of parts to whole, in either a work of art or a work of nature. In our appreciation of an art object, our understanding considers its various parts, allowing the free play of imagination to get a sense of the harmony of forms, a feeling of purposiveness in their arrangement; such feelings express those inarticulable ideas of beauty and allow the necessity and universality of the aesthetic judgment. Likewise, when the biologist assesses the traits of an organism, the same reflective procedure occurs: through an initial exploration of the parts, he formulates an idea of the whole—though a conscious and articulable one, an archetype—and thereby understands the organism's traits in relation to the whole. Indeed, the student of nature must, according to Kant, judge the structures investigated *as if* they came to exist by reason of the idea or archetype. But in this instance, the biologist makes only a heuristic assessment, and does not—cannot—presume that the idea at which he arrives has actually caused the structure. The scientist, according to Kant, ought to make *determinative* attributions only of mechanical causes, not of intentional causes, to explain natural phenomena.

In our scientific understanding of nature, according to the Kantian system, we apply categories like causality and substance determinatively to create, as it were, the phenomenal realm of mechanistically interacting natural objects. But in considering biological organisms, we must initially analyze the parts in reflective search of that organizing idea that might illuminate their relationships. But Kant suggested that we could conceive of another kind of intellect, one other than ours, which might move from the intuition of the whole to that of the constituents, instead of following our path from parts to whole. This would then be an *intellectus archetypus,* whose very idea would be creative. Concerning this Kantian notion, Goethe made a trenchant and many-layered observation:

> The author seems here, indeed, to refer to a divine understanding. Yet, if in the moral realm we are supposed to rise to a higher region and approach the primary Being through belief in God, virtue, and immortality,

then it also should be the same in the intellectual realm. We ought to be worthy, through the intuition of a continuously creative nature, of mental participation in its productivity. I myself had incessantly pushed, initially unconsciously and from an inner drive, to the primal image [*Urbildliche*] and type [*Typische*]. Fortune smiled on this effort and I was able to construct a representation in a natural way; so now nothing more can prevent me from boldly undertaking that "adventure of reason," as the grand old man from Königsberg himself has called it. (1817, 98–99)

In this passage and in the brief essay from which it comes, Goethe attempted to muscle into philosophical acceptance a thesis similar to one of Schelling—namely, that if moral experience required us to postulate God to make sense of that experience, then our experience of organisms should also require us to postulate an intellectual intuition to make sense of such experience.[25] But Goethe suggested that this would occur in two ways. First would be the intellectually intuitive action of nature—the assumption that nature herself, through a kind of instantiation of archetypal ideals, would create organisms according to such ideals. Here Goethe seems to allude to the Spinozistic notion of adequate ideas that themselves would be creative. The second construction that Goethe put on Kant's conception was that we also might share in this kind of intellectual intuition, presumably as the artist who created an aesthetic object and also as the scientist who penetrated the veil of nature to understand intuitively the archetypal unity underlying its variegated displays. Like Schelling, Goethe thus implied that if archetypal ideas were necessary for our experience of organic nature, then they had to be causal constituents of that experience—mentally creative of that experience. And there was the further implication of this analysis, namely, that in such mental creations we shared in nature's own generative power—indeed, that we become identified with nature in such activity. Goethe thus reaffirmed a Schellingian Spinozism: God, nature, and our intellect are one.

Goethe's final remark in the quotation above draws out the ultimate consequence of this ideal-realism. In the third *Critique*, Kant recognized that the variety of organic forms yet displayed "a common archetype" (*einem gemeinschaftlichen Urbilde*), and thus might be produced, as he put it, by "a common primal mother." This might lead to undertaking "a daring adventure of reason," namely, the belief that the earth had given birth to less purposive forms and these to more purposive, till the array of currently existing organisms appeared. Kant thought this transformational hypothesis would be logically possible if we initially assumed that the initiating cause of the series was itself organic. He yet rejected this evolutionary hypothesis because he did not think we had any empirical evidence of the generation of a more organized form from a less organized one (§80, V, 419–420). Schelling's theory of dynamic evolution, which Goethe accepted, postulated an organic foundation (i.e., absolute mind) for a transformational series; and by the time of *On Morphology,* researchers had accumulated fossil evidence of such transformations (e.g., the Megatherium). Goethe was thus ready, as he concluded, boldly to undertake that adventure of reason of which the Königsberg sage had spoken.

The evolutionary hypothesis as applied to nature reflected Goethe's own mental evolution: *On Morphology* tracked the gradual ascent of his morphological ideas, and those ideas gave rise to the transformational hypothesis he rather boldly embraced in the book.[26] The metaphysical foundation of these two evolutionary series—of the self and of nature—rested ultimately on the kind of ideal-realism for which Schelling had argued and which Goethe embraced.

NOTES

1. There is still a mystery as to whether Goethe actually made this discovery, so often attributed to him by many, including himself (for a reevaluation, see Richards 2002).

2. Goethe read the first *Critique* rather thoroughly, as his pencil scorings throughout the book indicate. Karl Vorländer (1923, especially 283–291), describes Goethe's marginalia. Sulpiz Boisserée notes on 3 October 1815 that Goethe mentioned having had a seminar with Reinhold on Kant (Boisserée 1978, 278).

3. Vorländer, in the fundamental treatment of Goethe's Kantianism, judged that the few notes the poet made directly from the *Critique of Pure Reason* indicated that he had not overestimated his penetration of the work (1923, 140–144). Reinhold's conception of the Kantian project would prove crucial for the later development of Fichte's and Schelling's idealism. When these two philosophers rejected so vehemently Kant's thing-in-itself, they were reacting mostly to Reinhold's version of that Kantian doctrine. The blunt formulations of Reinhold undoubtedly made Goethe's acceptance of the Kantian epistemology difficult, and allowed the poet to be more receptive to Schelling's later advance upon Kant (see Richards 2002).

4. Goethe sketched the ideas of the *Critique* that most struck him in his "Influence of More Recent Philosophy" (*Einwirkung der neueren Philosophie*) (1820, 96–97).

5. "This logical direction, which your mind was required to take in a reflective mode, comports ill with the aesthetic direction it takes when it creates. So you had one more task: just as you went from intuition to abstraction, so now you had to move in the reverse direction and turn concepts into intuitions and thoughts into feelings, since only in this way can genius produce anything."

6. The text is ambiguous as to whether the concept "lawfully free" (*gesetzmäßigfrei*) and the others in that string of adjectives should be applied to the mode of representation or

to the object represented (i.e., the type). Kant cultivated the notion that our moral and aesthetic representations were the result of bringing oneself freely under law. Goethe conceived of the archetype as embracing a lawlike pattern, but one that permitted the freedom of empirical expression. Insofar as Goethe attributed the same concept to the genius of nature as to that of man, perhaps he intended the ambiguity. More generally, Goethe liked the notion that freedom required constraint. Fichte, whose work Goethe knew quite well, had elevated the notion of representation [*Vorstellung*] to a prominent place in the epistemology of science, and he too insisted on the freedom of the act of representing.

7. Goethe frequently urged the idea that freedom was found only in restraint. His poem "Nature and Art" ends with the line "And only the law can give us freedom" (Goethe 1800, 780).

8. Fichte was quoted by Wilhelm von Humboldt to Friedrich Schiller (22 September 1794), in Goethe 1998, 1:564.

9. See Goethe's diary entry of 20 March 1797 in 1888, 2:62. Friedrich Schlegel's *Greeks and Romans* appeared in 1797 (though some of the essays had been printed earlier), bearing the preface that celebrated Schiller's essay.

10. Dietrich von Engelhardt has offered the most sensitive consideration of Goethe's relation to the science and philosophy of the romantics. But he argues: "Though Goethe's research into nature doubtlessly is close to metaphysical *Naturphilosophie* and romantic research into nature, it nonetheless is quite clearly distinct from these positions in its specific connections with aesthetics, philosophy, science, and biography" (Engelhardt 1998, 58). Walter Wetzels thinks Goethe rejected the presumed deductively a priori scientific methods of Schelling: "Goethe refused to embrace the approach advocated and practiced by Schelling, namely to proceed from preconceived general hypotheses about nature, and to deduce from them the properties of reality" (Wetzels 1987, 74). Of course, Goethe did reject such apriorism, but then so did Schelling. Nicholas Boyle attempts to show that Goethe, though initially friendly toward Schelling's *Naturphilosophie*, simply rejected the idealism of the young philosopher (Boyle 1991/2000, 2:667–668, 675–677).

11. After reading Henrik Steffens's 1806 *Outlines of the Philosophical Sciences* (*Grundzüge der philosophischen Wissenschaften*) Goethe observed to his friend Wolf (31 August 1806): "the preface to the little book has indeed a mellifluous edge, but we amateurs choke mightily on the contents" (1899, 187). Goethe complained to Sulpiz Boisserée about Schubert's Protestant "mysticism": "Thus the pitiable Schubert, with his attractive talent, attractive remarks, etc.—he plays now with death and seeks his healing in corruption—and indeed, he is already half corruption himself, that is, he quite literally has consumption" (Boisserée 1978, 232).

12. For instance, even after Schelling had turned more to investigate religious phenomena, Goethe read his works with avid curiosity, as he indicated to Jacobi (11 January 1808): "I owe [my attitude] to the more elevated standpoint to which philosophy has raised me. I have learned to value the Idealists. . . . Schelling's lecture has given me great joy. It sails in those regions in which we both like to tarry" (Goethe 1988, 3:62).

13. See Goethe's diary entries of 12, 13, and 16 November 1798 (1888, 2:222–223).

14. Goethe read Schelling's *Introduction* on September 23 and talked with him about it; and then from 2 to 5 October they read through the work together (see Goethe's diary: 1888, 2:261–263).

15. "A short while back he [Goethe] spent several weeks here. I was with him for a long time every day, and had to read aloud my work on *Naturphilosophie* and explained it to him. What a fluorescence of ideas these conversations have produced for me, you can well imagine" (Schelling to Carus, 9 November 1799, in Schelling 1962, 176–177).

16. Goethe's diary indicates that they met several times during Schelling's first year (1798) in Jena, and at the end of May they performed those optical experiments mentioned above (1888, 2:109).

17. "Ihr greifet rasch nach ungeformten Erden / Und wirket schöpfrisch jung / Dass sie belebt und stets belebter werden, / Im abgemess'nen Schwung." The exact date of the poem is uncertain, but it seems to have come from sometime in 1798–1799.

18. The essay is dated 28 April 1792 but did not receive its title until publication in 1823.

19. See Kant's third *Critique*, §46, V, 307–308. It is very likely that Schelling discussed his interpretation of Kantian genius with both Schiller and Goethe, with whom he met on frequent occasions during the period 1797–1800.

20. See Goethe's diaries (1888, 2:304–308) for the period of 5 September to 3 October 1800.

21. Goethe was in Jena from 12 to 26 December, and traveled back to Weimar with Schelling on the 26th. Among other diversions of the season, Goethe went with Schiller and Schelling to a masked ball at court (2 January 1801), though the three finally escaped to talk about aesthetics. Schelling stayed until 4 January, when Goethe had to take to his bed because of a severe catarrh. See Goethe's diaries (1888, 2:315–316; 3:1). Schelling wrote Goethe at the end of January to give thanks that the poet was feeling better. He then expressed his gratitude: "The recollection of the healing and happy stay in your house and under your gaze has not left me for an instant, and was, for me at this time, of infinite value" (Schelling 1908, 88).

22. "Natur und Kunst, sie scheinen sich zu fliehen, / Und haben sich, eh man es denkt, gefunden; / Der Widerwille ist auch mir verschwunden, / Und beide scheinen gleich mich anzuziehen."

23. Kant, it must be stressed, thought the regulative principle that nature should be comprehensible to us was a transcendental requirement for all determinative judgments about nature. Aesthetic judgment per se, however, which employed no conscious conception of its object—i.e., beauty—differed radically from determinative judgments about nature, which necessarily employed conscious concepts, concepts that should not be teleological.

24. "Wer Großes will, muss sich zusammenraffen; / In der Beschränkung zeigt sich erst der Meister, / Und das Gesetz nur kann uns Freiheit geben."

25. Though stated a bit vaguely, this was Kant's position in the second *Critique* and was reiterated in the third *Critique*. But Kant thought the postulate of God necessary as a practical matter—rather like a heuristic or regulative idea. The argument from moral experience could not be used as a proof of the existence of a transcendent entity. In the third *Critique* he made this point explicit in a footnote to his moral argument: "This moral argument should not be taken to provide an objectively valid demonstration of God's existence, not an argument that might prove to the skeptic that there is a God— rather that if he wishes to think consistently about morals, he must assume this proposition [that God exists] among the maxims of his practical reason" (§87, V, 450–451).

26. Most scholars of Goethe's biology reject the idea that he advanced anything like an evolutionary theory. I believe, however, that the evidence is quite persuasive that he in fact harbored evolutionary considerations of nature from a quite early period, at least from his intense interactions with Herder in the mid-1780s (see Richards 2002, chap. 11).

IV

KANT—*NATURPHILOSOPHIE*—ELECTROMAGNETISM

Michael Friedman

Robert Stauffer (1957) first gave prominence to the circumstance that Oersted's discovery of electromagnetism was intimately connected, at least in Oersted's own mind, with his rather deep philosophical engagement with both Kant's philosophy of science, as presented in the *Metaphysical Foundations of Natural Science* of 1786, and the further development of this philosophy within post-Kantian German *Naturphilosophie,* especially as represented by Schelling. Building on Stauffer's contribution, L. Pearce Williams (1965, 1966) developed a more general picture of the development of electromagnetic theory—especially in the work of Faraday but also placing particular emphasis on Oersted—as resting on a profound deviation from the Newtonian tradition due to *Naturphilosophie* and ultimately to Kant. For Williams, it was Kant's articulation of a so-called dynamical theory of matter in his treatise of 1786—a theory that represents matter as constituted out of the "fundamental forces" of attraction and repulsion rather than as a primitive, originally hard and impenetrable solid—that first opened up the possibility of a unified treatment of all the forces of nature, including, especially, the electric and magnetic forces.[1] And Oersted's work, in particular, is then seen as a direct expression of this fundamentally Kantian view.[2] Stauffer, for his part, makes a closely analogous argument to the effect that Oersted's "metaphysical faith" in "the unity of the powers of nature"—a faith nurtured and buttressed by his earlier philosophical work on Kant's *Metaphysical Foundations* and German *Naturphilosophie*—played a key role in motivating and sustaining Oersted's later experimental work, which culminated in the discovery of electromagnetism.[3]

Yet these contributions of Stauffer and Pearce Williams, as important and original as they are, remain on an excessively high level of generality. For a commitment to a unified treatment of all the forces of nature—a faith in "the unity of the powers of nature"—is far too abstract and unspecific to carry very much explanatory weight as a factor in Oersted's quite particular discoveries. For example, Newton, in the *Opticks,* had already sketched a general program for developing a unified account of nature (including electrical and magnetic phenomena, of course) on the basis of a small number of attractive and repulsive "Powers" or "active Principles," and it was precisely by virtue of this explanatory program, for Newton, that nature is "very comfortable to herself and very simple."[4] Or, to take a more directly relevant example, when Helmholtz, in his famous monograph of 1847, provided the first theoretical basis for the principle of the conservation of the total quantity of force (i.e., energy) in all the processes of nature in which it is converted from one form into another (a principle that became emblematic, in the mid-nineteenth century, of just the kind of "unity of the powers of nature" that Stauffer and Pearce Williams had in mind),[5] Helmholtz's own theoretical framework employed paradigmatically Newtonian central forces acting both in straight lines and immediately at a distance. Indeed, Helmholtz, more generally, is quite critical of German *Naturphilosophie* and sees his own work, in this respect, as a return to the more serious and sober scientific preoccupations of the original Kant.[6] Accordingly, while Helmholtz's 1847 monograph is clearly indebted to Kant's 1786 *Metaphysical Foundations*, it bears no traces at all of post-Kantian *Naturphilosophie*.

What is needed, then, is a more detailed account of Oersted's particular intellectual context, for it is precisely these details that distinguish Oersted's intellectual situation both from Newton's (and also Kant's) on the one side and from Helmholtz's on the other. And one of the most important of these details, from the present point of view, is the circumstance that the dynamical theory of matter favored by both Oersted in particular and *Naturphilosophie* more generally is fundamentally different from Kant's original theory. Indeed, as I myself have just intimated, Kant's theory, in most relevant respects, is far more Newtonian then theirs. It is especially misleading, therefore, when Pearce Williams portrays Kant as the great opponent of the Newtonian natural philosophy, for Kant's philosophy of nature, in most relevant respects, should rather be viewed as a culmination of the Newtonian tradition.[7] Whereas the dynamical theory of matter of the *Metaphysical Foundations* did serve, in fact, as a preamble or prolegomenon to the *Naturphilosophie* of Schelling and Oersted, it was only by radically reconceptualizing and transforming Kant's philosophy that the decisive step into *Naturphilosophie* was actually taken.

Fortunately, the main outlines of a more detailed and specific account of Oersted's particular intellectual situation—both philosophically and scientifically—have already been provided by Andrew Wilson, in his outstanding introduction to the Princeton edition of Oersted's (selected) scientific works (Jelved et al. 1998, xv–xl).[8] The principal points are these:

(1) Whereas Kant, in the *Metaphysical Foundations*, had declared that chemistry was not yet a science and therefore incapable of an a priori "metaphysical" foundation, one of the central ambitions of both Schelling and Oersted was precisely to provide such a foundation. This was to proceed by further articulating the fundamental forces of attraction and repulsion Kant had shown to be constitutive of all matter in general and as such, so that chemical forces, in particular, would thereby naturally emerge.

(2) New developments in electrochemistry provided the primary empirical impetus here, in so far as chemical forces could now be plausibly identified with electrical forces, and electrical forces in turn could be taken as a further articulation and development of the fundamental forces of attraction and repulsion in general. The electrochemical researches of Johann Ritter were particularly important in this regard, for Ritter had encountered Schelling's philosophy at Jena and was also a close friend and collaborator of Oersted's; and it was Ritter, in fact, who first introduced Oersted to this aspect of Schelling's *Naturphilosophie*.

(3) Galvanism or current electricity of course played a central role in the new electrochemistry, as exemplified, above all, in the electrolytic decomposition of water resulting in oxygen and hydrogen gasses. Oxygen and hydrogen were thereby associated with negative and positive electricity, respectively, and this suggested an especially close link between electrical forces and the fundamental chemical forces involved in combustion. Schelling, Ritter, and Oersted (along with many others, of course) took this as evidence of the electrical nature of chemical affinities quite generally.[9]

(4) Finally, the well-known parallels between electrical and magnetic forces suggested to Schelling, Ritter, and Oersted that magnetism, too, was essentially implicated in chemical interactions (including galvanism) and, more specifically, that what Schelling called the basic or original form of "the dynamical process" was further differentiated, at the level immediately following that of Kant's two fundamental forces of attraction and repulsion in general, into magnetism, electricity, and chemical forces (including galvanism). For Oersted, this suggested a testable link between magnetism and galvanism, in particular, and thereby stimulated the experimental researches that eventually resulted in his great discovery.

As Wilson admirably explains, the immediate intellectual context for Oersted's discovery therefore involved a very specific—and, as it turned out, especially timely—interaction between contemporary philosophical and scientific developments. Just when philosophy of nature in the Kantian tradition was ripe, as it were, for an extension of the dynamical theory of matter into chemistry, new discoveries in electrochemistry provided the needed fertile soil within which such a philosophy could then bear scientific fruit.[10] What I want to add here to Wilson's account is a bit more detail on the relationship between Kant's dynamical theory of matter and Schelling's, on the basis of which we might then be in a position to shed further light on both Oersted's philosophy of nature in particular and the relationship between philosophical speculation and scientific discovery more generally.

CONSTITUTIVE PRINCIPLES AND REGULATIVE USE OF REASON

Kant's philosophy of human knowledge and experience—including his more specific philosophy of (corporeal) nature developed in the *Metaphysical Foundations Natural Science*—is based on a number of fundamental distinctions. The most important is the distinction between the passive or receptive faculty of pure intuition or sensibility (involving the pure forms of sensible intuition, space and time) and the active or intellectual faculty of pure understanding (involving the categories or pure concepts of the understanding: substance, causality, community, and so on). It is this distinction that gives rise to the dichotomy between appearances (spatio-temporal objects given to our sensibility) and things in themselves (purely intellectual objects thought by the understanding alone), as well as the closely related distinction between *constitutive* a priori principles and merely *regulative* a priori principles.

Constitutive principles result from the application of purely intellectual representations to our spatio-temporal sensibility and yield necessary conditions for all objects of experience—conditions which therefore are necessarily realized in experience. The pure concepts or categories of substance, causality, and community, for example, are necessarily realized in our experience by a system of causally interacting conserved entities distributed in space and time—a system for which massive bodies interacting in accordance with Newtonian universal gravitation acting immediately at a distance consistently provided Kant with his primary model. Indeed, one of the main points of the *Metaphysical Foundations* is to explain how the general constitutive principles of experience of the *Critique of Pure Reason* are further specified or articulated to provide an a priori "metaphysical" foundation for precisely this Newtonian model.[11]

Yet Kant's general constitutive grounding of experience, even as extended into "the special metaphysical of corporeal nature" in the *Metaphysical Foundations*, leaves much of the natural world still unaccounted for. The *Metaphysical Foundations* provides a priori insight into only the most general properties and powers of all matter in general and as such (properties such as mass, gravity, impenetrability, and elasticity) and leaves even the property of cohesion for a physical and empirical rather than an a priori and metaphysical treatment.[12] Accordingly, Kant here assigns the problem of further specifying the general concept of matter as such into particular species and subspecies—the problem of what he calls "the specific variety of matter"—for the further development of empirical physics and chemistry; and so it is by no means surprising that Kant, as we have said, officially denies scientific status to chemistry and asserts, in the preface to the *Metaphysical Foundations,* that, at least at present, "chemistry can be nothing more than a systematic art or experimental doctrine, but never a proper science" (IV, 471).

How, then, are the more empirical sciences of nature to proceed? It is here that Kant invokes his famous doctrine of the regulative use of reason. The faculty of reason, in contradistinction to the faculty of understanding, generates a priori intellectual representations that cannot be fully realized in our human spatio-temporal experience. These include the ideas of God, Freedom, and Immortality, for example, and also, more relevant to our present concerns, the idea of the systematic unity of all empirical concepts and principles under the a priori constitutive concepts and principles already generated by the understanding. This idea of systematic unity guides our process of inquiry in the more empirical and inductive sciences, without constitutively constraining it, as we successively ascend from lower-level empirical concepts and principles toward higher-level concepts and principles. The goal of this process is an ideal complete empirical science of nature in which all empirical concepts and principles are constitutively grounded in the pure categories and principles of the understanding, but this is necessarily an ideal we can only successively approximate but never actually attain. And the paradigmatic application of the regulative use of reason, in the period of both the *Metaphysical Foundations* and the first *Critique,* is precisely to contemporary chemistry. Kant sees this chemistry—primarily Stahlian phlogistic chemistry as supplemented by the new discoveries in pneumatics but not yet including Lavoisier—as a purely empirical or experimental art guided by the regulative use of reason toward an entirely

unspecified and indeterminate future state of affairs in which the experimental results in question are finally grounded in the fundamental forces of
matter in a way that we are not yet in a position to anticipate.[13]

Moreover, as is well known, Kant, in the *Critique of Judgment*, extends
the doctrine of the regulative use of reason to what he now calls the faculty
of reflective judgment, and he now applies this faculty, in particular, to the
case of biology. The problem here, in a nutshell, is that all matter in general
and as such—all matter as the object of our outer senses in space—is essentially lifeless. This, in fact, is how Kant interprets the law of inertia, which
law, in turn, is itself constitutively grounded by a further specification of the
a priori principle of causality articulated in the first *Critique*.[14] Biology, the
study of life, can therefore never be a science in the strict sense for Kant; it
can never be constitutively grounded in the fundamental forces of matter.
The best we can do, in this case, is to extend the doctrine of the regulative
use of reason via the teleological idea of purposiveness (*Zweckmäßigkeit*)—
an idea that already arises for reflective judgment in general as it guides our
inductive ascent from particular to universal toward the ideal infinitely distant goal of a complete systematic unity of nature. And this idea can now be
applied to particular objects of nature or "natural products" (i.e., living
organisms) in so far as they are conceived, by reflective judgment, as themselves purposively organized. But such a mode of conception is in no way
constitutive of these objects themselves; it is rather a merely regulative
device for guiding our empirical inquiry into living organisms as far as it
may proceed.

SCHELLING'S DYNAMICAL THEORY OF MATTER

Post-Kantian German idealism—as successively articulated by Fichte,
Schelling, and Hegel—is characterized by a rejection of Kant's fundamental
distinctions, a rejection, that is, of "Kantian dualism." A prominent example
of such dualism, of course, is the distinction between appearances and things
in themselves and, more relevant to our present concerns, the closely related
distinction between constitutive and regulative principles. In particular, it
appears that Kant's doctrine of the regulative use of reason inevitably leaves
us with a quite intolerable skepticism concerning most of the phenomena
of nature. For only very few of these phenomena, as we have seen, are actually constitutively grounded, and, for the rest, we have at best the otherwise
entirely indeterminate hope that they might be constitutively grounded
some day. (In the case of biology, as we have seen, even this much hope
seems to be in vain.) It would appear, then, that the vast majority of natural

phenomena are not objectively (constitutively) grounded at all, and our claims to have rational or objective knowledge of nature are accordingly cast into doubt.[15] For the post-Kantian idealists, therefore, the very enterprise of transcendental philosophy—the attempt to give an a priori or rational foundation for the totality of our knowledge and experience—must be radically reconceived.

This radical reconceptualization of transcendental philosophy proceeds in outline as follows. Since, to begin with, we are also rejecting Kant's central distinction between a passive or receptive faculty of pure sensibility and an active or intellectual faculty of pure understanding, we must reject Kant's version of the distinction between understanding and reason as well; for the understanding, according to Kant, is the intellectual faculty applied to sensibility, whereas reason, in contradistinction to the understanding, is the same intellectual faculty considered independently of sensibility.[16] Here it issues in the ideas of God, Freedom, and Immortality and also, as we have seen, in the regulative use of reason under the idea of systematic unity: an infinite progressive process that approximates, but never fully attains, an infinitely distant goal. Moreover, this process, for Kant, also has an essentially "dialectical" character, in that it inevitably gives rise to the antinomies of pure reason (the idea of the world as either finite or infinite, and so on), which can only be resolved by a higher synthesis based on a thoroughgoing appreciation of the essentially incompletable nature of the dialectical process as such. The faculty of reason, considered in itself, is thus both infinitary and dialectical; and it is therefore very natural, once we have abandoned, once and for all, the distinction between active intellect and passive sensibility, to view the a priori constitution of knowledge and experience as an infinitary dialectical progression of the same kind.[17]

But how is such an infinitary dialectical progress of reason, which, for Kant, is a reflection of how we must (regulatively) view the objects of nature rather than how they (constitutively) are, to be reconceived as an a priori constitution of nature itself? It is precisely here that Schelling makes his decisive contribution. For Schelling, transcendental philosophy, the story of how human reason successively approximates to a more and more adequate picture of nature, has a necessary counterpart or dual, as it were, in *Naturphilosophie,* the story of how nature itself successively unfolds or dialectically evolves from the "dead" or inert matter considered in statics and mechanics, to the essentially dynamical forms of interaction considered in chemistry, and finally to the living or organic matter considered in biology. Since nature, on this view, dialectically unfolds or successively evolves in a way that precisely mirrors the evolution or development of our rational conception

of nature (and, of course, vice versa), it follows that there is no possible skeptical gap between nature itself and our conception of it, or, in Kantian
terminology, between the constitutive domain of the understanding and
the merely regulative domain of reason. *All* the phenomena of nature—
including, in particular, both chemical and biological phenomena—are
rationally or objectively grounded in the same way.

The key to Schelling's conception is a dialectical extension and elaboration of Kant's original dynamical theory of matter. From Schelling's
point of view, Kant's dynamical theory of the fundamental forces of attraction and repulsion necessary to all matter in general and as such (which
embraces, therefore, even the "dead" or inert matter considered in statics and
mechanics) has already introduced an essentially dialectical element into
nature, in so far as the dynamical constitution of matter in general proceeds
from the positive reality of expansive force (repulsion), through the negative
reality of contractive force (attraction), to the limitation or balance of the
two in a state of equilibrium.[18] But we now know, as Kant himself did not,
that chemistry can be dynamically grounded by a dialectical continuation of
this progression—as we proceed, more specifically, from the magnetic,
through the electrical, to the chemical (or galvanic) forms of the basic or
original dynamical process grounded in the fundamental forces of attraction
and repulsion. And, once we have gone this far, it is then a very short step
(particularly in view of the recently discovered parallel interconnections
among electrical, galvanic, and biological phenomena) to view biology, too,
as a further dialectical continuation of the same dynamical process.[19] Biology, too, can be a science, for all rational science, as Kant did not see, is
grounded in a single dynamical evolutionary dialectical progression. The
whole of nature, in this sense, is at once both rational and alive;[20] and this
means, in particular, that there actually is life—objectively, not merely regulatively—in even the very simplest forms of organized matter.[21]

At the center of this entrancing vision stands the new electrochemistry. For it is precisely here that we can unite the concept of matter in general as conceived in Kant's original dynamical theory (the "dead" matter
considered in statics and mechanics) with matter as conceived by *Naturphilosophie*—as an inexhaustible source of rational life. It is in precisely this
context that we can view chemistry, in Schelling's words, as a dialectical
"middle term" between mechanism, on the one side, and biological (ultimately rational) living purposiveness, on the other.[22] Indeed, even the inert
matter considered in statics and mechanics is already at least potentially
alive, since Kant's dynamical theory has shown that the fundamental forces
of attraction and repulsion are necessary to all matter in general and as such,

and we have just seen that the original or primary dynamical process governed by these forces must necessarily evolve or develop into first chemical and then biological forms of external nature. In the end, it is precisely by rejecting the fundamental Kantian contention that all matter in general and as such—all matter as the object of our outer senses—is essentially lifeless that Schelling, from his point of view, finally overcomes any possibility of a skeptical gap between our rational conception of nature and nature itself.[23]

It is also clear, as Wilson explains, that this grand *naturphilosophische* vision constituted an essential part of Oersted's intellectual development. Oersted began his career as a philosopher of nature with two dissertations on Kant's *Metaphysical Foundations* in 1799; yet even here his aim, unlike Kant's, was to secure an a priori "metaphysical" foundation for matter-theory and chemistry, including the fundamental property of cohesion.[24] At this time, however, Oersted was not yet acquainted either with the new developments in electrochemistry or with Schelling's radical transformation of Kant's original dynamical theory of matter intended to account for them.[25] When he did become acquainted with these developments soon thereafter, Oersted enthusiastically embraced Schelling's key idea of chemistry as an extension of general dynamics;[26] and he came to share Schelling's view, in particular, of how magnetic, electrical, and chemical (or galvanic) phenomena constitute what Schelling calls the "second potency" of the original dynamical process.[27] He also came to share, more generally, the broader philosophical vision of a dialectical or evolutionary development of nature as a whole in the realization or progressive unfolding of a single infinite rational life:

> A clearer perspective soon adds to this that there is nothing dead and rigid in nature, but that every thing exists only as a result of a development [*Entwicklung*], that this development proceeds according to laws, and that, therefore, the essence of every thing is based on the totality of the laws or on the unity of the laws, i.e., the higher law by which it has been brought forth. Every thing, however, must again be regarded as an active organ of a more comprehensive whole, which again belongs to a higher whole so that only the great All sets the limit of this progression. And thus the universe itself would be regarded as the totality of the developments, and its law would be the unity of all other laws. However, what finally gives the study of nature its highest meaning is the clear understanding that natural laws are identical to the laws of reason, so they are in their application identical to thoughts; the totality of the laws of a thing, regarded as its essence, is therefore an idea of nature, and the law or the essence of the universe is the totality [*Inbegriff*] of all ideas, identical with absolute

> reason. And so we see all of nature as the *appearance* [*Erscheinung*] of one
> infinite force and one infinite reason united, as the *revelation of God.*[28]

As the surrounding context makes clear, a primary motivation for this
broader vision was precisely the new discoveries in electrochemistry, as
interpreted, along with Schelling, as expressing the fundamental unity of
magnetic, electrical, and galvanic forces.[29] It is with perfect justice, therefore,
that Oersted himself describes his discovery of electromagnetism, in the
now well-known passage with which Stauffer begins his 1957 paper, by
asserting that he "was not so much led to this [i.e., the view that 'magneti-
cal effects are produced by the same powers as the electrical'], by the reasons commonly alleged for this opinion, as by the philosophical principle,
that all phenomena are produced by the same original power" (Jelved et al.
1998, 546; Oersted 1920, 2:356).[30]

KANT'S *OPUS POSTUMUM*

We can deepen our appreciation of the quite specific intellectual situation
faced by Schelling, Ritter, and Oersted by taking a brief look at how Kant
himself, very late in his career, attempted to extend his dynamical theory of
matter into chemistry. This attempt occurs in unpublished materials from
the years 1796 to 1803 collected into what is now known as the *Opus postumum,* and it involves, in Kant's own terminology, drafts and sketches of a
projected new work to be entitled *Transition from the Metaphysical Foundations of Natural Science to Physics.* Here we can see Kant responding to two
complementary pressures. On the one hand, there is a clear systematic need
to have something more to say about how the general concept of matter as
such is further specified, a priori, into various species and subspecies—the
problem, that is, of what Kant calls "the specific variety of matter." In the
Metaphysical Foundations, as we have seen, Kant had very little to say about
this, although he does present an outline of his conception of the foundations of contemporary chemistry (including such topics as cohesion, fluid-
ity, elasticity, and dissolution).[31] On the other hand, however, Kant, by 1795,
has completed the transition from the phlogistic chemistry of Stahl to the
new chemistry of Lavoisier, and he has now become convinced, contrary
to the doctrine of the *Metaphysical Foundations,* that chemistry has finally
entered the secure path of a science after all. There is a clear need, therefore,
to explain the a priori foundations of this new science; and the *Transition*
project, accordingly, is best conceived as an attempt further to develop transcendental philosophy so as to solve both of these problems at the same

time—to provide an a priori explanation of the possibility of the specific variety of matter that is simultaneously an a priori grounding of Lavoisier's new chemistry.[32]

The *Transition* project begins, naturally enough, from the idea that the two fundamental forces of attraction and repulsion constitutive of all matter in general and as such must be somehow further specified so as to account, at least in outline, for the more specific forces and powers exerted by specific types of matter—matter in the solid, liquid, or gaseous states, for example, or matter in the form of particular kinds of chemical substances. The preface to the *Metaphysical Foundations* had asserted that chemistry could not become a science properly speaking until forces of attraction and repulsion appropriate to specifically chemical interactions could actually be discovered (forces, for example, of the kind Newton speculates about in the *Opticks*); and, since such forces had not yet been found, chemistry was not yet a science.[33] Now, however, chemistry has become a science, but it has not done so, in the work of Lavoisier, by finally uncovering the specifically chemical forces. What Lavoisier has achieved, rather, is a fundamental conceptual reorganization of the subject—a new system of chemical classification based on the newly discovered central role of oxygen. The task of the *Transition* project, accordingly, is to show how this system of chemical classification itself receives an a priori grounding in transcendental philosophy.

This task, as one might imagine, is not an easy one, and Kant struggles mightily with it throughout a large and bewildering series of outlines, drafts, and sketches of his projected new work. Finally, in 1799, in drafts that came closer than any other part of the *Opus postumum* to actual publication, he presents the so-called Aether Deduction: an a priori proof that there is a universally distributed aether or caloric fluid, constituted by a perpetual interaction between the two fundamental forces of attraction and repulsion, filling all of space. This universally distributed aetherial medium is supposed to provide an a priori grounding for the central theoretical construct of Lavoisier's new chemistry—caloric or the imponderable matter of heat—and at the same time to serve, in a way that had long been familiar in eighteenth-century matter theory, as the medium or vehicle for light, electricity, and magnetism as well.[34] In this way, the totality of forces or powers of nature—including, above all, the specifically chemical forces—are, at least in principle, systematically unified in a single a priori representation, and in this sense the problem of the specific variety of matter has finally been solved.

The a priori representation in question—the representation of a universally distributed caloric fluid or aetherial medium filling all of space—is, by the standards of Kant's critical philosophy, an extremely peculiar one. In particular, it combines aspects of discursive or conceptual representation (what Kant calls "distributive" or "analytic universality") with the apparently entirely opposed characteristics of sensible or intuitive representation (what Kant calls "collective" or "synthetic universality"). As a continuum of forces providing a basis for further specification of the concept of matter in general, it is a discursive or conceptual representation; as a space-filling continuum, providing what Kant calls a perceptual "realization" of the pure intuition of space, it is a sensible or intuitive representation. Moreover, and by the same token, as an a priori principle for the further specification of the concept of matter in general it is a constitutive representation; as the ultimate ground for the systematic unity of all of the forces of matter it is a regulative representation. It is in this way, in fact, that the "top-down" constitutive procedure of the *Metaphysical Foundations* and the first *Critique* has a necessary intersection, as it were, with the "bottom-up" regulative procedure of the faculty of reflective judgement; and it is in precisely this way, accordingly, that the skeptical problems arising from the doctrine of the regulative use of reason that so vexed the post-Kantian idealists are finally resolved for Kant himself.[35]

We cannot delve more deeply into the *Opus postumum* here. But we have seen enough, I hope, to appreciate that Kant himself, at the very end of his career, was on the verge (both philosophically and scientifically) of the radically new conceptual situation faced by Schelling, Ritter, and Oersted. He had already considered the problem of extending the dynamical theory of matter into chemistry, and, at the same time, he had already subjected the fundamental distinction between constitutive principles and regulative principles to a radical reconceptualization. Nevertheless, the decisive step into *Naturphilosophie* is one that Kant did not and could not take. For, in the first place, the critical new developments in magnetism, electricity, and chemistry that provided the fertile empirical soil on which alone *Naturphilosophie* could take root played no role at all in Kant's thought. Kant, to the best of my knowledge, never engaged with even the electrostatic and magnetostatic work of Coulomb, to say nothing of the electrochemical researches arising from the Voltaic pile. The central idea of *Naturphilosophie* in its application to chemistry—that chemical forces are at bottom electrical in nature—never occurred to him; and, as a result, the idea that one could extend the dynamical theory of matter by conceiving magnetic, electrical, and galvanic forces as a further dialectical development of the original dynamical process governing the fundamental forces of attraction and repulsion was entirely foreign to Kant's own final attempt to solve the problem of the specific variety of matter.[36]

Moreover, and in the second place, although, as we have seen, there may be room, with considerable stretching and straining, to find a place in Kant's critical system for a representation that combines both constitutive and regulative aspects, there is no room at all for the grand *naturphilosophische* vision of nature as a whole as the realization or evolutionary development of a single divine infinite rational life. For this idea, of course, entails the total *Aufhebung* of all of Kant's most fundamental distinctions, along with the critical philosophy itself.[37] It is in this precise sense that Kant's own version of the dynamical theory of matter represents the prolegomenon—but only the prolegomenon—to the radically transformed version of the theory adopted by Schelling, Ritter, and Oersted. Seeing how close Kant actually came to their particular intellectual situation—but also how far he remained—helps us develop a better understanding of the extraordinary, and extraordinarily timely, confluence of scientific problems and philosophical ideas that framed their enterprise and thereby made it fruitful.

SCHELLING AND OERSTED

Our discussion of the relationship between Kant's dynamical theory of matter and Schelling's has tended to support, with a bit more specificity and detail, the original evaluations of Oersted's philosophical inspiration contributed by Stauffer and Pearce Williams, according to which Schelling's *Naturphilosophie,* in particular, provided essential philosophical motivation for Oersted's scientific discoveries. But these evaluations, not surprisingly, have since been challenged—first in a relatively measured and balanced way by Barry Gower (1973) and then in a very pointed and more contentious form by Timothy Shanahan (1989). The basis for Gower's skepticism is the conviction that Schelling's metaphysics is simply too abstruse and obscure to have exerted a substantial influence on the actual practice of science;[38] whereas the focus of Shanahan's attack, by contrast, is the idea that Schelling's speculative methodology is too rationalistic and aprioristic vis-à-vis empirical physics. For Shanahan, accordingly, the more sober and empirical approach of Kant's original dynamical theory, quite independently of its further elaboration by Schelling, provided by itself the decisive philosophical inspiration.[39] A consideration of these challenges, by way of conclusion, will, I think, prove useful, not so much for the sake of deciding whether Schelling's philosophy was or was not scientifically important, but rather as a means of shedding further light on the question of how philosophical speculation fruitfully interacts with scientific discovery more generally.

Let us begin with the more focused and pointed challenge raised by Shanahan, which raises empirically oriented doubts concerning Schelling's aprioristic methodology. In this connection, especially, it is appropriate to consider Oersted's own methodological remarks; and here it is clear, throughout his intellectual career, that Oersted's own view was that there is a necessary correspondence or parallelism between a purely aprioristic speculative deduction of natural phenomena on the one side and an empirical or experimental inquiry into these same phenomena on the other. This view is expressed, in the first instance, in Oersted's 1799 dissertations on philosophy of nature crafted under the explicit inspiration of Kant's *Metaphysical Foundations*—where, in particular, Oersted already argues for an even more aprioristic deduction on the speculative side than does Kant himself.[40] Moreover, the importance of such a fully a priori speculative deduction— now modeled explicitly on Schelling's *Naturphilosophie*—became ever more important to Oersted as his intellectual development progressed. In his paper "What Is Chemistry?" (1805), for example, Oersted contrasts "speculative" and "empirical" approaches to this science, where "[t]he latter merely comes up with scattered objects which invite reflection, and from this an arrangement into coherent elements emerges; the former seeks the first principle of everything, sees what constructions must result from it, and prefers to offer the most fundamental construction of science as its definition [of the science]." Thus, "[i]t is not at all necessary for chemistry to be only experimental"; on the contrary, "a universal construction is necessary to complete science, and it stands to reason that such a construction cannot be given through experience but can only be expected from speculation" (Jelved et al. 1998, 198–199; Oersted 1920, 3:114–115). As the context makes perfectly clear, the universal speculative construction envisioned here is precisely that originally attempted by Schelling.[41]

Perhaps Oersted's most striking and explicit statement of this dual methodology occurs in his 1811 "First Introduction to the General Doctrine of Nature" (*Naturlaere*):

> In our knowledge of nature we distinguish between something which comes more immediately from reason and something else which rather has its origin in the senses. The two are in the most intimate connection with each other. It is the essence of man to present reason in an organic body, not merely in one particular form, but in its self-contemplating [*selvbeskuende*] totality. His sensuous nature, in the most exact sense, can only be regarded as the embodiment of this reason. Therefore, the external sense organs already receive impressions in a manner which is in the most perfect harmony with it, and an unconscious reason in the internal

sense impresses its own stamp even more markedly on these various abilities. Imperceptibly, they thus approach the conscious reason which organizes and combines everything into even higher units which, step by
step, are finally transformed into the remarkable internal harmony of
the independent reason. Thus, *the science of experience* (empirical science)
comes into existence. Reason, on its side, is similar to the internal foundation and essence of nature. In a way, it contains the seeds of the entire
world and must develop them through its necessary self-contemplation
[*Selvbeskuelse*]. Consequently, it starts from the highest to which our spirit
can ascend, from the essence of beings, the origin of everything. In itself,
as a sign of this, it seeks out the various primary directions and through
them the origin of the essential fundamental forms in the eternal unity.
In its own laws it sees those of nature, in the variety of its own forms that
of the world, and thus it develops and creates from itself the great All. In
this way arises *speculative natural science,* which is also called *philosophy of
nature* [*Naturphilosophien*].

After remarking that "[o]n the empirical path, we are stopped by the enormous profusion of objects which our senses offer, in which, however, there
is no completeness," Oersted then concludes:

> [S]peculative natural science seems to lead us more immediately to our
> goal, but here we would do well to bear in mind that the reason which
> reveals itself in nature is infinite while ours, which must discover it there,
> is limited, trapped in finitudes. In innumerable sparks, reason spreads
> through mankind. Although a reflection of the whole in every individual,
> it has in each its distinctive direction which prevents it from spreading its
> light equally clearly and fully in all directions. Only recently shaped in its
> present form, speculative natural science will only approach significant
> perfection through the combined efforts of many thinkers.

And the context again makes it clear that the speculative natural science
which has been "only recently shaped in its present form" is Schelling's.[42]

What is most striking about this particular passage is that Oersted's
methodological ideal of a dual parallel development of both speculative and
empirical approaches is grounded, in turn, in a grand metaphysical vision of
rational mind and external nature as two complementary aspects of a single
total organic development. It is precisely because sensible nature, on the one
side, is a realization or externalization of an infinite rational spirit, whereas
rational mentality, on the other side, is the evolutionary culmination of an
infinite dialectical process by which sensible nature itself continually unfolds, that there is, for Oersted, a necessary harmony between speculative
and experimental inquiry. Since our external senses must be regarded as the

"embodiment" of our reason, and because sensible nature itself is the dialectical externalization of an infinite reason (to which our finite reason necessarily approximates), empirical natural science and speculative philosophy of nature must ultimately coincide. Oersted's dual methodology is itself grounded, in this sense, in the primary thesis of Schelling's metaphysics—in the claim, in Schelling's terms, that transcendental philosophy and *Naturphilosophie,* spirit within us and nature outside us, must ultimately coincide.[43]

But this now leads us back to Gower's question (note 38): In what precise sense is this grand—and, from our present point of view, rather wild—metaphysical vision actually implicated in Oersted's concrete scientific work? We have already made the essential point in our earlier discussion of Oersted's great discovery. For, as we have said (see pp. 58–60 above), at the center of this metaphysical vision stood the new electrochemistry. It was this new chemistry, interpreted, along with Schelling, as a further articulation and elaboration of general dynamics, that provided the crucial dialectical middle term between matter as conceived in Kant's original dynamical theory (as a balance or equilibrium of the two fundamental forces) and matter as conceived in *Naturphilosophie*—as an inexhaustible source of rational life. And it was Schelling's particular way of conceiving the new chemistry as such an extension of general dynamics—as a triadic dialectical "second potency" of the original dynamical process embracing magnetic, electrical, and chemical or galvanic phenomena—that provided Oersted with his model for the specific *kind* of unity of the fundamental forces of nature that guided all of his experimental investigations, including, above all, his discovery of electromagnetism.[44] Indeed, in the absence of Schelling's triadic model, there simply was very little reason, either theoretically or experimentally, to expect the relationship Oersted actually discovered in the first place; and this, no doubt, is why Oersted's discovery was initially received by the scientific community as such a surprise.[45]

It is precisely here, therefore, that Schelling's bold metaphysical speculations were fruitfully intertwined with the very latest results of empirical scientific research. Indeed, Schelling's speculative physics was intimately engaged with the new electrochemistry at its cutting edge, since, as we have seen, Schelling had become convinced of the fundamentally electrical nature of chemical forces already in the first (1797) edition of his *Ideas for a Philosophy of Nature,* and thus prior to the invention of the Voltaic pile and the consequent discovery of the decomposition of water by an electric current. When Schelling learned of the electrolytic decomposition of water through the work of Ritter, he was then perfectly positioned to seize upon this result and triumphantly announce a new dynamical synthesis of magnetic, electrical, and chemical or galvanic forces in the second (1803) edition.[46] So it is no wonder, in particular, that working experimental investigators at the forefront of the new electrochemistry—men like Ritter and Oersted himself—were also able to take positive philosophical inspiration from Schelling's metaphysical speculations.

As we have seen, the immediate context for Schelling's particular philosophical intervention into the ongoing practice of empirical scientific research were problems in Kant's original dynamical theory of matter. Kant had crafted his theory in intimate connection with the most fundamental principles of Newtonian physics (see notes 7, 11, 14, and 18 above), and Kant was increasingly conscious, at the same time, of the need to extend his constitutive grounding of Newtonian physics so as to embrace the new discoveries in matter theory and chemistry made at the end of the eighteenth century (see note 32 and pp. 60–61 above). In just this connection, however, Kant faced deep philosophical problems centering around his sharp distinction between constitutive principles and merely regulative principles, eventually leading to a final attempt, in the *Opus postumum,* to articulate a radically new type of a priori construction—a space-filling aetherial medium or caloric fluid—combining both regulative and constitutive aspects in a single unitary representation (see note 35 and pp. 61–62 above). Here, as we have seen, he was able, with some difficulty, to provide a kind of constitutive grounding for Lavoisier's recently articulated chemical system, but even this final effort fell considerably short of the new electrochemistry that became centrally important at the turn of the century. The timeliness, and consequent fruitfulness, of Schelling's particular intervention therefore consists in the way in which he responded to the deep philosophical problems created by Kant's sharp distinction between constitutive principles and regulative principles, on the one hand, while simultaneously engaging with the newest developments at the forefront of electrochemistry, on the other. Kant's own attempts further to develop his system so as to resolve both the philosophical problems in question and to accommodate recent scientific results had led him, as we have seen, right up to the verge of the radically new conceptual situation faced by Schelling, Ritter, and Oersted. Nevertheless, the decisive step of conceiving magnetic, electrical, and chemical or galvanic effects in terms of a triadic continuation of a dialectical process beginning with the fundamental forces of attraction and repulsion in general is one that Kant did not and could not take—and it is was just this crucial step that was then left to Schelling alone.[47]

It is clear, then, that the timeliness, and consequent fruitfulness, of a philosophical intervention into ongoing scientific research does not depend on the validity, by current standards, of the philosophical ideas in question—nor does it require that the theoretical schemata suggested by these philosophical ideas should themselves be correct according to subsequent scientific theorizing. In the present case, in particular, neither Schelling's bold metaphysical vision nor the triadic developmental dynamical model of magnetic, electrical, and chemical or galvanic phenomena managed to survive into the second half of the nineteenth century, let alone into the twentieth.[48] Yet Schelling's speculative physics did in fact fruitfully guide the experimental and theoretical work of the principal founders of electrochemistry and electromagnetism at the beginning of the nineteenth century.[49] And it was able to do this, in spite of the shortcoming we might perceive in it today, precisely because the new empirical situation at the turn of the century demanded the exploration of non-Newtonian physical ideas, while, at the same time, the Kantian philosophical system, at the very end of the eighteenth century, had pushed such Newtonian ideas to their outermost philosophical limits. A new philosophy of nature, like Schelling's, which, as we have said, intelligently responds simultaneously to both the deep tensions emerging in Kant's philosophical system and the new empirical results, was therefore, in this specific historical context, precisely what was then needed.[50] Oersted's spectacular discovery of electromagnetism provides us with perhaps our single most important confirmation.

NOTES

Copyright Michael Friedman. All rights reserved. Preliminary ideas for this paper were presented at the Dibner conference in the fall of 2000, and an earlier version was presented at a conference on Oersted at Harvard University in the spring of 2002. The paper will also appear in the proceedings of this conference. I am indebted to discussions at the Harvard conference, especially with Kenneth Caneva. I am also indebted to comments on earlier drafts by Frederick Beiser, Jordi Cat, and Allen Wood. Finally, my understanding of post-Kantian German idealism has been very substantially influenced by Frederick Beiser and Paul Franks—the former through his work on Schelling and *Naturphilosophie* (see notes 17 and 46 below), the latter through his work on Salomon Maimon's skepticism (see note 15 below). Translations from Kant and Schelling are my own. In quoting Oersted's works I follow the translations in Jelved et al. 1998, occasionally emended as indicated.

1. See Williams 1965, 62: "The reduction of all physical phenomena to attractive and repulsive forces was seductively simple. The different *kinds* of attraction and repulsion—electrical, magnetic, etc.—were the results of different conditions under which the two

basic forces manifested themselves. Behind these differences lay the essential *unity* of all forces."

2. See Williams 1965, 137: "The possibility that electricity and magnetism were but different modes of action of the underlying and fundamental forces of attraction and repulsion was a primary tenet of the Kantian dynamic philosophy. One of Kant's most ardent disciples, the young Dane Hans Christian Oersted, devoted twenty years of his life to making this possibility manifest."

3. See Stauffer 1957, 34: "To Oersted, his belief in the existence of a physical relation between electricity and magnetism seemed a logical corollary to this general principle of unity [of all the powers of nature]. His metaphysical faith in this relationship afforded a motive for his persistent experimentation in the field of electricity and magnetism."

4. See *Opticks*, Query 31 (Newton 1717/1730, 397): "And thus Nature will be very comfortable to her self and very simple, performing all the great Motions of the heavenly Bodies by the Attraction of Gravity which intercedes those Bodies, and almost all the small ones of their Particles by some other attractive and repelling Powers which intercede the Particles."

5. See, for example, the immediate continuation of the passage cited in note 1 above: "From this it followed logically that all forces of nature were convertible into one another; one need only find the proper conditions for accomplishing the conversion." The link between energy conservation and *Naturphilosophie* is also considered in a well-known paper by Thomas Kuhn (1959), which, with regard to electromagnetic forces in particular, follows Stauffer's account of Oersted.

6. It is for this reason that Helmholtz is generally taken as an early representative of nineteenth-century *neo*-Kantianism—a movement aiming to turn away from the "metaphysical" speculations of post-Kantian idealism (as represented, in particular, by Schelling's *Naturphilosophie*) to the original "epistemological" orientation of Kant himself. For discussion, see the introduction to Cassirer 1950. Helmholtz's celebrated address, "Über das Sehen des Menschen," delivered at the dedication of a monument to Kant in Königsberg in 1855, condemns the then-current rift between philosophy and natural science due, in Helmholtz's opinion, to the entirely speculative *Naturphilosophie* of Schelling and Hegel, and announces a new project of cooperation between the two disciplines in the spirit of Kant. This address then became a model for the neo-Kantian tradition.

7. Thus one central ambition of the dynamics of the *Metaphysical Foundations* is to defend the Newtonian attraction as a true and immediate action at a distance against both "all sophistries of a metaphysics that misunderstands itself" (viz., Leibnizean metaphysics) and Newton's own doubts, which, according to Kant, "set [Newton] at variance with himself" (IV, 514; the first quotation is at 523). Pearce Williams therefore commits a serious error when he portrays Kant's transcendental idealism about space as fundamentally inimical to Newtonian action at a distance: "Thus, *The Critique of Pure Reason,* if taken seriously as a fundamental critique of metaphysics, would mean the rejection out of hand of the very foundations of Newtonian physics. Particles could not act upon one another across empty space because empty space could not be known to the mind"

(1966, 34–35). What Williams misses here is that, whereas Kant indeed denies that empty space can be an object of experience, he still insists, in the *Metaphysical Foundations,* that "[t]he *attraction essential to all matter* is an immediate action of matter on other matter through empty space" (Dynamics, Proposition 7; IV, 512)—so that the action of gravitation, in particular, is entirely independent of any intervening matter that may fill the space between the attracting bodies (IV, 516): "this attraction is a penetrating force and acts *immediately* at a distance through that space, as an empty space, regardless of any matter lying in between." Williams (1973, 18–19) acknowledges that he has deliberately "misread" Kant—"as the *Naturphilosophen* misread him." However, while it is perfectly legitimate to depict Kant as the *Naturphilosophen* read him, it is even more important, I think, to become as clear as possible on the various ways in which they radically transformed him—and on the way in which, in particular, he is in fact much more Newtonian than they.

8. Another particularly helpful discussion is Gower 1973, although Gower himself paints a somewhat more skeptical picture of both *Naturphilosophie* in general and its relevance to Oersted's discovery in particular.

9. This same work, of course, was the basis for Davy's and Faraday's parallel electrochemical researches in England. As Pearce Williams (among others) has made abundantly clear, these researches were also intimately involved with German *Naturphilosophie* as mediated by the influence of Coleridge. But this specifically English context, together with Coleridge's particular version of *Naturphilosophie,* lies well beyond the scope of our present discussion.

10. Wilson also uses this quite specific intellectual situation, in the context of a detailed discussion of the development of Oersted's electrochemical experimental program, to provide a fundamental clarification of Oersted's theoretical ideas on the nature of electricity and magnetism—including the famous *conflictus*—a topic that I will have to leave aside here.

11. This idea is not particularly controversial today. For discussion, see Friedman 1992—although particular details of my interpretation are, of course, controversial. Compare note 7 above.

12. See IV, 518: "The *action* of the universal attraction immediately exerted by each matter on all matters, and at all distances, is called *gravitation*; the tendency to move in the direction of greater gravitation is *weight*. The action of the general repulsive force of the parts of every given matter is called its *original elasticity*. Hence this property and weight constitute the sole universal characteristics of matter, which are comprehensible a priori, the former internally, and the latter in external relations. For the possibility of matter itself rests on these two properties. *Cohesion,* if this is explicated as the mutual attraction of matter limited solely to the condition of contact, does not belong to the possibility of matter in general, and cannot therefore be cognized a priori as bound up with this. This property would therefore not be metaphysical but rather physical, and so would not belong to our present considerations."

13. For further discussion of Kant's conception of chemistry in this period, see Friedman 1992, chap. 5.

14. The statement of the law of inertia and accompanying proof is Proposition 3 of the mechanics chapter of the *Metaphysical Foundations*. The following remark explains the connection with lifelessness—viz., the nonexistence of any *internal* principle of change. See, in particular, the conclusion of this remark: "The possibility of a proper natural science rests entirely and completely on the law of inertia (along with that of the persistence of substance). The opposite of this, and thus also the death of all natural philosophy, would be *hylozoism*. From this very same concept of inertia, as mere *lifelessness,* it follows at once that it does not mean a *positive striving* to conserve its state. Only living beings are called inert in this latter sense, because they have a representation of another state, which they abhor, and against which they exert their power" (IV, 544). This rejection of "hylozoism" is clearly directed at Leibnizean natural philosophy.

15. My formulation of this skeptical problem is indebted to Franks 2003, although Franks himself does not emphasize, as I do, the distinction between constitutive and regulative principles—he instead formulates what I take to be essentially the same problem by means of a distinction between scientific judgments and everyday or "ordinary" judgments.

16. It does not follow, as Allen Wood has emphasized to me, that the post-Kantian idealists have no room for reinterpreting the distinction between understanding and reason in other terms—indeed, new versions of this distinction become central to their thought. The main point here, however, is that they have absolutely no room for Kant's particular version, which entails, in particular, a sharp distinction between constitutive principles (due to the understanding) and merely regulative principles (due to reason).

17. For Kant, by contrast, the constitutive grounding of experience by the understanding is necessarily finite (since there are only twelve categories), and only the contrasting regulative use of reason is infinite. Kant makes precisely this point, in fact, in the preface to the *Metaphysical Foundations*: "[J]ust as in the metaphysics of nature in general, here also the completeness of the metaphysics of corporeal nature can confidently be expected. The reason is that in metaphysics the object is only considered in accordance with the general laws of thought, whereas in other sciences it must be represented in accordance with data of intuition (pure as well as empirical), where the former, because here the object has to be compared always with *all* the necessary laws of thought, must yield a determinate number of cognitions that may be completely exhausted, but the latter, because they offer an infinite manifold of intuitions (pure or empirical), and thus an infinite manifold of objects of thought, never attain absolute completeness, but can always be extended to infinity, as in pure mathematics and empirical doctrine of nature" (IV, 473). For helpful discussions of post-Kantian idealism along these lines, see, e.g., Royce 1919 and, for a more advanced treatment, Cassirer 1920. For an important recent discussion, placing particular emphasis on the role of *Naturphilosophie* and the resulting "organic view of nature," see Beiser 2002.

18. This already represents a radical transformation and reinterpretation of Kant's original dynamical theory. To be sure, Kant's construction of matter in general out of the two fundamental forces of attraction and repulsion proceeds via the three categories of quality: reality, negation, and limitation. For Kant himself, however, in the necessarily finite constitutive grounding of experience provided by his metaphysics of nature (see note 17

above), these categories of quality, in turn, are necessarily subordinate to the categories of relation articulated in what he calls the analogies of experience: the categories of substance, causality, and community. In the *Metaphysical Foundations* these latter are realized by the laws of conservation of matter, inertia, and equality of action and reaction—which laws, as Kant understands them, entail the lifelessness of matter in general (see note 14). This overriding emphasis on what he calls the three "laws of mechanics" is central to Kant's much more Newtonian version of the dynamical theory (see notes 7 and 11 above).

19. For Schelling, this dialectical continuation takes a quite precise and specific form: Corresponding to the magnetic, electrical, and chemical (or galvanic) forms, we then have, as the "third potency" of the original dynamical process, the biological powers of reproduction, (nervous) irritability, and sensibility (see Schelling 1803, supplement to book I, chap. 6).

20. See Schelling's *Ideas for a Philosophy of Nature* (1803, 54; 1988, 40): "Finally, if we comprehend nature as a single whole, then *mechanism,* i.e., a past-directed series of causes and effects, and *purposiveness* [*Zweckmäßigkeit*], i.e., independence of mechanism, simultaneity of causes and effects, stand opposed to one another. In so far as we now unite these two extremes, an idea of a purposiveness of *the whole* arises in us—nature becomes a circle that returns into itself, a self-enclosed system."

21. See Schelling 1803, 46–47; 1988, 35: "This philosophy must admit, therefore, that there is a graduated development [*Stufenfolge*] of life in nature. Even in mere organized matter there is *life,* but only life of a limited kind. This idea is so old, and has been preserved until now in the most varied forms up to the present day—(already in the most ancient times the whole world was [regarded as] penetrated by a living principle, called the world-soul, and Leibniz's later period gave every plant its soul)—that one can well surmise in advance that some ground for this natural belief must lie in the human spirit itself. And it is in fact so. The entire mystery surrounding the problem of the origin of organized bodies rests on the circumstance that in these things necessity and contingency are united in the most intimate way. *Necessity,* because their *existence* is already purposive, not only (as in the case of the work of art) their form; *contingency,* because this purposiveness is nonetheless only actual for an intuiting and reflecting being."

22. See Schelling 1803, 187; 1988, 149: "Therefore, already in the chemical properties of matter there actually lie the first, although still completely undeveloped seeds of a future system of nature, which can unfold into the most varied forms and structures, up to the point where creative nature appears to return back into itself. Thus, at the same time, further investigations are marked out, up to the point where the necessary and the contingent, the mechanical and the free, separate from one another. Chemical phenomena constitute the middle term between the two. It is this far, then, that the principles of attraction and repulsion actually lead, as soon as one considers them as principles of a *universal system of nature.*"

23. Schelling is here self-consciously returning to precisely the Leibnizean "hylozoism" Kant explicitly rejects (compare notes 14, 21, and 22 above). This essentially biological or organic conception of nature then implies the overcoming of all skepticism in the sense that the closing of the circle mentioned in note 20 above (embracing both mech-

anism and teleology) means that transcendental philosophy and *Naturphilosophie*—spirit and nature—are ultimately identical, in so far as nature itself gives rise to both life in general and conscious or rational life in particular. See Schelling 1803, 56; 1988, 42: "Nature should be the visible spirit, spirit the invisible nature. It is *here*, therefore, in the absolute identity of spirit *within* us and nature *outside* us, that the problem of how a nature outside us is possible must be solved."

24. Compare note 12 above. Accordingly, Oersted here envisions a far-reaching alteration of the very structure of the *Metaphysical Foundations,* so as to include a theory of motion on the one side and a (separate) theory of matter on the other. This already represents a fundamental (and quite self-conscious) divergence from Kant's own procedure, according to which "natural science [throughout] is either a pure or applied *doctrine of motion*" (IV, 476).

25. Oersted does refer to the first edition of Schelling's *Ideas,* published in 1797. But this, of course, was prior to the invention of the Voltaic pile and the consequent discovery of the electrolytic decomposition of water by Nicholson and Carlisle in 1800. Schelling himself had already become convinced of a close link between electrical forces and chemical forces (especially combustion) on the basis of electrochemical phenomena involving static electricity (sparking)—including the researches of Priestley and Cavendish, as well as early work on the decomposition of water by sparking of the Dutch chemists Van Troostwijk and Deiman (see Schelling 1803, book I, chap. 4).

26. It appears, as already suggested, that Oersted was initiated into much of this new material by Ritter; see, e.g., Christensen 1995. Ritter himself performed very early experiments with the Voltaic pile and in particular discovered the electrolytic decomposition of water independently of Nicholson and Carlisle. (Ritter also originated the electrochemical series linking the affinities of metals for oxygen with the metals employed in the Voltaic pile.) Schelling reports on this work of Ritter's in the second (1803) edition of the *Ideas* (supplement to book I, chap. 3), and it was this edition in particular that then exerted a decisive influence on the further development of Oersted's philosophy of nature.

27. Again, Schelling is much clearer about this crucial piece of the story in the second (1803) edition. On this view, magnetic forces present the basic or elementary form of the cohesion of bodies, expressed in a one-dimensional line (between two magnetic poles); electrical forces act at the two-dimensional surfaces of bodies determined by such cohesion (in a distribution of charge manifesting static electricity); and chemical or galvanic forces act within the three-dimensional volumes of such bodies (as their parts move relative to one another under the influence of affinities and other chemical processes). See Schelling 1803, 148–152, 164–166, 176, 225, 256, 274–275, 338–342; 1988, 115–118, 128–129, 137, 180, 204–205, 219, 268–271. Compare Oersted 1920, 3:103–105, 110, 113–116; 1:330–332; 3:165–166; 2:39–40; 2:146–149; Jelved et al. 1998, 190–191, 195, 197–199, 252–253, 291, 312, 378–379. For a detailed account of how Oersted's theoretical and experimental work takes off from the second edition of Schelling's *Ideas,* culminating in the grand *naturphilosophische* vision considered immediately below, see Wilson's discussion in Jelved et al. 1998, xxix–xxxix.

28. Jelved et al. 1998, 384, translation emended; Oersted 1920, 2:156–157. This occurs near the conclusion of the 1812 paper entitled "The Chemical Laws of Nature Obtained through Recent Discoveries."

29. Compare note 27 above. It is several pages earlier in his 1812 paper that Oersted famously suggests that "[t]he mode of action [*Wirkungsform*] in the circuit, or the galvanic [mode], stands between the pure electrical and the magnetic [modes] in that its forces are bound far more than in the former and far less than in the latter," so that "one should experimentally investigate [*versuchen*] whether one can produce some effect on the magnet as a magnet in one of the states in which electricity occurs as very bound [i.e., by galvanic or current electricity]" (Jelved et al. 1998, 378–379, translation emended; Oersted 1920, 2:147–148).

30. Immediately thereafter Oersted refers to the passage from 1812 cited in note 29 above. Skeptical doubts about Oersted's later portrayal of his discovery—as expressed, for example, by Gower (1973, 345–346)—are therefore (in view of what we have just seen) quite unwarranted.

31. This discussion occurs in the General Remark to Dynamics. Kant here introduces it as follows: "Instead of a sufficient explanation of the possibility of matter and its specific variety on the basis of these fundamental forces [of attraction and repulsion], which I cannot provide, I will present completely, so I hope, the moments to which its specific variety must collectively be reducible (although even so not conceivable in regard to its possibility)" (IV, 525).

32. For further discussion of the evolution of Kant's conception of chemistry between the years 1786 and 1795, together with a reading of the *Transition* project along these general lines, see again Friedman 1992, chap. 5.

33. See *Metaphysical Foundations*: "So long, therefore, as there is still for chemical actions of matters on one another no concept to be discovered that can be constructed, that is, no law of the approach or withdrawal of the parts of matter can be specified according to which, perhaps in proportion to their density or the like, their motions and all the consequences thereof can be made intuitive and presented a priori in space (a demand that will only with great difficulty ever be fulfilled), then chemistry can be nothing more than a systematic art or experimental doctrine, but never a proper science, because its principles are merely empirical, and allow of no a priori presentation in intuition. Consequently, they do not in the least make the principles of chemical appearances conceivable with respect to their possibility, for they are not receptive to the application of mathematics" (IV, 470–471).

34. Kant had consistently believed in the existence of such an aetherial medium throughout his long career (including, in particular, in the period of the first *Critique* and the *Metaphysical Foundations*). What is new, in the *Opus postumum*, is the idea that the existence of this medium can now be proved a priori. The central role of the matter of heat in Lavoisier's chemistry provided Kant with new grounds for optimism that such a proof is (and must be) possible.

35. This reading of the aether-deduction, as developed by Friedman (1992, chap. 5, § IV), is quite controversial. For an alternative discussion, including some criticisms of

my reading, see Förster 2000. Another important aspect of the *Opus postumum*, to which Förster pays particular attention, is a reconceptualization of the place of biology and teleology in the critical system.

36. By contrast, precisely this idea, as we have seen, lies at the heart of Schelling's approach. See his *Ideas*: "*[A]ll particular or specific determinations of matter have their basis in the differing relation of bodies to magnetism, electricity, and the chemical process*" (1803, 256; 1988, 205).

37. The representation of God—of an infinite rational being—is paradigmatic, for Kant, of a thing in itself lying entirely beyond all human knowledge, at least from a purely theoretical point of view. For the post-Kantian idealists, by contrast, the very distinction between theoretical and practical points of view is also necessarily called into question.

38. See, e.g., Gower 1973, 302: "This preoccupation with metaphysics provides a reason for skepticism concerning the impact of *Naturphilosophie* upon early nineteenth-century science. It is difficult to believe that anyone whose energies were primarily devoted to science could have employed with any confidence the immensely complex metaphysics of Schelling and others. The idiom in which the philosophical spokesmen for the movement chose to express their views is notorious for its virtually impenetrable abstractness. To this extent, *Naturphilosophie* is not comparable with, say, Cartesianism or Leibnizeanism where the metaphysical component is recognizably relevant to the development of science."

39. See, e.g., Shanahan 1989, 303–304: "Kant held that pure reason alone is unable to produce a science of nature, and that therefore careful empirical investigations are necessary. In this last respect Kant differed sharply from certain other 'dealers in the *a priori*' who assumed that empirical research was unnecessary, because pure reason was capable of arriving at a complete knowledge of nature—Schelling, in particular, seemed to suppose this. Oersted's attitude was much closer to Kant's than to Schelling's. Like Kant, Oersted was both more knowledgeable about and more respectful of the actual achievements of physical scientists than were the majority of *Naturphilosophen,* Schelling included. . . . The *Naturphilosophen* were mainly impressed by how *much* physical knowledge can be arrived at by the use of pure reason. Kant, and Oersted, tended to focus on how *little* could be achieved by pure reason alone, and thus regarded experimental investigations more highly." Christensen makes a point of endorsing Shanahan's conclusion about the relative importance of Kant's and Schelling's influence on Oersted (1995, 184–185).

40. See Jelved et al. 1998, 46–47; Oersted 1920, 1:35–36: "In order to achieve completeness in our knowledge of nature, we must start from two extremes, from experience and from the intellect itself. The former method is regressive, beginning with composite facts and resolving these until it arrives at the most simple; the latter is progressive and thus begins with the simplest and progresses toward the most composite. Consequently, the former method must conclude with natural laws which it has abstracted from experience, while the latter must begin with principles, and gradually, as it develops more and more, it also becomes ever more detailed. . . . When the empiricist in his regression toward the general laws of nature meets the metaphysician in his progression, science will reach its perfection." Kant, according to Oersted, has taken the necessary first steps along

the path of such a metaphysics; yet, as Oersted points out at the end of this dissertation, Kant did not fully satisfy the demands of a priori metaphysics in so far as he began with an explicitly empirical concept (the empirical concept of matter). See Jelved et al. 1998, 76; Oersted 1920, 1:76: "By taking empirical concepts as a basis and inferring the natural laws from them, one imparts to the natural laws thus proved only hypothetical validity instead of the rigorous generality which they should have. According to critical philosophy, all natural laws ought to be deduced from the nature of our cognition, which Kant has developed so excellently in his *Critique of Pure Reason,* and I believe I have proved that this can be done by deducing them all a priori and taking only what that book has proved for my basis. Therefore I did not hesitate to deviate from Kant's letter in order to follow the spirit of critical philosophy." Shanahan is particularly misleading, therefore, when he (unfavorably) contrasts Schelling's purely aprioristic ideal with Kant's reliance on "the empirically given concept of matter" in his *Metaphysical Foundations*— for it is precisely this latter that Oersted explicitly rejected (1989, 297).

41. On the preceding page Oersted outlines Schelling's further articulation of the fundamental forces of attraction and repulsion in general required by the science of chemistry in accordance with the three dimensions of space: magnetism acting in a *line,* electrical effects on *surfaces,* and chemical effects (in this case heat) acting "equally freely in *all directions* in a body" (compare note 27 above). On the succeeding page Oersted concludes his discussion of a projected complete science of nature, in which experiment and speculation are perfectly balanced, with the words: "It is well-known that Schelling, through speculation, has produced an attempt which, as such, is of incalculable value, but the combined efforts of a great number of blessed geniuses are probably required for the accomplishment of this task" (Jelved et al. 1998, 199; Oersted 1920, 3:116).

42. See Jelved et al. 1998, 288–289, translation emended; Oersted 1920, 3:162–163. In the following pages Oersted again puts forward the characteristic view that chemistry is now a branch of general dynamics, where the two original fundamental forces manifest themselves in "electric, galvanic, and magnetic effects" (Jelved et al. 1998, 291; Oersted 1920, 3:166); and a comparison of the passage quoted in the text above with the closely related one cited in note 41 strongly suggests that Schelling represents the initial decisive contribution to speculative physics that now needs to "approach significant perfection through the combined efforts of many thinkers." This is confirmed near the end of the 1811 paper where Oersted describes "[t]he progress of philosophy in the eighteenth century" (Jelved et al. 1998, 305; Oersted 1920, 3:185): "The perspicacity of Immanuel Kant liberated it from the atomistic system, which, though of speculative origin, was made the basis of experimental physics. F. W. J. Schelling created a new philosophy of nature [*Naturphilosophie*], the study of which must be important to the empirical student of nature and must both inspire many new ideas in him and also prompt him to re-examination of much that was previously considered unquestionable." (The idea that Kant liberated natural philosophy from atomism is one of the main thrusts of Oersted's 1799 dissertations on the subject.)

43. Oersted's dual methodology is thereby grounded, from his own perspective, in Schelling's definitive overcoming of all skepticism; see note 23 above. To see even more explicitly the connection between this point and the grand *naturphilosophische* vision expressed on pages 59–60 above, compare the following passage from a discussion of the

history of chemistry in 1807: "Not just chemistry but all human knowledge has always, although with varying clarity, intervened in the essence of things. This has always developed through a continually renewed struggle, which, however, has resolved itself in perfect harmony. And it is not just science, not just human nature, it is all of nature that develops according to these laws. . . . In short, the development of the earth is precisely like that of the human spirit. This harmony between nature and spirit is hardly accidental. The further we progress, the more perfect you will find it, and you all the more readily agree with me in assuming that both natures are shoots from a common root. . . . We have glimpsed a higher physics, where the development of science itself, along with all its apparent contradictions, itself belongs to the doctrine of nature. It shows us that everything in the great whole has grown from a common root and will develop into a common life" (Jelved et al. 1998, 259–260, translation emended; Oersted 1920, 1:342–343).

44. See notes 27, 29, 30, 41, and 42 above. The question of the specific kind of unity envisaged here is crucial, for, as we have emphasized, it is precisely this that distinguishes Oersted, Ritter, and Schelling from Kant's original dynamical theory, on the one hand, and from other conceptions of the unity of nature, on the other; see notes 4, 5, and 7 above, together with the paragraphs to which they are appended. The need for considering the precise kind of unity involved has been especially well emphasized by Kenneth Caneva in a very rich and sophisticated discussion of the issue. See, for example: "One of [*Naturphilosophie*'s] most important characteristics, rightly emphasized by virtually all commentators, is the belief in the unity, the interconnectedness, of the phenomena of nature. Having said that, however, one must immediately recognize not only its vagueness, but also the fact that there were other, very different routes to the same end—in particular, via various aether theories. When one explores the terms in which that unity was to be achieved, one gets closer to notions especially characteristic of *Naturphilosophie*—here, in particular, the concepts of *Steigerung* and *Stufenfolge* and the elaborate tripartite schemas that embraced broad classes of phenomena in hierarchically situated analogical relationships" (Caneva 1997, 40). Caneva then emphasizes, in my view perfectly correctly, the importance of such "triadic periodicity"—as explicitly taken from Schelling—in the work of both Ritter and Oersted.

45. Of course the obvious similarities and analogies between electrical and magnetic attractions and repulsions were well known since antiquity, and it was also well known that static electricity produces magnetic effects (e.g., in thunderstorms). The difficulty, however, as Oersted himself points out, is that "electrified and magnetized bodies [appeared to] have no attractive or repulsive effect on one another as a consequence of their condition" (Jelved et al. 1998, 378; Oersted 1920, 2:146). Moreover, there was as yet very little reason for anticipating magnetic effects from chemical or galvanic phenomena. To be sure, Ritter had done some experimental work in magneto-chemistry—but, aside from being rather shaky, this work was guided by the very same triadic schema (i.e., Schelling's) as was Oersted's. As Oersted himself explains (notes 29 and 30 above), it appears that he arrived at his decisive experiment by beginning from Schelling's triadic schema and then adding the idea that electrical forces are much more "bound" (i.e., less manifest as static electricity or charge) in a galvanic circuit than they are in a static distribution of charge. From this point of view, current electricity is much closer, as it were, to magnetism, and so it is precisely here that we should seek the desired "attractive or repulsive effect."

46. See notes 25, 26, and 27 above. The supplement to book I, chapter 3 of *Ideas* mentioned in note 26 concerns "the History of the Decomposition of Water," and it begins with the words: "A more non-sensical undertaking could hardly be imagined than to attempt to outline a universal theory of nature from particular experiments; nevertheless, the whole of French chemistry is nothing but such an attempt—but it could hardly [be expected], as well, that the overwhelming value of higher prospects, directed upon the whole, over those based on particularities, would have been finally so admirably confirmed, as in precisely the history of this doctrine, especially in that part of it that concerns the nature of water" (1803, 119; 1988, 93). Thus, although it is certainly true that Schelling was no empiricist (and there is also no doubt that Schelling's speculative flights often led him astray), Shanahan is completely off the mark when he suggests (note 39 above) that, for Schelling, "empirical research was unnecessary" and insinuates that Schelling was neither "knowledgeable about" nor "respectful of" concrete scientific work. Schelling's attitude toward speculation and experiment was not so different from Oersted's: The former is necessary to organize the results of the latter into a coherent and scientific unified totality on the basis of "higher prospects, directed upon the whole." For discussion of Schelling's similarly dual methodology—whereby a priori deduction must be verified by the empirical facts, which it serves, in turn, to systematize—see Beiser 2002, § IV.4.6.

47. See note 36 and p. 62 above. Thus, once again, there was simply no way—as both Shanahan and Christensen appear to assume (note 39)—that Kant's original dynamical theory of the fundamental forces of attraction and repulsion in general could, by itself, have suggested the specific *kind* of unity of the various forces and powers of nature that was central, in particular, to the new electrochemistry. Here Pearce Williams hits the nail directly on the head when he writes (1973, 15): "Kant's original attractive and repulsive forces were adopted and transformed by Schelling. What Schelling added was the substitution of development for equilibrium between conflicting forces. Thus when equal and opposing forces met the result for Schelling was not stable equilibrium, but development into a higher conflict. It was in Schelling that the famous triad of thesis, antithesis, and synthesis first appeared in the guise of dynamic, conflicting forces. The world is thus the scene of ceaseless activity and dynamic progression."

48. It is largely for these reasons, it seems, that Gower, in particular, remains skeptical of the scientific relevance of Schelling's ideas (see note 38 above). At one place, for example, he refers to Schelling's "bizarre theory that nature and self are ultimately identical" (1973, 313); at another he asserts that "there is little of philosophical value to be gained by an examination of [the] arguments and theories [of Schellingean metaphysics]" (1973, 352). He also complains that "[Oersted's] dynamical theories of physical action, with their emphasis upon the interaction of polar forces, contributed little to the creation of a conceptual framework with which the energy conservation principles could emerge and be understood" (1973, 349). Nevertheless, as I have suggested (note 8 above), Gower's discussion is generally quite balanced and helpful, and, as a whole, it conveys a remarkably good sense of how Schelling's ideas concerning "polarity"—or what Caneva calls "triadic periodicity" (note 44 above)—in fact fruitfully guided the work of both Ritter and Oersted.

49. Aside from the work of Ritter and Oersted, as we pointed out above (note 9), this also includes such seminal figures as Davy and Faraday. And, although later figures, such as Helmholtz, for example, self-consciously rejected the pan-organic conception of nature originating in *Naturphilosophie* (see note 6 and p. 52 above), the question of the relationship of the new electrochemistry and electrodynamics to the life sciences—in Helmholtz's case, to sensory and motor physiology—continued to be a central driving force. Indeed, this is still true of Maxwell's electrodynamical theorizing—which, as in the case of Helmholtz, continued to draw on nineteenth-century physiology (see Cat 2001). Only with the triumph of such slogans as "Maxwell's theory is Maxwell's equations" at the beginning of the twentieth century did this originally biological problematic finally completely disappear.

50. The crux of the matter, of course, is that Schelling's *Naturphilosophie* was in fact an intelligent, perceptive, and appropriate response to both the tensions in Kant's system and the new empirical results. Viewed in this particular context, I believe, there is indeed much of "philosophical value" to be gained by studying it. For a more general attempt—focusing on quite different examples—to elucidate the crucial interaction between philosophical ideas and scientific problems along these lines, see Friedman 2001.

V

Extending Kant: The Origins and Nature of Jakob Friedrich Fries's Philosophy of Science

Frederick Gregory

On 13 August 1829, seventeen-year-old Ernst Apelt wrote to Jakob Fries in Jena seeking advice about the best way to pursue a career in philosophy. He wrote that he had become familiar with many philosophers and had come to the conclusion that only critical philosophy possessed the proper method. Yet after reading Kant's *Critique of Pure Reason* and Fries's *New Critique of Reason,* he found himself full of doubts about what clearly was an essential issue: What relation obtains between appearances and things-in-themselves?

> The thing-in-itself is that which appears to us. Is this, then, divinity itself or an extra-divine being whose holy grounding is divinity? Further, is this absolute being an intelligence? How can an intelligence appear as matter? If intelligence were the absolute thing, would not matter be mere appearance and would this not lead to spiritualism? Is the thing-in-itself individual or general, i.e. is there a thing-in-itself corresponding to every appearance? Or is the thing-in-itself only a general expression for something free of all accidentals (which therefore exists only in thought)? (Fries 1997, 4)

Recognizing in Apelt the same intellectually precocious and curious teenager he himself had been four decades earlier, Fries chose to reply. He urged caution, adding that Apelt must not be content with doubt but must learn how to distinguish what he knew from what he did not know. Fries conceded that Apelt had put his finger on a central and difficult matter, one, he claimed, that was more difficult to understand in Kant than in his own system. He noted that he and Apelt would have to talk these things over in person to get them straight. Nevertheless, Fries was willing to express his bottom-line position. Yes, each object of appearance did have a corresponding thing-in-itself. His view, Fries wrote, amounted to transcendental spiritualism joined together with empirical dualism (1997, 6).

In his second letter to Fries, Apelt was a bit more forthcoming about the origin of his interest in philosophy, perhaps in response to some general advice about the proper preparation for studying philosophy that Fries had included in his letter. Fries had noted that if Apelt was very serious about philosophy, "then, as Plato advised, study mathematics above all and then physics as well. Only he who possesses knowledge of these can acquire really sharp and healthy philosophical judgment" (1997, 5). In a letter dated 13 January 1830, Apelt was eager to assure Fries that natural science and mathematics were high on his list of favorite subjects. He explained that from his early youth onward the heavens had made a powerful impression on him and that this had led him in turn to the study of mathematics (Fries 1997, 7). But his first existential encounter with larger questions had occurred while contemplating the wonders of astronomy. Years later he recalled very specifically the captivating experience of reading Johann Elert Bode's *Introduction to Knowledge of the Starry Skies* (1823).[1] Here he was brought face to face with the immensity of the universe, with questions of the infinity of time and space. "I stood before the great riddle of philosophy without knowing it. . . . A dark feeling told me that I could find the answer to my questions only in metaphysics."[2]

Apelt's claim, in his recollections, that he had read Fichte's metaphysics around this time and found it unsatisfying is not corroborated in the early letters to Fries. His reply to Fries rehearses his prior curriculum of study, most of which was carried out privately, apart from the regimen of the Gymnasium. Fichte is not mentioned, but his account hints at why Fichte might not have addressed his need and why he turned to someone like Fries. "It seems to me moreover appropriate for a youthful disposition not to set itself up in sharp opposition to nature since it receives its first impressions from nature and is not loosed from these mother's arms until reflection is awakened in it (Fries 1997, 8).

In light of these sentiments, one might ask why it was Fries and not Schelling who supplied what a youth like Apelt sought. Schelling's system, after all, took nature seriously and was occupied with the very concerns Apelt had raised in his first letter to Fries. One might reply simply that Schelling was out of vogue, that among natural scientists, at least, disenchantment with Schelling's *Naturphilosophie* was widespread, although not complete. Why, then, did Fries rather than Hegel hold such appeal for Apelt? Around 1830, when the exchange with Fries occurred, Hegel was the most celebrated philosopher in Germany, and he too had written a work on the philosophy of nature. Answers to the question why it was specifically Fries who filled Apelt's need go to the central place that nature, mathematics, and natural science occupied in Fries's thought. To understand how this dimension of Fries's position came about, we would do well to become familiar with the extraordinary path Fries trod as a youth.

EDUCATIONAL BACKGROUND

Jakob Friedrich Fries had an unusual heritage. On his father's side he descended from military nobility. We are told in his son-in-law's biographical account that his grandfather, who became a merchant, had to drop the "von" from his name because of a decline in the family's fortunes (Henke 1867, 3). Peter Conrad Fries, Jakob's father, persuaded his parents to permit him to attend the seminary at Maulbronn, where Johannes Kepler had studied as a boy more than a century earlier. After eventually receiving his doctorate, Peter Fries became a Protestant cleric. He came upon the writings of Nikolaus Ludwig von Zinzendorf, who in 1722 had established a refuge on his estate for a small Protestant group from Moravia known as the Unity of the Brethren. By 1727 the refugees had established a small village of three hundred, dubbed Herrnhut because it was to be "under the Lord's care" (Burckhardt 1893/1897, 1:36–37). Fries's mother, Christiane Sophie Jäschke, was born in Herrnhut in 1738. Her grandfather had been one of the original small group to come out of Moravia.

Fries was born in 1773. When he and a younger brother were five and three years old, respectively, they were placed by their parents in the Unity school for boys in Niesky (Silesia). After Jakob was promoted into the *Pädagogium,* the educational level for those over twelve years in age, he discovered mathematics. He had never enjoyed arithmetic and calculation, but when a substitute teacher introduced the class to geometry, Jakob responded immediately. Mathematics was an area in which one could procure new books without censure, and Fries took advantage of it. He went through geometry, trigonometry, and the basics of analysis, quickly leaving all the others, including the instructors, behind. When, at age fourteen, he told one of his teachers how much he appreciated the certainty and clarity of mathematics, he received the reply that God's existence was equally certain and clear, since God was the ground out of which the world was derived. Jakob received this answer in polite silence but asked himself what, then, was the ground of God? If God was his own ground, why could the world not be likewise? This, Fries recalled, opened for him the world of philosophy (Henke 1867, 17).[3]

Although he also took courses in philosophy, the best philosophy teacher, Karl Bernhard Garve, was now teaching in the sect's theological seminary.[4] Fries did not take well to the philosophy he was taught. He was repeatedly brought back to a question he could not dismiss: Could not the objective validity of our ideas in some way lend itself to proof comparable to the proofs of geometry? As he completed the course of study at the *Pädagogium,* he possessed a favorable perception of mathematics but a decidedly skeptical view of philosophy.

But two fundamental lessons had come through to him from his Herrnhuter life: the important things in life are sure truths, not arbitrary sentiments, and the source of this truth is transcendent, not of our making. Unlike his teachers, Fries was beginning to locate the certainty he sought in the rigorous nature of mathematical reasoning, not in the explicated conclusions of pietistic religion. It was not long before he began to ask himself a second question: How much of any conclusion, whether a religious doctrine or some other proposition, is of our making and how much is not? He would come back to an appreciation of religion later in his life, but not until he satisfied himself about what could be known with confidence.

Fries had known but one way of life since childhood, so that in spite of his suspicion of philosophy and his religious doubt, he really had no choice, on completion of his secondary training, than to go to the seminary to pursue the career in the *Gemeinde* for which he was being trained. In the seminary Fries's doubts only increased. His teacher Garve insisted on imparting to his students an appreciation of the current issues raging in philosophy. The Moravian Conference of Elders and others were not enamored of what they called "the new philosophy that has sprung up of late in all German universities." It was the time of Kant and of his popularization through the lectures of Karl Reinhold in Jena. Other names were also becoming known, names like Jacobi and Fichte. Unlike Garve, the Brethren were suspicious of the merits of the movement, but they tolerated Garve's philosophical lectures because they were convinced that Kant's system was merely another phase through which university faculty were passing, just as the Wolffian system Kant criticized apparently had been (Henke 1867, 21).

The three years in seminary were, in Fries's later recollection, the most beautiful of his life. He needed no outside incentive in his studies, but Garve's enthusiasm for the latest philosophy was an additional prod. Garve taught his students philosophy, true, but he also taught them how to philosophize. His own favorite was not Kant, about whom he had reservations, but Reinhold. Garve had corresponded with Reinhold by the time Fries came to him and must have appreciated the natural interest that this young seminarian brought to his studies.[5] Fries, however, found Reinhold's claims of proof to be woefully inadequate when compared, he tells us, to the precision of the mathematical reasoning to which he had become accustomed (Henke 1867, 26). Eager to read Kant's own works for himself, a difficult prospect, since the Niesky authorities did not wish to provide them, Fries got hold of Kant's *Prolegomena to any Future Metaphysics* and his prize essay on natural theology and morals and discovered there Kant's wonderful precision of argument. He followed Kant without objection through the metaphysics of nature, although he began to be uncomfortable with aspects of Kant's "dialectical exposition" (Henke 1867, 26).

Fries next got hold of and studied in the seminary Kant's critiques and his book on religion with his friend Karl von Zezschwitz. If we are to take his recollection from 1837 at its word, he had sensed even then that in the *Critique of Pure Reason* Kant had not investigated the psychological basis on which our whole capacity for knowledge rests, and that both it and the *Critique of Practical Reason,* taken together, lacked an appropriate psychological foundation. He expected, he tells us, that he would find these missing pieces in Kant's *Critique of Judgment.* While the introduction to this work did attempt to address the deficiency, it was cursory and unsatisfactory; hence Fries took on the task of carrying out these investigations himself. He recalled taking long walks in the woods around Niesky, planning how he would complete what he viewed as a necessary extension of Kant's program, a task he identified as a "propadeutic of general psychology" (Henke 1867, 26–27).

As a result, Fries's basic philosophical stance was set remarkably early in his life; the essentials were in place even before he left Niesky for Leipzig in October 1795, a youth of twenty-two, ostensibly to study law at the university.[6] By the time he left Leipzig for Jena a year later, the intellectual challenge to which he wished to devote his life was becoming clearer to him. At Leipzig any enthusiasm he had had for the study of law waned quickly. He stole time from his other studies to read philosophy, especially Fichte, on his own. In January he wrote to his friend from Herrnhuter days, Samuel Christlieb Reichel, who had remained with the Herrnhuter, that he was hard at work on his propadeutic of a general theory of the soul (Henke 1867, 41).[7] No doubt he portrayed this task with greater clarity in retrospect than it could possibly have had at the time; nevertheless, Fries claimed that he had come to see his reworking of Kant as having three phases. First, he had to demonstrate that the critique of reason, which according to Kant was a propaedeutic for philosophy, was as such a branch of empirical psychology, in which the objects of observation were the contents of the human mind. A critique of reason was therefore really general psychology—later he called it philosophical anthropology. A second task was to make precise the relationship of general psychology to empirical psychology and also to what was called rational psychology. Once this had been completed the third task could be addressed, namely, the writing of a new, anthropological critique of reason (Henke 1867, 44–45). The implications of all this for his understanding of natural science, which are dealt with below, emerged during the next decade and found practical exposition in his later works dealing directly with scientific subjects. Fries therefore saw his own contribution as making important technical corrections to Kant's critical philosophical system and working out its implications for the study of nature. But, although he set out to improve upon the master, in spirit Fries remained a Kantian all his life.

How are we to understand Fries's bold confidence that he could extend Kant's philosophical treatment of nature and natural science? One explanation is that it was due to the zeal of youth; Josiah Royce has described the general preoccupation of philosophers of this era "first of all to transcend Kant, and secondly to transcend everybody else" (1967, 171). One could also argue that Fries realized how many of the advances in mathematics since Newton's time found no place in German idealistic philosophy and that the failure of all systems of philosophy, including Kant's, to convincingly hold together the expanding realms of mathematical and empirical knowledge had led to a crisis of scientific rationality that Fries sought to address (Pulte, this volume).

Without denying the merits of this argument, I wish to consider a different kind of explanation, one that derives from what Thomas Söderqvist has called the dimension of "existential biography" (1996), by which Söderqvist means the emotional motivations that affect individuals at a deep and basic level. I wish to suggest that Fries possessed fundamental dispositions that informed the development of his attitudes about the significance and relationship of empirical knowledge and mathematics. These basic dispositions were due to deeply held convictions inherited from pietistic religion that had impressed both Kant and Fries.

Both Kant and Fries believed that in our response to the human condition, we must identify that in it which is certain. Both believed that understanding God's truth is a human responsibility that can be achieved only within limits. For the Herrnhuter, the meaning of these fundamental beliefs was visible in their confidence that the central tenets of their communal life were sure, and in their deferential surrender to an almighty God who, unlike them, knew no limitations on his perfection. Kant's and Fries's embrace of certainty can be seen in their appreciation of the significance of mathematics for philosophy; and their acknowledgment of imperfection is evident in their conviction that while God's truth can be approached through understanding, it can never be exhaustively captured by this means. Fries in particular, surrounded by other philosophers who were trying to "transcend Kant," never mistook the ennoblement of humanity for its deification.

RECONSIDERING CHEMISTRY AS SCIENCE

As we have seen from his own testimony, Fries was among the many thinkers who challenged themselves to bring the same clarity, precision, and certainty to philosophical reasoning that they had discovered and appreciated in mathematics. Fries also soon became convinced that his greatest challenge was the incorporation into his philosophy of the clarity and immediacy of empirical observation, while acknowledging that its truth value is rarely if ever immune to doubt. Fries was among the first to combine with his respect for mathematics a recognition of the fundamental yet problematical role of empirical observation in the matter of constructing scientific theories. In the words of Karl Popper, "the problem of the basis of experience has troubled few thinkers so deeply as Fries" (1968, 93).[8]

It was, of course, his encounter with Kant's thought, which dovetailed so nicely with the spiritually conditioned intellectual needs he inherited from the Herrnhuter, that resulted in his unique understanding of scientific knowledge. His agreement with Kant, as he put it to Apelt near the end of his career, that behind each object of appearance there was a corresponding thing-in-itself, meant that he would never be able to follow Fichte in his attempt to correct Kant's "mistake." His patience quickly wore thin already in Leipzig, at Fichte's constant flirtation with what Fries regarded as nonsense. "One third of all his books," he wrote to Reichel in the spring of 1796, "are the words I, Not-I, suppose, oppose—nothing but horribly indefinite expressions, yet twisted with infinite cleverness into an entire philosophy" (Henke 1867, 42). On coming to Jena he gave Fichte a second chance by attending his lectures, but by May 1797 he was done with him.[9]

Schelling was another matter. Fries did not read Schelling's early works on the philosophy of nature until after 1800, for in the fall of 1797 he took a position, out of financial need, as a tutor to the children of a family in Switzerland. While in Switzerland he pursued his program of improving Kant's philosophy, began a critique of Fichte, read Newton's *Principia* and Gilbert's treatise on mathematics (*Dissertatio de Mathesi Prima*), and wrote two works for a *Privatdozent* named Alexander Scherer, a man he had gotten to know in Jena who was editing a new journal in chemistry.[10]

In his "Attempt at a Critique of Richter's Stoichiometry" (*Versuch einer Kritik der Richterischen Stöchyometrie*), which appeared eventually in Scherer's journal in 1801, Fries began to make clear how his respect for empirical observation and mathematics led him to an understanding of natural science different from Kant's. Kant had concluded that some inquiries into nature's workings could not be regarded as genuinely scientific because their results did not rest on a priori axioms of knowledge. The indication that an investigation of nature was genuinely scientific, Kant wrote in the *Metaphysical Foundations,* was evident only to the extent that one encountered mathematics in it (1786, 470). Chemistry, for example, could never be a genuine science, because even its most general results rested on a merely empirical foundation (1786, 471). If one agreed with Kant, then, one was not disposed to try to quantify the results of chemical experiments by searching for mathematical equations that accurately captured chemical phenomena.

In a later exposition, in 1822, of his understanding of natural science, Fries would explain why one should always "remain in contact with mathematical theories," regardless how individuated and removed from a constitutive theory the data may appear to be (1822, 610). Here he simply made clear that he did not agree with Kant about chemistry; rather, he endorsed the attempts of Jeremias Benjamin Richter to quantify chemical relations. Fries's "Attempt" was a review of two works of Richter, the three volumes of his *Foundations of Stoichiometry* (*Anfangsgründe der Stöchyometrie*) and the first eight parts of his *On the More Recent Subjects of Chemistry* (*Über die neueren Gegenstände der Chemie*), all of which had appeared between 1793 and 1797. Richter was primarily concerned to uncover the quantitative ratios by which different acids neutralized a given base, or of different bases that saturated a given acid. He had noticed that the ratios were independent of the neutralizing substance. By establishing patterns in specific known cases, he sought to uncover arithmetical and geometric series that would allow him to predict combinations not yet known.[11]

Fries of course applauded Richter's tireless efforts to seek out mathematical relationships that described how chemical constituents combined. But he was also critical of Richter. Richter's fundamental error, according to Fries, was that he had never asked, let alone answered, the necessary preliminary question: To what extent can mathematics be applied to chemistry? Richter's arithmetical and geometrical series were, in fact, only practical solutions, since they were not sought with pure theory in mind. Fries set out to determine what a pure theory of stoichiometry might look like. It would, he concluded, contain a mathematical and a dynamical part. The mathematical part would consider the quantifiable relations of mass, volume, and density displayed in every chemical reaction. The dynamical part would define the units of intensity and the magnitudes of force present (1801, 19–21). Fries's own discourse on pure stoichiometry consisted of his providing algebraic proofs of theorems that were consistent with the manner in which our minds conceive fundamental parameters such as volume, mass, density, and force (see Gregory 1984).

Fries had written this critique for a journal of chemistry with an eye toward a possible career as a professor of chemistry, a prospect made more urgent by the death of his mother in 1799. In a letter to Reichel he resolved to pursue a more permanent vocation than his tutoring offered, prompting his determination to return to the university to take up chemistry or philosophy (Henke 1867, 63–64). While his work on chemistry is an early indication of his fascination with natural science, it also reveals that Fries was more taken with the philosophical implications of empirical science than with its practical side. Throughout his career he never displayed an interest in conducting experiments of his own. In this he was unlike slightly younger contemporaries, Johann Wilhelm Ritter and Hans Christian Oersted, who were meeting in Jena when Fries's critique of Richter was published and who also took a philosophical interest in natural science. Whether or not Fries realized it, his heart lay in pure philosophy. Furthermore, the option to pursue a career in chemistry received a setback when he discovered, on his return from Switzerland, that he could no longer work with Scherer, who had taken a position in industry. So he committed himself to study philosophy. He set for himself the task of doing what he felt Kant had never satisfactorily completed—writing a work on the relationship between philosophy and natural science.

When Fries composed the critique of Richter, he was unaware of Schelling. Only on returning from Switzerland did he discover that Schelling had already published four new works on the philosophy of nature while he had been away. I have dealt with the relationship of Fries and Schelling elsewhere (Gregory 1983a). Suffice it here to repeat an observation about the initial impact Schelling's work had on many. It was received by numerous individuals as a welcome relief from the empty musings to which philosophy had recently surrendered itself. Gotthilf Heinrich Schubert, for example, turned to Schelling's work with relish because it opposed what Schubert called "the Kantian-Fichtean egoism—which I hate" (Bonwetsch 1918, 10–11). Even Fries himself confessed to a momentary infatuation with Schelling's work; hence Fries regarded Schelling as a major obstacle in his path, a formidable rival who not only had beat him to publishing, but whose particular interpretation had struck a nerve in German academe.[12]

Soon Fries came to the conclusion that, after all was said and done, Schelling had opted for Fichte's solution to the problem of the thing-in-itself. Schelling had ignored Kant's caveats about the limits of reason, as if the cleft between the thing-in-itself and the thing-of-perception could be transcended. Indeed, Fries would soon conclude that it was a common feature of most post-Kantian philosophers to think that their philosophy could get at the essence of things by way of an eternal unity. They all confused the representative view of the world that humans create with the idea of an ideal divine understanding, thus turning their human reason into divine reason (Bonsiepen 1997, 328; compare Fries 1828–1831, 4:17).[13]

RESTRUCTURING KANT'S CRITIQUE/RESTRUCTURING THE KANTIAN DEDUCTION

Fries undertook his philosophical exposition of natural science in the context of his attempt to rework the starting point of Kant's system. The main deficiency of Kant's critique of reason, according to Fries, was that he was mistaken about its beginnings. What Kant did in the *Critique* was to begin with the principle of the possibility of experience and claim to deduce from it a system of basic propositions (*Grundsätze*) of pure reason.

> How can he want to prove the law of causality from [the principle of the possibility of experience] when experience is grounded in the interaction of our sensual and intellectual powers of knowledge? Or more clearly, how will he prove the law of possibility in general from the law of the possibility of experience? That would break all rules of philosophical knowledge since the general law would follow from a single instance of it. (1828–1831, 4:25)

No wonder, Fries pointed out, those who had come after Kant had tried to improve on this starting point—Fichte with his Ich, Reinhold with his "thinking as thinking," Schelling with his Absolute (1828–1831, 4:17).

Clearly, for Fries, experience was the unqualified starting point, a replacement for possibility in general. But the content of experience could not be further proven; it could only be exhibited (*aufgewiesen*), and this went for the content of inner experience as well as outer, for the perception of the content of our minds as well as for that of the external world.[14] We avoid circular reasoning because we do not claim to have a proof (Bonsiepen 1997, 328). In the following passage about the agreement of our ideas with their objects, one must be careful to realize that Fries is not talking here about a specific idea of a given object but about the general assertion that there is an agreement of our ideas with their objects:

> The agreement of our ideas with the being of their objects is something that the human mind can never subject to a mediate proof; rather, it is

only the immediate presupposition of every knowing reason that is valid solely through the power of its trust in itself. In the nexus of human thoughts it is neither given, received, nor altered in a mediate fashion. Intuition by itself is its own testimony of truth. Only to the degree I trust the intuition do I know something of the being of real objects. Just as immediately valid for us are the metaphysical basic truths that come to consciousness immediately for us in the feeling of truth. The truth about which humans argue, with respect to which they can err and doubt, is never this transcendental truth of the agreement of idea [*Vorstellung*] and object. That is rather the empirical truth of consciousness which requires the correct comparison of mediated representations with the immediate. (1828–1831, 4:xxvii–xxviii)

Fries goes on to explain how, in the absence of a proof, we assure ourselves of our a priori possibilities of knowledge through inner perception, and to evaluate as well the status of knowledge gained through outer perception. In the former case, where the results of inner perception are concerned, it is the act of making judgments that exposes a priori laws that lie within our reason, laws that we use even before we are conscious that we are doing so. Through reflection, however, we can become "conscious again" (*wieder bewußt*) of the presence of these laws. An oft-cited passage from Fries's *New Critique of Reason* makes this clear: "When I say, for example, every substance persists, every change has a cause. . . . I acknowledge laws of nature. . . . But these very laws, of which I become conscious again in the judgment, must lie in my reason as immediate knowledge. It's just that I need the judgment in order to become conscious of them" (1828–1831, 4:341–342).

How is knowledge gained from outer experience? Complete knowledge results from the working together of two separate ways of knowing: intuition and thinking. Through intuition we acquire ideas and through thinking these ideas are unified and connected. The mind possesses a capacity to acquire ideas (*Vorstellungen*) through a sensibility (*Empfindlichkeit*) called sense (*Sinn*). These ideas consist of sense intuitions (*sinnliche Anschauungen*) that we receive in the act of sensation and that are accompanied by ideas of space and time. According to Fries, it is important to note that the idea is already present in the sense intuition; it is not something added afterward. Further, Fries insists that the sense intuition should not be regarded, as it is with Kant, as causally linked to the object.

> Let us take an example. I see a green tree before me and through sensation attain knowledge of it. If now I inquire how this occurs I receive the answer according to the usual relation: the tree affects my eye, by which

> I receive the sensation of green and, because the sensation must have a cause, I reason to the tree as the affecting agent and as the cause of the sensation of green. And some add with Fichte that when I name the tree green or the sugar sweet it is quite falsely expressed—we ourselves are really the green and sweet. I hold however: God protect us from a sweetened mind. I believe the entire explanation to be completely false. The tree is green and the sugar sweet or no one is. And when I contemplate the tree [*den Baum anschaue*] I see immediately in the sensation something green outside me without inquiring at all about a cause of my sensation. . . . Were I to contemplate the thing outside me. . . . as the cause of my sensation I would in that case not come to the tree but either to the light or to a mere motion in my eye, etc. The matter is much simpler when we stay with the observation itself. The main thing here to note is, namely, that in the sensation an intuition of something outside me. . . . is contained from the outset and that the idea of an object. . . . is not added afterwards, but is completely present already from the beginning. (1828–1831, 4:87–88)[15]

Sense intuitions come to us involuntarily in sensation and contribute to the course of ideas (*Gedankenlauf*) we possess. Thinking makes use of our understanding (*Verstand*), which contributes the voluntary ideas of reflection to the course of ideas. All past ideas are stored in our memory and new ones are constantly being experienced. Fries argues that, since our internal mental life is not a chaos of representations, something independent of external impressions must be present that regulates inner mental activity. This is what he calls the power of imagination (*Einbildungskraft*), which acts in different modes.

Beyond the two components by which knowledge originates in the mind, intuition and thinking, are yet other inner states and alterations of ideas that have to do with the presence, exchanging, and mutual play of ideas in our inner world. This is the realm of the imagination. It belongs neither to intuition nor to thinking; rather, it involves a mechanism of inner changes where the ideas already present in the mind further influence one another. The imagination exists between intuition and thinking because it is a broad field of inner activity partly dependent on voluntary determinations and partly not. (But even in its involuntary ideas imagination does not belong to sense in any real significance.) Here internal modifications and exchanges of ideas occur that are not given in immediate sensations of sense but result from an inner play of our ideas. This is the area of imagination (1828–1831, 4:132–134).

Imagination exists in both a reproductive and a productive mode. Reproductive imagination is a capacity of the memory to combine ideas in accordance with the laws of association. Productive imagination brings a totally new element to knowledge in mathematical intuition. Intuitions are given in sensation as isolated (*vereinzelt*), but the ideas of space and time lie at the basis of each as original and necessary forms of sensibility and, as Kant showed, bring unity and connection into the intuition (1828–1831, 4:173, 175). Although the idea of the object comes with the intuition, the construction of objects of sense perception in space and time does not; it comes by virtue of the power of the imagination. We are able to handle ideas of figure and duration just as mathematics uses them. In applying these ideas to things outside us, our geometrical sketches seem to belong to sense and not to our imagination. But all our ideas of shape, size, and distance are also used voluntarily. They are dependent on voluntary reflection that is not given with the sensation (1828–1831, 4:179).

Fries now faces head on a contradiction he seems to have created. He claims on the one hand that this use of the productive imagination is dependent on voluntary reflection. But if this is so, then the productive imagination does not give us intuitions of a world outside us but only ideas of a world we have made up. His resolution of the matter involves his own unique Kantian tightrope walk between what he calls "one-sided empiricism" and "one-sided rationalism," represented in his mind by Condillac and Schelling, respectively (1828–1831, 4:186).

The phrase "productive imagination" is ambiguous, according to Fries, depending on whether one refers by it to the capacity of merely having mathematical intuitions of the spatial connection of things or to our awareness that we have them. The law of all spatial connection is embedded in our possession of mathematical intuitions of space and time and of the spatial connection of things. Our use of these laws is but the application of a living geometry that we constantly make without thinking about it. In this activity of the mind one does not encounter voluntary choice. The world is as it is, and in our immediate sense perception of it we apprehend it according to laws that are part of the fabric of reality. But, through reflection, we can become aware of the mathematical relationships that lie within our reason, and this involves voluntary deliberation and reflection. Just as judgments help us to become conscious of the law of causality we are using without being aware of it, so reflection helps us both to uncover the living geometry in which our reason is embedded and to extend this use by clarifying its implications. Fries argues that the whole of mathematical relationships lies darkly in our reason and that we bring implications of this whole out into the light again through reflection (at which, obviously, some are more gifted than others). This voluntary activity of reflection does not give us new knowledge, since it is already in our reason (1828–1831, 4:179). The contradiction is resolved because, while we do not create the world outside us, we need voluntary reflection to become fully aware of the knowledge we already have of it.

As already noted, Fries, unlike Kant, argues that the objects of experience are not causally connected to our ideas. His system provides only an exhibition (*Aufweisung*) of the contents of our reason, not a deduction or proof. The basis for his claim that our ideas describe the real world outside us lies in the self-confidence of reason described above.

Of course all this refers to what Fries called "constitutive pure theory," because it emerged from an analysis of the constitution of the mind in its relationship to the world outside. Fries used this phrase in *The Mathematical Philosophy of Nature* (1822), which is devoted to a detailed analysis of mathematics and of the implications of a mathematical representation of natural process (1822, 610). But what about those aspects of natural science that did not appear to lend themselves to mathematical description, those that Kant had regarded as merely empirical sciences? Fries's position was that even in those branches of natural science where a constitutive theory seems unattainable and whose results are arrived at only through induction and hypothesis, the scientist cannot escape pure theory completely. "Empirical natural science must not tear itself free from pure theory and operate on its own good fortune. It should rather always remain in contact with mathematical theories by borrowing from them the highest leading maxims, without which no healthy scientific induction can be successful, and by applying the inductive method at every stage in accordance with these higher conditions" (1822, 610).

The Relevance of Fries's Moravian Past

Before concluding this sketch of Fries's position, I should like to point out one additional reason why it is helpful to associate his philosophical stance, in which mathematics and natural science occupy so central a place, with his Moravian past. I have suggested that the certainty Fries found in mathematics resonated with the need for certainty so prominent among the Herrnhuter. But others who did not share Fries's religious outlook had expressed the same desire to find a way to import into philosophy the self-evident quality they had discovered in mathematics—one thinks most obviously of Hobbes. What convinces me that there is a link between Fries's pietistic background and his eventual position is the way he brings together mathematical certainty and our sensual experience of the world. What results is the special combination of confidence and humility that marked the Herrnhuter disposition.

The confidence is very visible in Fries's declaration of his so-called *Wahrheitsgefühl*, the claim that the agreement of our ideas with the being of their objects must be assumed and is true solely through reason's trust in itself, as is the case with the validity of the metaphysical basic truths of which we become conscious (again) through judgment. In both cases one has something resembling a clear declaration of faith.

The humility becomes evident in Fries's recognition that his initial hope as a youth—that he could demonstrate the objective validity of our ideas through a proof comparable to the proofs of mathematics—was impossible. He had been taught as a student that some things must simply be accepted on faith, an assertion that at the time provoked doubt and resistance in his mind. Now he realized that the certainty of mathematical proof could not in fact be transferred to our knowledge of the world. One could not transfer mathematical certainty to the "declarations of faith" implicit in the *Wahrheitsgefühl*.

All this becomes clearer in Fries's discussion of empirical natural science. Fries contends for the first time among philosophers, in the view of Popper and Lakatos, that one cannot prove a statement by appeal to a perceptual experience (Popper 1968, 105n3; Lakatos 1970, 99n3).[16] In his view, proof has to do only with what he calls mediate knowledge, the logical chain of reasoning that occurs when moving between premises and conclusion. Proof is purely an objective restatement of the premises; it can say nothing at all about the status of the premises.

Fries has harsh words for the idea that one can prove a statement by appeal to a perception, or, in Fries's words, "that a given piece of knowledge is objectively firmly grounded if one demonstrates it, i.e., if one has reduced it to intuition." Fries calls this belief "the misleading principle of all speculation, through which a false confidence in proof is maintained and through which the false hope of achieving something for the grounding of our knowledge through intellectual intuition is preserved" (1828–1831, 4:352). Deriving a statement from a perception amounts to creating a mystical union of dissimilar things.

Fries believes he has avoided this mistake by not regarding the idea of sensation as causally linked to the object. If they were related in this way, then it would be reasonable, presumably, to attempt to include perception in the same class as statements. Kant had it right, Fries tells us: "It is always just a matter of how human reason knows and cognates, not an immediate matter of how things are in themselves." But Kant was wrong to insist that the idea of the object causally makes the object possible. By "wanting to explain the objectivity of the sense intuition through the causal efficacy of that which brings it about [*des Afficirenden*]," he brings the sense perception into the realm of statements and becomes susceptible to "the misleading principle of all speculation" (1828–1831, 4:355).

But what difference does this make? One implication of Fries's "anthropological" revision of Kant's position has to do with the possibility of error in our thinking. Because we only know something of the being of real objects to the degree that we trust the intuition, it must be the case that the immediate knowledge of the senses is never subject to error. But our judgments are inevitably subject to error, primarily because our construction of objects of sense intuition usually involves temporal and spatial imprecision: "In many cases the intuition is not completed, or we mistake ourselves totally. From this we are able to find that the idea depends on our self activity and on our judgment" (1828–1831, 4:181). In his earlier works Fries was quite clear about the source of error.

> The senses do not lie. Even here it is always the understanding that makes the error. Optical illusion rests on the activity of the productive imagination which is subject to the will, and which is itself occasioned by previous erroneous judgments of the understanding. . . . The only erroneous factor in our mental representations. . . . is erroneous judgment. Thus error in no wise touches the immediate cognition of reason that lies at the basis of all judgments. (1805, 28–29)

The skeptic confuses things by holding that the immediate knowledge of perception is open to error, but "the one who sees that the striving for truth belongs only to reflection will never think of his reason as surrendered to uncertainty" (1828–1831, 4:405).

About the mediated conclusions of natural science, then, Fries has great humility. They are in constant need of reexamination and correction. And yet what Fries has accomplished is to win a place for empirical knowledge that stands at the top of the intellectual capacities of humankind. In the words of his follower Leonard Nelson,

> The rule that all metaphysical knowledge is formal in character means that it is not possible by any logic however sophisticated to extract from the pure metaphysical forms a body of knowledge which would outbid empirical knowledge and which, if such a thing were possible, would really suppress and replace it. . . . The principles of metaphysics are in fact purely regulative principles underlying the empirical use of our knowledge and serving as guiding maxims in various fields of experience. . . . There is no single closed system in which the totality of our knowledge could be combined. Our knowledge has several distinct and separate entrances. Nothing, therefore, is so important for the

understanding of critical metaphysics as an insight into the principle, enunciated by Fries, that truth splits up into different world views. (Nelson 1971, 2:247–248)

Although Nelson himself characterizes them differently from Fries, the worldviews he has in mind go back, of course, to what Fries early in his career characterized as worlds of knowledge, belief, and aesthetic sense.[17] It can be no accident that Fries's ultimate position is in the final analysis compatible with the Herrnhuter disposition to be confident with certainty of divine truth in spite of the limitations on what can be known. This ultimate position, after all, never surrendered reason to uncertainty, but deferred to a final divine truth that could only in part be an object of cognition.

The knowledge of natural science stood at the crux of Fries's answer to the challenge his background had imparted to him. It best revealed the answer to the question Fries had used to challenge the young Ernst Apelt: to ascertain what could be known and what could not. Some four years before Fries wrote to Apelt that he would have to learn mathematics and physics if he wanted to study philosophy seriously, Fries extolled the virtues of natural science in a more personal manner in the textbook on science that he had written to use with lectures.

> Among scholars one of the happiest lots falls to the serious natural scientist. For if in human life order and opinion may be revised, for him the unchangeable truth always remains constant. In this way his science is by far the richest and one of the most alive of all.... The joy of accomplishment stays more with the natural scientist than with others and [it] increases for him daily with new discoveries more than for others, here in the celestial breadth of astronomy, there in microscopic observation of the smallest things on earth, wherever the infinite fullness of life keeps transporting us into new astonishment. (1826a, 9)[18]

When Matthias Schleiden chose to remember Fries in his biographical sketch of him, he entitled his work "Jakob Fries, the Philosopher of the Natural Scientists" (1857). When Fries described himself for the young Apelt, he too emphasized his fundamental respect for empirical knowledge by characterizing his position as "empirical dualism," referring to his insistence that one must begin with observation of the outer world of nature and the inner world of the mind. In that letter of 1830, however, we also hear echoes from his spiritual past. Joined to empirical dualism was what he called "transcendental spiritualism" (1997, 6), no better exemplification of which can be cited than the declaration of faith called his *Wahrheitsgefühl*. In the end, Fries was content to proclaim his faith in reason boldly, bracketing the certainty he encountered in a manner that both supported his confidence about what could be known and yet required humility and awe in the face of what lay beyond cognition.

NOTES

1. Bode lived from 1747 to 1826. Which edition of this work Apelt used is impossible to determine.

2. As quoted in Apelt's *Autobiography*, in the editor's introduction to Fries 1997, 15. Apelt's sketch appeared in 1862 in Hermann Schäffer's *Erinnerungsblätter der Mathematischen Gesellschaft zu Jena*, 19–31 (see Fries 1997, 13n17).

3. In 1837 Fries's son-in-law persuaded him to write down his recollections and drew on them and on letters available to him to produce the biographical volume (Henke 1867).

4. Because of the repeated moves, Fries's location during his schooling can be confusing. Up to the age of five he was with his parents in Barby. At age five he was sent to the boys' school in Niesky, where he remained until the school was moved back to Barby. Here Fries completed the remaining three years at the boys' school before he entered the seminary, which now was in Niesky. At age nineteen he finished the upper level of the boys' school in Barby and went to the seminary in Niesky for the next three years, until 1795, when he left the Herrnhuter to go to the university in Leipzig. His instruction in philosophy at Barby between age sixteen and nineteen was conducted without Garve, who was then no longer teaching in the boys school. Fries was taught philosophy by Garve, however, when he went to the seminary from 1792 to 1795.

5. In his letter to Reinhold of 14 August 1789, Garve asked how Reinhold could confine himself to someone else's system as much as he did. It is here that we learn of Garve's dissatisfaction with Kant, although he deferred to Reinhold's claims about Kant with the excuse that perhaps he just did not understand Kant deeply enough (Reinhold 1825, 346–348).

6. Why Fries opted to study law is not clear. One reason might be that of the three higher faculties in the German university, law was the most popular. On the dominant place of the faculty of law in eighteenth-century German universities, see Hammerstein 1985, 312, 318, 324, and Heilbron 1979, 140.

7. On Reichel, who remained a loyal member and rose through the ranks of the Brethren, see Hamilton 1900, 341, 390.

8. Popper also may have been attracted to Fries because of his appreciation of falsifiability. In the *Mathematische Naturphilosophie* (discussed below), Fries included a criterion of falsification among his list of rules for the experimental natural scientist: "One should make no presupposition that cannot be definitely contradicted [*widerlegt*] by experience" (1822, 21). Fries credits this insight to Heinrich Link in his *On the Philosophy of Nature* (*Über Naturphilosophie*) of 1806. But, as the editors of Fries's complete works point out,

Link had only said there that "one must assume nothing in natural science that is not, or could not be, an object of experience." See the introduction to 1822, 35n49.

9. See Fries's letter to Karl von Zezschwitz of 23 May 1797 in Henke 1867, 47.

10. The two articles for Alexander Scherer's *Archiv für die Theoretische Chemie* dealt with an investigation of the nature of heat and light and a critique of Richter's chemistry. The latter reveals the stimulating effect of empirical natural science on Fries's thought and is discussed below.

11. On Richter, see Partington 1951/1953, and Rocke 1984, 7–10.

12. Fries conceded early on (1802) that he had something of a forerunner in Schelling. See his "Investigations on the Nature of Heat and Light" (*Untersuchungen über die Natur der Wärme und des Lichtes*) (1802, 150, 179–181). Later he recalled thinking of Schelling's considerable talent as a challenge for his own abilities. Compare the quotation (in 1822, 20) from his "Aufzeichnungen an Mirbt" (*Notes on Mirbt*) (this autobiographical sketch was written to E. S. Mirbt to cover matters not contained in the sketch he had written for Henke and may be found in the Fries Archive in Düsseldorf). For other positive expressions of Fries's view of Schelling, see Gregory 1983a, 148, 156.

13. This refers to volume 1 of the second edition (1828) of Fries's *New or Anthropological Critique of Reason*. Since Fries pronounced himself satisfied with the original volume 1 (which appeared in 1807 simply as *New Critique of Reason*), he changed very little of it in the second edition.

14. Fries rejected Kant's claim that there are but two possibilities of knowledge of an object: either that experience makes the concept possible or that concepts make experience possible. Kant, accepting the second of these options, rejected Leibniz's third alternative of a preestablished harmony between an object and its idea. Fries, who in his *Handbook of Psychical Anthropology* did not shy away from identifying himself as one of "us Leibnizians" (1837/1839, 2:xi–xii), wished to accept what Wolfgang Bonsiepen has called "an original coordination" (*Zuordnung*) of concept and object that we have to presuppose as transcendental truth resting in the self-trust of reason not susceptible of further inquiry (Bonsiepen 1997, 329). More on the implications of this below.

15. Fries's reference to Leibniz is again relevant, for when Leibniz spoke of a preestablished or predetermined capacity of the mind, he was never able, according to Fries, to explain the temporal origin of this knowledge. This was because it is established in our reason independently of sense stimulation; it comes to us with consciousness, not as something that arises later (1837/1839, 1:ix).

16. This raises the issue of what Popper calls "psychologism" (1968, 94) and whether Fries is guilty of it, a subject that would take us too far afield. Of course much depends on what is meant by psychologism. Examples of the disagreement include the views of Popper and Lakatos, who tend to minimize psychologism in Fries's thought (and in their own), as compared with standard encyclopedia entries, which have concluded that he is justly criticized for confusing a mental act with its logical content. See Mourelatos 1967,

254; Abbagnano 1967, 520; Wood 1998, 798. Most recently Malachi Haim Hacohen has examined the pros and cons of the matter (2000, 228–229).

17. I have sketched the interrelationships of these world views in Gregory 1990, esp. 73–78.

18. For an interesting conjecture about how Fries's 1826 *Textbook of Natural Science* (*Lehrbuch der Naturlehre*) (1826a) may have influenced Robert Mayer, see Caneva 1993. I have discussed Caneva's treatment of Fries in Gregory 1997, 149–151.

VI

Kant, Fries, and the Expanding Universe of Science

Helmut Pulte

The relation between science and philosophy in the first half of the nineteenth century in Germany was characterized by a significant tension: Science became the prevailing signature of culture, and philosophy—at that time dominated by the systems of speculative idealism—lost its authority in matters of scientific rationality. Quite the contrary, philosophy itself became increasingly the target of "scientistic" criticism, that is, it was accused of not (or of no longer) being able to judge what rationality meant in the different discourses of science and of not obeying scientific standards in its *own* discourse (see, for example, Schnädelbach 1983, 88). The growing alienation and even hostility between science and philosophy later in the century led to the formation of a philosophy of science that was relatively isolated from "school philosophy" and was promoted by scientists themselves (as, for example, by Ernst Mach, Hermann von Helmholtz, Ludwig Boltzmann, and Heinrich Hertz).

Jakob Friedrich Fries (1773–1843)[1] was one of the few philosophers and scientists in the first half of the nineteenth century who perceived this development early on and tried to keep philosophy and science together on the basis of a somehow "dynamized" Kantianism. Though the reception of his philosophy of science suffered from unfavorable historiographical, biographical, and political circumstances not to be discussed here (see Pulte 1999a), his approach to the philosophy of science deserves special attention, as it reflects and integrates post-Kantian developments in mathematics and the natural sciences without giving up Kant's principal aim of a transcendental foundation for all scientific knowledge. Fries's commitment to Kant is best summarized at the end of an unpublished letter from 1832:

> Despite all this I remain a Kantian, because in the history of philosophy, what will be estimated more than any of our new findings is Kant's distinction of analytic and synthetic judgments, the fundamental question of how synthetic judgments a priori are possible, the discovery of a transcendental guideline and the system of categories and ideas, the discovery

of pure intuition, and finally the implementation of the doctrines in his critiques.[2]

Kant's transcendental philosophy has been taken as a turning point (not only, but also) in the *history* of the philosophy of science and of the philosophy of nature. As such it serves as an important landmark still in current systematic discussions. The question regarding the significance of Fries's philosophy therefore amounts to a question also about the relevance of all the work that originates in Kant. I propose to deal with his approach as a continuation of Kant's doctrine, which was motivated by the scientific achievements of his time; it can therefore be labeled as a "scientifically adequate" attempt to carry on Kant's approach.

To this end, Fries had to extend Kant's narrow definition of "science proper" as it is highlighted in the introduction to the *Metaphysical Foundations of Natural Science*. Kant's three necessary conditions for "science proper"—mathematicity, apodicticity and sytematicity—are closely related, though not reducible to each other. First, his claim that "in every doctrine of nature only so much science proper can be found as there is mathematics in it" (IV, 470) does not mean, of course, that any use of mathematics within a natural doctrine turns it into science. Second, not any apodictic doctrine is science, as is shown by metaphysics. And third, not any systematically organized doctrine is a proper science for Kant—though it can be science and even "rational science" (IV, 468)—as in the case of chemistry. Rather, proper science, according to Kant, needs a pure part in which the apodictic certainty of its first principles is founded and the possibility of physical objects is guaranteed by a construction of its concept in pure intuition (IV, 469–470). This is the basic idea underlying Kant's concept of "science proper." Its range "shrinks" even further with Kant's elaboration of this concept in the subsequent parts of the *Metaphysical Foundations*. Here it becomes an apodictic and systematic natural science that aims at an explanation of all natural phenomena by the interaction of corporeal masses according to fundamental attractive and repulsive forces. In the end, their mathematical construction is the kind of mathematization of nature that Kant asks for.

Kant's program of the *Metaphysical Foundations* met increasing resistance in the first decades of the nineteenth century.[3] One important reason was that his understanding of "science proper" excluded important new areas of research, especially within chemistry and biology. Even those "new sciences" that made extensive use of mathematics (and in so far followed Kant's ideal) did not reach the type of mathematization Kant was asking for.

And even within Kant's "model sciences," that is, mathematics and mathematical physics, certain developments—such as the rise of algebraic analysis within pure mathematics, of analytical mechanics within mathematical physics, or of the calculus of probability[4]—meant a challenge for those philosophers and scientists who in principle shared Kant's understanding of science.

At first sight, Fries's approach can be characterized as a twofold extension of Kant's strict and rigid understanding of science. First, he develops a *methodology* of science that gives scientific meaning to Kant's synthetic principles a priori in those areas where their constitutive character is by no means obvious. Second, he weakens in an "empiricist" direction Kant's demand that science has to form a *system*, that is, he weakens it in a way that allows the formation of different empirical theories (as sciences) without giving up the idea of a system of all scientific knowledge (as a regulative *ideal*).

This essay aims at a survey of Fries's philosophy of science with special attention to his extension of Kant's understanding of science in relation to scientific development in general. To suit this purpose, details of the history of the different sciences in question are omitted throughout. I first discuss in some detail Fries's "methodological transformation" of Kant's approach. I then illustrate this transformation with some examples from mathematical physics, chemistry and biology. Some concluding remarks on Fries's philosophy of (pure) mathematics are meant to show that it, too, can be characterized by Fries's predominant aim to keep together Kantian philosophy of science and the actual development of science.

FROM SCIENCE TO THE SCIENCES: "SYSTEM" AND "THEORIES" IN FRIES'S PHILOSOPHY OF SCIENCE

Fries devoted a substantial part of his large philosophical oeuvre to methodological and foundational problems of the natural sciences.[5] I will concentrate on one aspect that seems most significant with respect to his extension of Kant's understanding of science, that is, Fries's separation of "theories" from "system" and its attendant methodology.

According to Kant, systematicity is a necessary prerequisite for a body of knowledge to become a science: "Any doctrine, if it is to be a system, that is, a whole of knowledge ordered by principles, is called science" (IV, 467). It is well known that, with respect to natural science, Kant addressed the problem of systematic unity from two different directions. First, he approached it from the "bottom up," where empirical laws are successively brought under more general laws by our reflective judgment and where a logically conceived unity of all laws is presupposed as a regulative ideal of our reason. Second, he considered it from the "top down," where more and more special empirical laws are subsumed under the a priori laws of our understanding as they were "deduced" by Kant in his first *Critique* and specified in the *Metaphysical Foundations* (see, for example, Friedman 1992, 48–49, 242–264). It was shown elsewhere that the first approach is deeply rooted in Kant's precritical physico-theology—Kant himself later refers to a "subjective" and a "formal" teleology (V, 193)[6]—and that it seems philosophically insufficient according to Kant's own standards, in so far as it cannot explain the necessity of the special laws without which they are, for Kant, no proper laws at all but mere Humean regularities (see Pulte 1999b, 306–327): Reflective judgment is not constitutive. The second approach, on the other hand, seems insufficient in so far as it does not show how the great variety of empirical phenomena that refer to different kinds of matter are to be brought under a few very general concepts and laws: Principles of the understanding are not "immediate." Both approaches taken together raise the problem of how they are and how they can be coordinated so as to realize the ideal of a systematic whole of our scientific knowledge.

FRIES'S FRAMING OF THE ARGUMENT

Fries comes to this problem at an early stage of his career, and he locates it in one of the most serious defects that he finds in Kant's whole theoretical philosophy: Kant did not separate understanding and reason with sufficient clarity and he therefore mixed up knowledge (by our understanding) and belief (by our reason) at several important points of his transcendental argument. Fries's remedy is to sharply demarcate a so-called "natural world view" of the understanding and an "ideal world view" of reason as two different types of judgment about reality on equal footing (1828–1831, 5:310–324), and to introduce a mediating faculty called *Ahndung* or presentiment.[7] Without going into the subtleties of this modification of Kant's theoretical system (see Elsenhans 1906, 1:335–345), Fries's general argument can be summed up and focused with respect to philosophy of science in three steps.

First, Fries states that Kant's weak demarcation of the faculties of understanding and reason results in a "confusion of theory and idea" (1828–1831, 5:333). A distinction between theory[8] and idea is nevertheless absolutely necessary to circumscribe the legitimate claims of scientific knowledge and separate them from the excessive claims made by the ideal worldview: Science, belonging to the natural world view, emerges in the shape of *theory*. The necessary distinction between theory and idea also implies a demand for the differentiation of two kinds of regulatives that Kant often mixes up: "ideal regulatives" referring to ideas (reason) and "heuristic maxims," referring to theory (understanding) (1828–1831, 5:313ff.).[9]

Second, according to Fries, Kant had declared ideas to be only regulative, but in fact he also used them as constitutive. Fries thereby offers a new interpretation, or rather puts the Kantian notions "constitutive" and "regulative" in concrete terms with respect to philosophy of science. A principle is called constitutive "if, as soon as it is given, it decides the case of its application for itself so that the *subsuming judgment* is able to develop from it science in theoretical form; a principle is called *regulative*, on the other hand, if the *reflecting judgment* has to first seek out for it a case of application and its constitutive purpose" (1828–1831, 5:311). For the present it could be stated that constitutive principles enable theory, while regulative principles enable generalizations. It is important to note that in the case of Fries, as opposed to the case of Kant, this distinction arises relative to particular theories. And even within such particular theories it is not absolute: As we shall see later, regulative principles can become constitutive. According to Fries, Kant did not realize the "potentially" constitutive function of certain regulative principles. By endowing ideas with a regulative function for judgment, Kant implicitly allowed them to function as "physical regulatives" and thus as constitutive of experience, "after he had initially denied them all claims to constitutiveness" (1828–1831, 5:346).

Third, this problem can be removed in light of the first step and the split within Kant's subjective formal teleology, that is, by a distinction between two kinds of regulatives that are not differentiated in Kant's use of ideas: According to Fries, regulatives *of* theories and regulatives *in* theories have to be distinguished. Ideal regulatives contain general definitions about aims and forms of theories and serve mainly to separate theories from ideas; they are not constitutive and cannot become so. Heuristic maxims, however, are regulatives in theories; their function is to subordinate the special (particular empirical facts, particular empirical laws) under the general (particular empirical laws, laws of higher level); they play a leading role for induction. Fries wants to apply his thesis regarding regulative and constitutive principles exclusively to these: Heuristic maxims, that is, maxims of the systematizing understanding, can become constitutive for experience. As he indicates and as will be examined in more detail by an analysis of his understanding of "theory," the heuristic maxims operate on actual given experience, while the ideal regulatives operate on all possible experience—a difference that results from Fries's separation of natural and ideal world-view (1828–1831, 5:332). From the start Fries thus places Kant's problem of coordinating the bottom-up and the top-down approaches in the context of the *natural worldview*, as only in this context it can become a subject for the philosophy of science. The *ideal worldview*, on the other hand, does not refer to a given manifold of experience but to the whole of possible experience, which is not accessible to real science and therefore can have no impact on the philosophy of science.

ONE SYSTEM, VARIOUS THEORIES

Kant's subjective formal teleology is "global" in character, that is, it refers to the whole system of nature or the whole system of possible experience. For Fries, this all-embracing notion of "system" can be relevant only to the ideal worldview. However, the relevant "unit of knowledge" for the natural worldview is "theory": "We therefore demand theory in its strictest meaning from the natural world view of things; but just in its opposition to the ideal view" (1828–1831, 5:345). Fries defines theory as "a science in which facts are recognized in their subordination under general laws and their connections are explained by these" (1837, 541). It is crucial in this respect that the unity embodied by a theory can be given neither through experience itself nor through philosophy, because its necessary principles cannot say anything about a particular fact (1837, 551). Theories are characterized rather by mathematical unity. Only pure intuition includes particular facts and general rules, so that only mathematics can contrive the connection of both: "If at all we therefore achieve theory and explanation only through mathematics" (1837, 551).

Fries draws two important conclusions from this: On the one hand, theory can only explain such empirical facts that can be subsumed under the same notions of magnitude (*Größenbegriffe*). On the other hand, it "follows that there should be as many theoretical beginnings in our cognition as there are different qualities. Of these there are, however, various ones in the doctrine of nature [*Naturlehre*], so that any theoretical task in our cognition is limited; the theories of our science cannot be unified in a system, there are instead as many individual theories separated from each other as there are separate qualities" (1837, 552).

This means, in more concrete terms, that different qualities (like sound or heat) can define (at least provisionally) different theories (like acoustics or a theory of heat). However, Fries's "pure doctrine of motion" (*Reine Bewegungslehre*)—an elaboration and extension of Kant's *Metaphysical*

Foundations[10]—retains an exceptional, "towering" role, as its laws are valid for any objects of outward experience, whatever their qualities are; it is the constitutive theory par excellence (1822, 3). But this heritage of Kant's "top down-systematization" is methodologically transformed by Fries in a characteristic manner: As the a priori laws of motion, as given in Fries's phoronomy,[11] are valid for all physical objects, they form what I will later call "a priori anchors" for the development of heuristic maxims of the different theories (such as acoustics, the theory of heat, etc.) in question. As such, however, they do not determine the empirical content of the more special theories, but are to be understood merely as heuristic guides for these theories. Though we shall always try to reduce sensory given qualities to fundamental properties of matter, force, and movement, any actual theory has to accept sensory qualities as given and thus starts its mathematical development with the notions of magnitudes belonging to them: "no outward quality like color, sound, heat, smell etc. can be explained as such, but each alone is the principle of a theory in which the gradual differences are reduced to its most simple relations" (1837, 595, cf. 551–552). Regardless of his general Kantian orientation, Fries here expresses an "empiricist concession" that is rooted in his detailed knowledge of and intimacy with the scientific practice of his time and the differentiation of particular *theoretical* subdisciplines of physics that traditional mechanism could no longer hold together.

We therefore find a pluralism of theories with Fries that clearly goes beyond the scope of Kant's concept of system. This point is decisive for understanding the difference between Kant's subjective formal teleology and Fries's heuristic maxims, because these maxims correspond to concrete theories that need to have a limited range of experience (1828–1831, 5:345). In this sense the maxims always refer to a "really given manifold" (the "reality" of which is the practice of the scientific development of theories) and not to "any somehow imaginable [*irgend zu gebende*] manifold," that is, not on an all-out system of experience inaccessible to science (1828–1831, 5:323).

This restriction of maxims, taken by itself, does not solve Kant's problem, but it points the way to a solution: Kant's problem does not so much reveal a defect of empirical theory building, but rather poses a problem for empirical methodology. Even a theory that has constitutive, that is, mathematical, principles (1837, 551) is in need of such a methodology, because the "deductive range" of such principles is most often limited:[12]

> In each mathematical system we can actually develop the system from the
> highest principles in forward direction by putting together each complex

[*Komplexion*] out of its elements; but with these developments we always reach only a certain point where the composition of the complexes will be too large. Here we follow the reverse way of observation, regard the complex as a whole and just try to organize the complexes at large by an involution without completing the evolution down to the last detail. The latter method of induction demands a development of constitutive laws as precisely as possible in order to obtain certain heuristic principles; however, it remains indispensable in its own sphere as all theoretical compositions always treat only general laws without finding the way to a particular story. (1828–1831, 5:312–313)

The experience of incompleteness of each actually performed "development" of a theory expresses the impossibility in principle to complete such a "development," which is a consequence of Fries's restriction to the natural worldview. The quotation reveals, however, that this restriction does not relax the demands on the formation of scientific knowledge, but in a way strengthens them: It is a matter of getting theory and experience into a kind of "dynamical balance" in order to gain (as far as possible) complete scientific explanations—a problem of balancing deduction and induction, constitutive laws and heuristic maxims.

Now, Fries's picture of theory formation is roughly this:[13] Theory starts genetically, as does all our knowledge, with experience and proceeds by means of induction and speculation to general concepts, rules, and classifications, at best up to constitutive principles. This process is not linear: Rules already gained have to be reconsidered in regard to particular cases and serve as guidelines for further generalization on their part. These guidelines therefore have an "anchor point" in prior experience. Speculation provides a second "anchor point," which is a priori: It demonstrates by means of mathematical and philosophical abstraction which general laws are possible at all for a certain field of experience and in what way these laws relate to constitutive theories (that may already exist). For example, in the theory of gravitation experience shows that an attractive and central force between single masses exists, and only experience can find out its degree with respect to mass (empirical fixation); mathematical and metaphysical abstraction, however, define the form of the law of gravitation (a priori fixation; compare 1822, 400–401, 443–499). The guidelines thus "fixed" twice are nothing else than Fries's heuristic maxims, that is, maxims of the systematizing understanding. They regulate the further formation of theory in form of a "rational induction" as opposed to an unguided (in a way "blind") induction. At best, they lead to the discovery of constitutive principles for the theory in question.

This regressive procedure describes only the genetic development up to a constitutive theory. However, the ideal case of such a theory that Fries recognizes in celestial mechanics is the exception (1822, 345). In various fields, for example in natural history, there are theories that he himself— rather misleadingly—describes as "regulative": Their laws are nothing more than generalizations, more or less probable, though it should be noted that even their heuristic maxims have to be directed to the most general constitutive principles of phoronomy (1837, 596). Fries therefore talks about "two different ways" that may serve to develop "theoretical science" (1828–1831, 5:595):

> First we gain *constitutive theories* following the progressive method of the subsuming judgment and then [we gain] regulative theories following the regressive method of the reflecting judgment. In their presentation the constitutive theories proceed systematically from their principles, they therefore demand a principle that allows developments on its own accord, that is, it demands a precisely defined mathematical task. . . . *Regulative theories* first require induction as the method of invention in order to proceed from facts to general laws which here are to be asserted as principles of the theory. (1828–1831, 5:595–596)

These two methods of theory formation—"bottom up" and "top down"— highlight Fries's methodological dissolution of Kant's problem: If (and only if) it succeeds—usually by an interaction of both methods—in reaching a complete constitutive theory, the laws of this theory can be subsumed in a logical, deductive system. Only then does it make sense to talk about the "necessity" of special laws, which was Kant's cardinal problem from his precritical period on (see Pulte 1999b, 314–327). For Fries, however, there is neither a guarantee nor an absolute requirement to demonstrate the necessity of the laws of a theory—no guarantee, as constitutive principles are scarce, and no absolute requirement, as all human science is natural science, and all natural science must be restricted according to the natural worldview: "We presuppose as known that in human convictions this whole [natural] *science must remain separated from the belief in eternal truth*, though it is subordinated to belief" (Fries 1822, 1).

THEORY AND UNITY OF EXPERIENCE

Fries's methodological considerations bring him to another remarkable conclusion: If heuristic maxims as guidelines of rational induction reveal "a priori-anchorage," and if, furthermore, theory in general develops as an interaction of regressive and progressive methods, it is not at all possible to differentiate sharply between the merely regulative and constitutive functions of these maxims: Heuristic maxims of rational induction therefore must have constitutive contents (1828–1831, 5:311).[14] Based on this conclusion, Kant is reproached for "mixing and confusing theory and idea" (1828–1831, 5:333). As Kant does not differentiate clearly enough between natural and ideal worldviews, he consequently makes no difference between heuristic (that is, "systematizing") maxims of the understanding and the regulative ideals of reason, which are both involved in the regulative use of ideas. Though Kant did not want to admit this, the regulative use of ideas goes beyond the mere regulation of experience. As Fries remarks:

> In the most general case this mistake reveals itself in the use of the ideas of soul, the world and the deity, which even Kant falsely recognizes as physical regulatives after he had first denied them any claims to constitutive character. Here he did not understand, however, the nature of the systematizing maxims, otherwise he would have understood that, when applied, each regulative maxim for the natural view of things is only different in degree from the constitutive law and is actually a yet unknown constitutive law at the bottom of the theory.... (1828–1831, 5:346)

By declaring theory to provide the proper unity of experience, Fries of course reinterprets Kant's terms "regulative" and "constitutive": For him, unity of experience proves a meaningful aim only in relation to a certain theory, which means that regulatives as well as constitutives can be specific only to theory. One might call this Fries's principle of localizing by rendering empirical.

With this principle, Fries also intensifies a problem of the philosophy of science—and offers a respectable methodological solution—which remained unsolved with Kant. This is the problem of the relation between the theoretical unification of experience by general laws and the constitution of experience (in the peculiar sense of gaining objective experience of particular facts by science). Kant's claim for a unity of experience without constitution of experience, his subjective formal teleology of the "as if," is hardly satisfactory in this respect. In contrast, Fries's position avoids this kind of teleology and, in a way, appears "modern": Theoretical unification and the constitution of scientific experience are, according to his view, two sides of the same coin.

KANT'S METAPHYSICAL FOUNDATIONS AND FRIES'S MATHEMATICAL PHILOSOPHY OF NATURE

The criticism of Kant's philosophy of science that was sketched in the previous section might obscure the fact that Fries's approach is first and foremost aimed at an elaboration and, so to speak, at an "updating" of Kant's *Metaphysical Foundations*. At the same time, Fries sharply rejects the speculative strand of German *Naturphilosophie* as it appears in the works of Fichte, Hegel, and, above all, Schelling.[15] Among Fries's works, his *Mathematical Philosophy of Nature* is most significant in both regards (1822, v–vi, 1–3, 31–32, 397–398, 507–509). It would go beyond the scope of this paper to provide a detailed comparison of the *Mathematical Philosophy of Nature* and the *Metaphysical Foundations*. Instead, I will confine myself to some general observations about several branches of the natural sciences and of mathematics as treated in Fries's work. In this I am guided by two aims: First, I would like to use some examples from the "special" sciences to illustrate and underscore Fries's methodological reflections as presented in the previous section. Second, I will hint at some of the amendations and improvements of the *Metaphysical Foundations* that were offered by Fries and that may be representative of his approach in general. There can be no doubt that Kant's ingenious attempt to provide a transcendental foundation for the scientific knowledge of his time not only reflects the spirit of his time with respect to the extension of "science proper" but also reveals serious gaps within the domain of what was actually accepted as "science proper," though these gaps have been most often ignored in the German reception of the *Metaphysical Foundations* up to now.[16] A look at Fries's *Mathematical Philosophy of Nature* may contribute to a more complex picture.

Fries's principal work concerning the philosophy of science is divided into two parts: I will deal later with the first part, on the "philosophy of pure mathematics."[17] The structure of the second part, on "pure theory of motion," already shows that it is guided by Kant's *Metaphysical Foundations* but that Kant's work by no means determines Fries's approach to the philosophy of the different sciences: (1) "phoronomy," (2) "foundations of dynamics," (3) "foundations of mechanics," (4) "foundations of stoichiology" (*Stöchiologie*) or "foundations of the doctrine of the kinds and compositon of masses," (5) "foundations of morphology," and (6) "foundations of phenomenology" (1822, ix–x). Fries obviously accepts Kant's *Metaphysical Foundations* as the starting point of his investigations (1–3, 6), but not as sufficient (4, 5) (1822, 411–412).

MATHEMATICAL PHYSICS

Kant's *Metaphysical Foundations* are synthetic not only in an epistemological sense (synthesis of a priori concepts) or a methodological, especially Newtonian sense ("proved" explanations of phenomena and special laws by deduction from principles), but also in a traditional mathematical sense (relying on Euclidean geometry). The analytical tradition of mechanics goes back to the late seventeenth century, and achievements like the principle of least action, the principle of virtual velocities, and various forms of conservation laws—especially, of course, the conservation of mechanical energy for a large class of mechanical systems—might have shown Kant that conceptual foundations of mechanics fundamentally different from Newton's may well have been possible. And yet, this strand of mathematical physics is totally absent from the conceptual analysis of his *Metaphysical Foundations*. Its philosophical relevance was not acknowledged in Kant's critical period at all.[18]

Fries by contrast appreciated this development in the foundations of mathematical physics manifest especially in the works of Leonhard Euler, Pierre Louis Moreau de Maupertuis, Jean le Rond d'Alembert, Joseph Louis Lagrange, and Simon Denis Poisson. Both in his phoronomy (1) and in his mechanics (3), Fries refers to their approaches as alternative, that is, essentially non-Newtonian frameworks of mechanics. The *Mathematical Philosophy of Nature* is in fact probably the only German work in the first half of the nineteenth century in which this divergence of different attempts at the foundation of mathematical physics is reflected at all as a philosophical problem and in which an integration is proposed.

This proposed integration follows Fries's methodological reflections as described earlier: The "constitutive" or "direct" principles of the pure doctrine of motion are, by and large, Newton's laws of motion. Newton's second law is added to Kant's "legislative framework" of the *Metaphysical Foundations* as a conventional stipulation prior to any empirical observations about motion—a priori not in the sense of "condition of the possibility of experience" but in the sense of "necessary to judge given experience," or in more concrete terms: to deal properly with forces and motion (Fries 1822, 402–403; see König and Geldsetzer 1979, 26*). The principles of analytical mechanics, on the other hand, are "indirect"; they are results of "bottom-up-approaches" for systematizing mechanical experience before constitutive principles were found, and they are still useful when applied to mechanical systems with unknown interactions of forces (1822, 399–400, 404–405).[19] Thus Fries stresses the heuristic relevance of these principles according to the regressive method of theory formation: "All theory here starts from experience, but experience does not teach us the laws of motion, but requires us to search for these laws and determine the applications of pure laws to particular phenomena. So, the treatment of particular experiences at first always leads to indirect methods, where not all laws of the acting forces are known" (1822, 404).

Though the constitutive "Newtonian" laws are necessary in order to develop the pure theory of motion progressively and in a "synthesizing" manner, the "integrals of motion" and variational principles of analytical mechanics remain important as heuristic devices and instruments of applying the pure theory to intricate mechanical problems. In general, and with respect to the enormous rise of mathematical physics in the late eighteenth and early nineteenth centuries, Fries stresses the creative and formative role of mathematics for the natural sciences: "For, this science [the pure theory of motion] is actually the armory of all those hypotheses from which later explanations are drawn that have success in experience. Most of it concerns mathematical developments, the basic concepts, however, are of philosophical nature, and should this be successfully communicated to experts of natural science [*Naturkundigen*], we would gain quite a lot for the discipline of hypothesis" (1822, 10).

Chemistry

According to Kant's well-known dictum, it is likely that "chemistry can become nothing more than a systematic art or experimental doctrine, but never science proper" (IV, 471). This expresses neither his lack of appreciation nor his lack of interest in chemistry.[20] It rather highlights the fact that Kant saw no possibility of giving chemistry an a priori foundation that would meet the standards laid down in the *Metaphysical Foundations*, that is, an a priori foundation beyond chemistry's merely empirical generalizations, which lead to empirical rules instead of laws and to regulative ideals instead of fundamental concepts. Though the "chemical revolution" dramatically changed the character of chemistry during Kant's lifetime and especially succeeded in reaching important quantitative laws through the research of A. L. Lavoisier, L. J. Proust, J. Dalton, J. J. Berzelius, J. B. Richter, and others, this was, of course, not sufficient according to Kant's foundational claims.

Fries first discusses the problem of mathematizing chemistry in his *Criticism of Richter's Stoichiometry* (1801). Jeremias Benjamin Richter was a former student of Kant who somehow trivialized his teacher's demand for *mathematical* foundations of natural science (Carrier 1990, 200–201). Though Fries welcomes Richter's attempt to make chemistry a mathematical science (Fries 1801, 135), he firmly criticizes his realization of this aim. His criticism concentrates on two points (1801, 49): First, Richter gives no systematic presentation of stoichiometry but only a rhapsody (1801, 19, 25)—or, to use Kant's phrase, an "aggregate"—because he does not sufficiently reflect the metaphysical foundations of his science. Second, and even more important, Richter does not recognize that mathematics in natural sciences—aiming as it does for a foundation—cannot be applied to arbitrary experience but must be used to construct a priori concepts that make possible the experience relevant to the science in question (1801, 9–10, 13–18, 22–23, 48–49, 88–89). By mistake, Richter applies mathematics to the "art of chemical experimenting," whereas he should have applied it to gain a "theory of chemistry" as a subsystem of the "physical sciences" that is in need of both metaphysical and mathematical principles in order to be accepted as a science (1801, 16–17, compare 18–19, 89, 118). And in order to reach a pure theory of chemistry, Fries argues, proper mathematical principles must enter at the level of dynamics in Kant's sense (1801, 14–16). As Richter does not recognize the importance of dynamics for his "Kantian project," and especially underestimates the complexity of forces acting in chemical compounds, the quantitative regularities he finds in his "mass rows" (*Massenreihen*) can at best be compared to Kepler's laws of planetary motion, for which a Newton had yet to come (1801, 121–122, compare 17–19, 116).[21]

In the part of his *Mathematical Philosophy of Nature* devoted to "stoichiology" (1822, 540–571),[22] Fries tries to develop a dynamical foundation of chemistry: "The kinds of masses must not be separated according to mechanics . . . but according to dynamics, that is, according to the different relations of their fundamental forces. So the concept of *substance* (in the chemical meaning of the word) becomes meaningful to natural philosophy" (1822, 540). In his theory of fundamental forces, Fries adopts Kant's double dichotomy of attractive and repulsive forces on the one hand, penetrating forces (*durchdringende Kräfte*) and contact forces (*Flächenkräfte*) on the other hand. But contrary to Kant, Fries takes all four kinds of fundamental forces into account: attraction and repulsion at a distance, attraction and repulsion in contact (1822, 543–547, compare 451–453, 620–622). Though Fries's elaboration of this considerable deviation from Kant's dynamics cannot be discussed here, it should be noted that, according to Fries, the "gap" between the essentially mathematical level (1822, 451–452, 621–622) of constructing proper forces and the level of chemical phenomena cannot be bridged without leaving space to conjectures and hypothesis: "The future development of science will decide if hypotheses of this kind are useful or not. In any case all processes of gravitation as well as all phlogistic and chemical processes have to be explained by universal penetrating forces and contact forces" (1822, 571). Chemistry may thus not become a proper science according to Kantian standards, but is definitely a science according to Fries's "methodological extension," because it can be developed in the form of theory.

BIOLOGY

In his "foundations of morphology" (1822, 572–600) Fries also transcends Kant's realm of "science proper." Part of it is a "theory of morphotic processes" (*Theorie der morphotischen Prozesse*) or of "natural drives" (*Naturtriebe*) (1822, 584–585), as he calls it in a rather misleading manner. The designation "natural drives" is misleading, because it suggests an animistic or even anthropomorphic understanding of organic processes that Fries seeks to avoid and that he criticizes throughout his philosophy of biology (see Fries 1813, 394–400).

Morphology has to do with the forms of those interactions of physical bodies which cannot sufficiently be explained by fundamental forces alone (1822, 581). It is not restricted to organic processes, but is relevant already for a constitutive theory of mathematical physics, like celestial mechanics. In order to explain the movements of planets along conic sections, for example, the law of gravitation is not sufficient but must be accompanied by considerations of the configuration of the system or, to use mathematical terms, by the consideration of initial and boundary conditions. The aim of morphology is a mathematical classification of the different types of these conditions in order to distinguish different forms of physical interaction under the same fundamental forces. As far as they are relevant to a causal explanation of physical interactions beyond the fundamental forces, these conditions are designated by Fries's unfortunate notion of "natural drives" (1822, 582).

Now, the mathematical philosophy of nature must construct its different kinds mathematically. In the case of living plants or animals the accomplishment of this program may create immense mathematical and empirical problems. It will be essential, however, that one never introduce "an unexplainable fundamental force for certain substances, namely organic matter," but that instead one always strive for "an explanation in terms of a law that governs a certain kind of interaction in the world of physical bodies" (1822, 583). Fries thus rejects vitalism, but also the use of a material or "objective teleology," in order to explain organic processes. This kind of teleology was criticized but not always avoided by Kant. In contrast, according to Fries's twofold (that is, progressive and regressive) way of developing theory, this kind of teleology can always be used as a heuristic device in the regressive approach. In other words, this teleology can be used in order to reach scientific explanations by fundamental forces and morphotic structures, but always has to be excluded in the progressive approach, that is, as an explanation in its own right (1822, 597–598).[23] Without going into the details of Fries's methodology of biology, one might say that the appearance of a "Newton of the blade of grass," which seemed impossible to Kant (§75, V, 400) was no mere utopia to Fries but seemed reachable one day by the application of his direct and indirect approach. Fries's adherent Matthias Jacob Schleiden, botanist and one of the founders of modern physiology, later made abundant and successful use of this methodology in biology (Schleiden 1989; see Charpa 1988 and 1999).

PURE MATHEMATICS

Of course, Fries accepts not only Kant's premise that mathematics is decisive for reaching a proper understanding of natural phenomena, but also his premise that mathematics is of philosophical interest in its own right. Therefore it is not by accident that the whole first half of Fries's *Mathematical Philosophy of Nature* deals with "philosophy of pure mathematics" (1822, 33–395). Though this subject is actually beyond the scope of this paper,[24] some remarks about its character may show that its development fits the general objective of Fries's philosophy of science, that is, to "modernize" Kant's approach in light of actual scientific developments.

For internal as well as external reasons (especially the rise of neohumanism), German mathematics in the first decades of the nineteenth century was strongly oriented toward "pure" mathematics. This pure mathematics was sharply distinguished from sensory experience and aimed at rigor beyond questionable intuitive foundations. Therefore, arithmetic and algebra, rather than geometry or mechanics,[25] become models of mathematical research. The growing autonomy, abstractness, and "symbolladenness" of mathematics leads to doubts about Kant's understanding of mathematical concepts as mere constructions in space and time.

Fries seems to be not only the first German-speaking philosopher who explicitly asked for a philosophy of mathematics as a metatheory of pure mathematics (see Pulte 1999a, 74), but also the first to work out such a metatheory as a "complete system of mathematical forms" (1822, 50).

According to him, the two principal problems of this metatheory are the origin of mathematical knowledge and the foundational claims of mathematics in the context of all human convictions (1822, 48). With respect to the new developments within mathematics, which Kant did not reflect, two characteristics of his approach are worth mentioning, namely, his introduction of "syntactics" and his modification of Kant's understanding of mathematical apodicticity.

First, Fries clearly differentiates between a "syntactics or theory of combination [*Kombinationslehre*] as a theory of the pure laws of arrangement of given parts" and the "theory of numbers, arithmetic, which is based on the idea of wholeness composed of homogenous parts" (1822, 65). Arithmetic is more restricted than syntactics in so far as it composes its objects (that is, numbers) from a special syntactical postulate (homogeneity), though our productive imagination allows for other forms of composition (1822, 68). One can undoubtedly trace back to the works of Carl Friedrich Hindenburg and his so-called "combinatorial school" Fries's view that arithmetic aims at a measuring determination of magnitudes by concepts of pure intuition and is preceded by a regulating syntactics that is interested in the construction of the "most general mathematical concepts" and is not based in intuition.[26] This view takes up and develops Euler's and Lagrange's algebraical foundation of analysis:

> That syntactics is in principle independent of arithmetic is decided among us since Hindenburg. The task of syntactics is putting in order, the task of arithmetic is measuring. To syntactics belongs no separate purely imaginative [and] fixed sequence; but only the peculiar operation of productive imagination, that is, putting in order. Therefore syntactics has no axioms, but only postulates. In contrast, arithmetic borrows its postulates from syntactics, but has its separate fixed sequence of the larger and smaller and separate axioms for this. (Fries 1822, 68)

In Fries's philosophy of mathematics syntactics becomes a second basic discipline next to (and in a way prior to) arithmetic. The first creates more "qualitative" mathematical concepts (one might think of B. Riemann's later concept of an *n*-dimensional manifold), while the second creates more quantitative concepts (such as numbers and magnitudes).

It fits into this context that in Fries we encounter second (and more generally) a separation between pure intuition and mathematical apodicticity. Admittedly, mathematical knowledge that is different from philosophical knowledge is not given to us by thinking, but "already by itself in clear intuition. To realize, however, its universality and necessity I need thinking"

(1837, 417). It is therefore right to say that Kant's "apodicticity dualism" of intuition and thinking is replaced by an "apodicticity monism" of thinking in Fries's philosophy of mathematics (see Ende 1973, 35). For Fries, all apodictic knowledge is "discursive, philosophical knowledge as well as mathematical knowledge" (1837, 412).

This thesis only seems to signify a restriction of the Kantian meaning of apodicticity: By substituting the productive imagination as foundational authority for pure intuition, Fries actually opens the field of mathematical apodicticity to such propositions that have no foundation in Kant's pure intuition. He thereby takes into account the general development of mathematics in his time, which is characterized by an increasing abstraction and self-reference of its laws and by the complexity of its structures.

Fries did not (and, for several philosophical reasons, could not) extend his originality to the foundations of geometry and therefore remained strongly in favor of one (and only one) axiomatic system of geometry, that is, Euclid's (1822, 355–380; see Gregory 1983a, König and Geldsetzer 1979, 63★–69★). Nevertheless, his philosophy—and especially his philosophy of mathematics—found strong supporters among mathematicians. C. F. Gauß, for example, praised his work as exceptional and lucid in times of growing philosophical obscurity (König and Geldsetzer 1979, 39★–40★).

CONCLUSION

My outline may have shown that Fries's rather limited impact on later philosophy of science stands in remarkable contrast to his actual achievements in this area. It may have also indicated at least one important reason for this discrepancy: Fries's approach aims at establishing an autonomous philosophical metascience that develops in close contact with science. Philosophy of science can neither replace scientific research nor become superfluous owing to scientific developments—both areas are, on the contrary, complementary and interacting. In a way, however, this model was too modern to be successful in his time. While German academic philosophy and its historiography stuck to the idea of the predominance of philosophical speculation over empirical research, most practicing scientists turned away from German "school philosophy" and considered science and its history from the point of view of naive positivism. Neither view could perceive and appreciate Fries's peculiar approach and his achievements. Neo-Kantianism, however, could have done so, but frequently lost sight of a respectable part of its (potential) history when it followed nearly unanimously O. Liebmann's slogan "back to Kant." Among other reasons, this historical development contributed to the neglect of the philosophy of science in the tradition of Fries and his adherents (Ernst Friedrich Apelt, Matthias Jacob Schleiden, Oscar Xaver Schlömilch, Leonard Nelson, and others) up to now. But as Ernst Cassirer put it:

> It is his [Fries's] and his pupil Apelt's decisive merit that they . . . related the fundamental question of philosophy again to the "fact of science" and thereby brought it back on a strictly scientific ground. . . . what Fries and Apelt did for the elaboration of Kant's doctrine of synthetic principles, what they did especially for the understanding of particular fundamental concepts and fundamental methods, remains valid and has to be accepted also by him who rejects Fries's "anthropological" criticism as a foundation of philosophy. (Cassirer 1923, 482–483)

NOTES

1. For biographical information, see Frederick Gregory's chapter in this volume; Mourelatos 1967 gives a short but informative overview. The standard biography on Fries is still Henke 1937. The last volume of Fries's complete works (*Sämtliche Schriften*) (1967–), however, will contain rich additional material on his life and work. Glasmacher (1989) provides a valuable bibliography on Fries and his school up to 1988.

2. J. F. Fries to an unknown recipient, 21 September 1832 (letter no. 1177, to appear in the final volume—volume 29—of Fries's complete works).

3. I do not take into account here Kant's *Opus postumum* (especially his *Transition from the Metaphysical Foundations of Natural Science to Physics*) as it was largely unknown at the time. For Kant's later philosophy of science, see Friedman 1992, 213–241.

4. I return to the first two examples below. For the calculus of probability, see Fries 1842.

5. The bibliography includes his most important contributions to the natural sciences and to philosophy of science. Useful presentations of this part of his work can be found in Amir-Arjomand 1990, Hermann 2000, and, above all, König and Geldsetzer 1979.

6. I will take up and shorten Kant's paraphrase given in this passage and use the term "subjective formal teleology." See also Kant's first Critique (A620/B648ff).

7. Ahndung literally means "presentiment" but is used by Fries also in the meaning of "aesthetic sense." See Fries 1805, 601–755. In the following analysis I draw on Pulte 1999b, 330ff.

8. As a proper understanding of Fries's notion of "theory" depends to a certain extent on the frame described here, I will postpone a discussion of it to the next section.

9. The latter are also called "maxims of the sytematizing understanding" (e.g., 1828–1831, 5:323). Both ideal regulatives and heuristic maxims can be considered fission products of Kant's subjective formal teleology.

10. See Fries 1822, 397–690, where the pure doctrine of motion is treated in a sense close to Kant's phoronomy (in the first chapter of the second part, 397–442); see also the section "Mathematical Physics" below.

11. "This science is in a way the philosophy of applied mathematics. The pure doctrine of motion ... is mathematics applied to metaphysical knowledge; it contains the system of the whole [and] complete scientific knowledge of man" (1822, 397; cf. 3, 10).

12. Fries regards celestial mechanics as an exception; I will come back to this point.

13. For the following summary, compare Fries 1828–1831, 5:325–332, and 1837, 426–433.

14. Compare the section "Fries's framing of the argument" above. Fries's consideration can be illustrated as follows: A heuristic maxim serves to generalize a hypothesis about a field of experience that conforms to certain a priori constraints, because speculation supplies a structural framework of conditions that have to be obeyed in the construction of a hypothesis. If one of the hypotheses can be confirmed by eliminative induction, it is constitutive in so far as it contains new cases of application that were not considered before. Newton's law of gravitation was framed as a hypothesis with respect to the system earth-moon. In the sense described above it becomes constitutive with respect to other systems (like sun-earth).

15. A subtle analysis of this antagonism is given in Bonsiepen 1997; see also Gregory 1983b and 1989.

16. I would like to refer to Kant's omission of Newton's second law of motion in his attempt to give a foundation of mathematical physics. Kant does not try to give an a priori derivation of this law—which would be crucial for a foundation of rational mechanics in general. Moreover, this point is not discussed in a number of books devoted to Kant's *Metaphysical Foundations*. See, for example, Gloy 1976, Plaass 1965, and Schäfer 1966. For a reasonable analysis of this point (and conflicting interpretations), see, on the other hand, Pollok 2001, 387–388.

17. Fries 1822, 33–395; see the section "Pure Mathematics" below.

18. However, for the role of teleology (and especially the principle of least action) in his precritical period, see Buchdahl 1969, Waschkies 1987, and Pulte 1999b.

19. In some cases, as in the theory of capillarity, for example, the "indirect method" even seems indispensable in order to find the correct laws of the interacting forces; see Fries 1822, 408.

20. On the contrary, Kant's work—especially the *Opus postumum*—underpins his strong interest in foundational questions of chemistry; see Carrier 1990 and Friedman 1992, 264–290.

21. In later works, Fries was much more favorably disposed toward Richter's stoichiometry, obviously because he saw how difficult it would be to modify Kant's dynamics according to the demands of quantitative chemistry; see, for example, Fries 1822, 644, 654, and 1826a, 15–16, 52, 248–249; compare also Henke 1937, 49.

22. Fries uses the notion "stoichiology" as a synonym for "chemistry." See Fries 1826a, 15.

23. In Kant's application of teleological arguments in the realm of organic processes, Fries finds important evidence for his thesis that Kant did not sufficiently distinguish ideas (where teleology may be used) and theories (where teleology must be forbidden), that is, the ideal and the natural worldview (see section "One System, Various Theories" above, also Pulte 1999b, 327–329).

24. Several aspects of Fries's philosophy of mathematics are discussed in König and Geldsetzer 1979, 36*–69*, Gregory 1983a, and Schubring 1990.

25. It seems worth mentioning, however, that in the field of mechanics the orientation to "pure" mathematics leads to a conventional interpretation of mechanical principles half a century before Poincaré transferred his conventionalism from geometry to mechanics; see Pulte 2003, chaps. 5 and 6.

26. Hindenburg's school and its relevance for early nineteenth-century German mathematics is discussed in some detail in Jahnke 1990, 161–232.

VII

Kant, Helmholtz, and the Meaning of Empiricism

Robert DiSalle

The advent of non-Euclidean geometry has always been recognized as one of the fundamental challenges to Kant's theory of spatial intuition. If non-Euclidean geometries are possible, then it is at least questionable whether Euclid's postulates have the "necessity and universality" that Kant attributed to them as aspects of the form of outer intuition. Through most of the nineteenth century, of course, one could regard those geometries as mere conceptual possibilities and maintain, as many Kantians continued to do, that the actual space of our experience remains necessarily Euclidean. But by the turn of the century, whether space is Euclidean was widely regarded as an empirical question, to be settled by measurement. And the theory of relativity, especially general relativity, eventually demonstrated the applicability of non-Euclidean geometries to the physical world. In the early twentieth century, consequently, an important neo-Kantian aim was to generalize the theory of spatial intuition in order to render it compatible with the new science of physical geometry.

The other fundamental challenge was closely related to the empiricist challenge in its origins and motivations, but opposed in principle. For Poincaré and his followers, including, most notably, Einstein and the logical empiricists, reflection on the empirical foundations of geometry led naturally to conventionalism. For an empirical decision could not be made among mathematically possible geometries that have, in themselves, no empirical content; their empirical content is determined by some connection between an abstract geometrical object and an observable object or process—for example, the familiar association between Euclidean straight lines and the paths of light rays. And such a connection cannot be established by empirical measurement, but only by a stipulation that first makes empirical measurement possible. On the one hand, the logical empiricist discussions of convention, correspondence rules, and coordinative definitions are all twentieth-century reformulations of an essentially Kantian notion, namely, that knowledge begins with a framework of constitutive principles that first makes objective knowledge possible. Geometry, in particular, seemed to the logical empiricists to exemplify such an a priori constitutive framework—a framework within which objective measurement is possible but whose principles themselves cannot be justified by—because they are the presuppositions of—geometrical measurement. On the other hand, if these constitutive principles are essentially mere stipulations, then they represent not a synthetic but an analytic a priori. In fact they are, according to the logical empiricists, constitutive of linguistic frameworks rather than of "possible experience," and given the multiplicity of such frameworks, a decision among them can only be made on pragmatic grounds.

In both of these post-Kantian developments, Helmholtz played an essential role. Before Helmholtz, the empiricist alternative to Kant was a naive one, based on the idea that geometrical postulates are arrived at by straightforward induction (cf., e.g., Mill 1843). Helmholtz saw a subtler connection between geometry and experience, not directly of the postulates of geometry but of underlying physical facts. In particular, our experience of the "free mobility of rigid bodies" and our acquaintance with the paths of light rays provide the basis for our confidence in Euclid's postulates. For the postulates assert the possibility of certain constructions in space, and our intuitive conviction that those constructions are possible derives from our gradually acquired expectations about the behavior of rigid bodies and light rays. The far-reaching implication that Helmholtz drew is that, on the very basis on which our confidence in Euclidean geometry rests, a variety of non-Euclidean geometries may be constructed. For the constructive principles derived from rigid motion and light propagation are too general to favor Euclidean geometry over all other homogeneous geometries; precisely which homogeneous geometry characterizes physical space must be decided by experiments involving rigid bodies and light rays. It follows that non-Euclidean geometries are not merely mathematically possible but are "imaginable" in precisely the way that Euclid's geometry is—that is, they are possible objects of spatial intuition.

From Helmholtz's relatively sophisticated empiricism, however, the step to conventionalism is a small one. Poincaré's arguments for conventionalism began, in fact, precisely where Helmholtz's empiricist arguments ended. For if the laws of rigid motion and light propagation are, as Helmholtz said, "facts that lie at the foundation of geometry" (cf. the title of Helmholtz 1868b), their associations with geometrical concepts appear to be something other than factual claims; statements like "light rays travel on straight lines" and "lengths are congruent that coincide with the same rigid body" do not provide information about light rays and rigid bodies but only state *what we mean by* "straight" and "congruent." The conventionalist perspective, then, is only a slight modification of that of Helmholtz: The fundamental fact underlying geometry is that light rays and approximately rigid bodies provide particularly convenient *definitions* of straightness and length. As it turns out, precisely what distinguished Helmholtz's empiricism from its more naive predecessors—his appreciation of the physical meaning of our concepts of congruence and straightness—invited its reinterpretation as conventionalism.

Helmholtz's empiricism, then, has always been in a somewhat ambiguous position with respect to the apriorist alternatives of Kantianism and conventionalism. With respect to the nativist theory of spatial perception, Helmholtz's empiricist position was fairly straightforward; the question whether spatial distances are gradually learned or immediately detected by the sensory apparatus had philosophical ramifications, but was eventually amenable to straightforward empirical tests. Whether the founding principles of geometry are supposed to be empirical facts, as opposed to a priori constitutive principles of one sort or another, is harder to discern from Helmholtz's writings. An answer to this question is required for any clear account of the relationship between Helmholtz and Kant, as well as for a proper appreciation of the relevance of Helmholtz for later philosophy and science. In considering the question, I will suggest that the unclearness surrounding the relationship between Helmholtz and Kant has much to do with a deeper philosophical uncertainty, persisting into the present century, about the precise status of those principles that Kant called "synthetic a priori," that Helmholtz called "the facts that lie at the foundation of geometry," and that Einstein and the positivists characterized as "stipulations."

KANT AND HELMHOLTZ

It is well known that Helmholtz's interest in the foundations of geometry arose from his studies of spatial perception. The empiricist motivations of both studies have a particular aim in common: to investigate how much of our developed notion of space, geometrical or perceptual, can be derived from simple empirical principles. Helmholtz, understandably, saw this investigation as opposing the belief in an irreducible spatial intuition. In order to understand his view in opposition to Kant, however, it is important to distinguish the uses of intuition by Kant and by the nativist school in perceptual psychology—two closely related ways of thinking whose differences Helmholtz did not always express or perhaps comprehend clearly.[1]

For these are completely divergent ways of understanding the claim that intuition acquaints us with the properties of space. The nativist account (cf. especially Hering 1861) is based on the notion that we have direct qualitative sensations of the three-dimensional detailed space; the empiricist theory, advocated by Helmholtz but developed already by Berkeley and others, held that our direct experience is only two-dimensional, corresponding to the two-dimensional surface of the retina, and that three-dimensional space arises from a gradually acquired system of associations and inferences. According to Kant, however, our intuitive notion includes knowledge of its global Euclidean geometrical structure: We are acquainted with space as "an infinite given magnitude," concerning which we know, for example, that two lines cannot enclose a space. On the nativist view it is conceivable, if not likely, that the intuitive picture could be a poor approximation to the mathematical description. But Kant's view will not countenance this possibility: The intuitive picture and the mathematical picture are founded on the same kind of constructive process—not on a direct awareness by visual inspection of mental images, but on the "successive synthesis of the productive imagination in the generation of figures" (B 155). That is, the means by which we are acquainted with the properties of space are identical to the means by which mathematics constructs elementary figures and secures the truth of geometrical propositions in constructive proofs.

Considerations of this sort motivate a "logical" interpretation of Kant's use of intuition, an interpretation recently defended and refined by Friedman (1992). On the contrasting "phenomenological" view, the affinity with nativism appears more plausible: Intuition acquaints us immediately and directly with those properties of space that cannot be articulated by the logical analysis of the concept of space. For space is a "singular intuition" that contains infinitely many particular spaces "in" it, rather than a general concept that contains infinitely many instances "under" it. On the logical interpretation, however, Kant recognized that crucial mathematical deductions—especially those involving principles of infinity and continuity—could not be carried out on purely logical grounds, given the monadic logic available in Kant's time. Therefore such arguments therefore had to appeal to constructive principles whose only warrant comes from intuition. This view acknowledges Kant's insight into the limitations of eighteenth-century logic, and his diagnosis of a central error in the rationalist tradition, namely, the belief that mathematical knowledge rested on purely logical foundations and was independent of any sensory content; that degree of independence could be achieved only in the nineteenth century, when propositions involving continuity and infinity could be proved by means of modern polyadic logic.

On the "phenomenological" interpretation of Kant's theory, however, the focus on the logical use of intuition ignores the crucial question about how we come to understand the content of geometrical ideas. It would seem, moreover, that some element of the phenomenological picture is needed to understand the theory of intuition in connection with some of Kant's larger philosophical concerns—in particular his concern with "how pure mathematics is possible as a science." Friedman's view answers one aspect of this question: How is it possible for geometry to be a deductive science when, given the state of formal logic at the time, key propositions are incapable of rigorous proof? In such cases, only the direct appeal to intuition justifies the propositions at all. But we know that there is another aspect to the question, rooted in Kant's efforts to diagnose the failure of dogmatic metaphysics and to see in the exact sciences a possible model for success—in other words, a crucial aspect of this question is its relation to the question how *metaphysics* is possible as a science.

This concern began well before Kant arrived at his critical philosophy, and was clearly articulated first in his "Prize Essay" (II) in a comparison of the methods of the moral and the natural sciences. There it appears that what is most characteristic of the exact sciences is not their method of proof but their method of defining fundamental concepts: Metaphysics must define concepts by analysis, and there is always some uncertainty about whether the analysis has captured everything that truly belongs to a concept such as that of God or the soul; the fruitless disputes of metaphysicians have to do not with any disagreement about the methods of proof but with irreconcilable differences concerning fundamental concepts. In the exact sciences, by contrast, concepts are always constructed by procedures that are transparent to intuition. In this case there can be no doubt whether the definition captures what the concept is supposed to contain, since the concept first comes into being by the constructive definition. The crucial role of intuition in furnishing these constructions explains why metaphysics cannot define constructively: The objects of metaphysical understanding are not possible objects of intuition. Because intuition places no constraints on the objects of metaphysics, any synthetic definition uses essentially an arbitrary combination of concepts, one of the most prominent examples of which (for Kant) was Leibniz's definition of a monad (II, 277). Analytic definitions, meanwhile, are constrained by the task of capturing some supposedly preexisting meaning, which is why analysis is the appropriate method for definition in metaphysics—even if the task is correspondingly more difficult than that of defining mathematical concepts, and less likely to end in agreement. It would seem, then, that the logical use of intuition, in Friedman's sense, has to be supplemented by its constructive use in first singling out the object of mathematics from all possible objects of logical discourse—to single out what is mathematically possible from what is merely logically possible.

From a larger perspective, however, this difficulty with the logical interpretation does not suggest a resort to the phenomenological alternative, but an extended understanding of the role of construction. For Kant, the construction of fundamental definitions and the constructive proof of propositions are inseparable: In the case of a triangle, for example, it is the method by which we construct a triangle that secures our knowledge of those synthetic propositions about it, such as that its internal angles sum to two right angles, that are not contained in the bare concept of "triangle." And, as Friedman (2000b) points out, our acquaintance with the global properties of space, especially its infinitude, cannot be the product of an immediate quasi-sensory awareness but only arises from our ability to iterate indefinitely, in the "productive imagination," any step in any construction. (It is worth recalling that in Plato's *Meno,* when Socrates elicits geometrical knowledge from an untutored boy, the boy begins by agreeing to the iterability of basic Euclidean constructions, in particular the extendibility of the sides of a square to produce a square of arbitrary size.) Contrary to the phenomenological interpretation, then, to prove geometrical propositions and to justify geometrical postulates are not two distinct problems. Indeed, this fact is reflected in Euclid's postulates themselves, which—with the exception of the parallel postulate—only define the general conditions under which the classical constructions, with compass and straightedge, are possible. The first three postulates only assert the possibility of the simplest classical constructions of straight lines and circles; the fourth asserts the equality of all right angles, establishing the uniformity required for the constructive proof of general propositions involving right angles.

It is precisely the peculiar characteristics of the parallel postulate that lead from Kant's view to Helmholtz's. For the postulate asserts not a general condition on construction or constructive reasoning but the result of a particular kind of construction: that (in Euclid's formulation) two lines transverse to a third line will intersect at most on one side of the third line, and only on that side where the internal angles that they form sum to less than two right angles. This circumstance obviously encouraged the belief, which persisted until the nineteenth century, that the parallel postulate should be derivable from the others by classical construction. But as Gauss and others eventually recognized, this postulate asserted a global feature of space that was not implicit in the constructive methods of Euclidean geometry, so that the conditions defined in the other postulates could apply to a space in which the parallel postulate failed. And as Helmholtz pointed out more clearly than anyone had before him, it followed that one could give an intuitive picture of a non-Euclidean space as one in which, quite simply, the construction of lines would fail to produce Euclidean results—for example, in which lines perpendicular to a given line could be produced until they intersect. He accomplished this by giving a precise analysis of what it means to picture a geometrical possibility: "By the much misused expression 'to imagine,' or 'to be able to think of how something happens,' I understand that one could depict the series of sense-impressions which one would have if such a thing happened in an individual case. I do not see how one could understand anything else by it without abandoning the whole sense of the expression" (Helmholtz 1870, 8/248).

What had seemed impossible was to "visualize" a curved space in the sense of the phenomenological interpretation—that is, to have an immediate awareness of the space and its non-Euclidean structure. But Helmholtz showed that what makes space seem Euclidean is a *series* of sense-impressions, in particular the result of iterating basic constructive steps, and that the very same kind of experience could acquaint us with the structure of a non-Euclidean space. Moreover, he showed that the constructive principles expressed by Euclid's postulates have an empirical origin in a more general principle, the principle of the free mobility of rigid bodies. So our sense that we can perform Euclidean constructions in "the successive synthesis of the productive imagination," which to Kant was the basis for our intuitive belief in Euclid's geometry, turns out to be based on expectations that we develop in perceptual experience, as we learn to coordinate our lines of sight with the motions of our bodies and other objects that we suppose to be rigid. Helmholtz's derivation of the general form of the Pythagorean metric from the axiom of free mobility reaffirms an important part of Kant's view, namely, that the visual perception of space and the geometry of space have a common basis. But if that basis is nothing more than an empirical fact that might have been otherwise, then the postulates of geometry have no claim to necessity.

FACT AND CONVENTION IN HELMHOLTZ'S ACCOUNT OF GEOMETRY

From the beginning, however, the status of the principle of free mobility has not been as obvious as Helmholtz seemed to think, and the question has been raised whether a reduction of geometry to this principle really achieves any empiricist aim. Again, relative to the nativist theory of perception, its success is fairly clear: If our conception of the three-dimensional space around us really does arise from our experience with the (assumed) rigid motion of our bodies and of retinal images, then it is at the very least superfluous to assume that that conception is immediately given to us by our sensory apparatus. But the Kantian apriorist view is entirely different: It asserts that there are synthetic principles that are presupposed by all empirical claims about space—that constitute the conditions of the possibility of our experience of space. Such principles are therefore not empirical facts but formal and transcendental principles. The question for Helmholtz is whether, in showing that Euclid's axioms are not such principles, he merely derived them from another principle that is itself transcendental.

Helmholtz raised this question himself when he suggested that a contemporary Kantian, in order to defend some remnant of Kant's position, might take the principle of the rigid body to be a "transcendental concept, which is formed independently of actual experiences and to which these need not necessarily correspond" (Helmholtz 1870, 30/264). In that case the postulates of geometry could not be tested by the behavior of rigid bodies, for they would provide the criteria according to which bodies could be said to be rigid. But this remark could be turned back upon Helmholtz himself, however, insofar as he takes the concept of rigid body to be a condition of the possibility of geometry. Dingler raised this objection in the 1920s, and it became the basis for his "constructivist" account of Euclidean geometry, in which the latter is granted an a priori status because of its indispensable role in the constitution of our measuring instruments. Euclidean geometry, in other words, is the basis on which we construct those objects that represent for us the rigid displacements (cf. Dingler 1934). According to Torretti (1978, 168), such reasoning shows that the notion of rigidity "is not an ordinary scientific notion"; geometry would be impossible without it, and thus "a whole field of experience, which provides the basis for physical science, would fail to exist." The notion of rigid body must therefore be regarded, if Helmholtz is right, as "a concept constitutive of physical experience, that is, as a transcendental concept in the proper Kantian sense" (1978, 68). Torretti adds the qualification that what is made possible by the notion of rigid body is not the organization of our sensory intuitions but the manufacture of our measuring instruments, in accord with Dingler's suggestion.[2] But perhaps this is granting too much; given Helmholtz's claim that rigid motion plays the same constitutive role in our ordinary conception of space that it plays in the geometry of spaces of constant curvature, his accord with Kant seems even more significant.

Helmholtz's empiricism on this point is expressed in his conviction that mechanics provides the proper understanding of the nature of rigid bodies, and so mechanics gives some content to the claim that certain bodies really are rigid. It would then be simply a fact that there are rigid bodies, recognizable by empirical criteria. And therefore the condition of the possibility of geometry, and of our experience of space in general, would be not any transcendental principle but the simple fact that, in the world as far as we know it, there are sufficiently rigid bodies to enable us to carry out spatial measurements (cf. DiSalle 1993). For the Kantian, by contrast—as understood by Helmholtz—admitting that the notion of rigid body is a transcendental notion amounts to taking away the factual content of geometry. So, according to Helmholtz, this purported defense of Kant would effectively eliminate the synthetic a priori, as the principles of geometry would merely follow from the definition of rigid body. In that case the principles are analytic, lacking the synthetic content that only the appropriate mechanical principles would provide.

From Helmholtz's point of view, then, the principle of rigid motion turns out to be crucial to making sense of Kant's idea that geometry is synthetic, and that our intuitive knowledge of geometry is based on "successive synthesis" in the construction of figures. Our acquaintance with the special domain of geometrical objects, among the general class of logically possible objects, turns out to depend on our acquaintance with fundamental physical regularities. Yet if Helmholtz's foundation for geometry is therefore not a synthetic a priori principle, as Dingler's interpretation would suggest, the question remains whether it is any kind of synthetic principle. It is Helmholtz's own remarks that first place this in doubt: In the introduction to his 1868b paper deriving the consequences of the principle of free mobility, he notes that "in geometry we deal constantly with ideal structures, whose corporeal portrayal in the actual world is always only an approximation to what the concept demands, and we only decide whether a body is fixed, its sides flat and its edges straight, by means of the very propositions whose factual correctness the examination is supposed to show" (Helmholtz 1868b, 639).

There are analogous statements in Helmholtz's "epistemological analysis of counting and measurement" (1887). Here his broad purpose was to show that arithmetic, like geometry, can be shown to rest on a few general empirical principles, including basic psychological facts—in this case, the presupposed psychological operation of counting itself. "The first axiom—'If two magnitudes are both alike with a third, they are equal to each other'—is therefore not a law with objective significance; it only determines which physical relations we may recognize as equality" (1887, 380). More generally, he writes, "we should not wonder if the axioms of addition are verified in the course of nature, since we recognize as addition only those physical connections which satisfy the axioms of addition" (1887, 384). That the fundamental axioms of arithmetic have a nonempirical, seemingly analytic status was obviously one of the traditional opinions that Kant's idea of the synthetic a priori was meant to overthrow; that the fundamental idea of geometry should have such a status seems to go against Helmholtz as well as Kant, undermining the claim that free mobility is the "fact" underlying geometry. Helmholtz's claim that it is a physical regularity that makes geometry possible would have to mean no more than that the definition of rigid body, as expressed implicitly in the postulates of geometry, is a particularly convenient one in the world as we experience it. If geometry, like arithmetic, merely explores the consequences of a certain set of fundamental concepts, if its application to the world consists not in providing empirical evidence for its basic principles, but merely in identifying instances of its fundamental concepts, then Helmholtz's empiricism and the Kantian synthetic a priori would appear to be in the same difficulty.

From the point of view of Poincaré, the logical empiricists (e.g., Schlick 1917; Reichenbach 1957), and later twentieth-century commentators (e.g., Coffa 1991), the resolution of this difficulty is obvious. The constitutive principles that Helmholtz is appealing to are not transcendental principles but conventional stipulations: They are constitutive not of possible experience but of the meanings of geometrical terms. Helmholtz's remarks simply show that the principle of free mobility provides a conventional definition of length, on the basis of which objective measurement is then possible. This explains Helmholtz's ambiguity concerning the apparent necessity of the principle, which seems at the same time to be entirely empirical in character. As Coffa expressed this view, "convention, semantically interpreted, is merely the opposite side of necessity. In the range of meanings, what appears conventional from the outside is what appears necessary from the inside" (Coffa 1991, 139). Helmholtz, on this view, had identified the necessary principle at the foundation of ordinary geometry and had some inkling of its definitional character, but simply failed to complete the step outside the framework of ordinary geometry. Thus he stopped short of recognizing the principle of free mobility as a convention. This last step was left for Poincaré to take, and the logical empiricists, following his lead, came to understand the principle as nothing more or less than the "coordinative definition" of length, that is, as associating, by an arbitrary stipulation, the geometrical concept of length with a particular class of physical phenomena. That such a definition is conventional is demonstrated by the resulting relativity of geometry: Any geometry can be made to describe our experience correctly, provided only that we make a suitable stipulation about how congruence and straightness are to be defined. Helmholtz thought that one could test whether space is Euclidean, on a large scale, by taking the parallax of stars and measuring the angle sum of the resulting triangle (Helmholtz 1870, 22–23/258); we could just as well consider this a test, instead, of whether light rays provide a convenient definition of straightness. If the angle sum differs from two right angles, according to Poincaré's now familiar argument, we are free to decide whether to adopt a new geometry or a new definition. That Poincaré erred in assuming that physicists would prefer the second option—according to the broad philosophical consensus in the aftermath of general relativity—does not by itself vitiate the general principle that something in the nature of a decision is required.

In light of such considerations, modern commentators have found the notion of the relativity of geometry to be not merely suggested by, but already implicit in, Helmholtz's writings. Such a view appears to be encouraged by Helmholtz's much-discussed thought experiment with the mirrored sphere (1870, 24–25/259–260). Imagine that we and our measuring instruments, which reveal to us the approximately Euclidean character of our space, observe our reflections in a spherical mirror. Then our own rigid motions and straight lines of sight will appear on the mirror as distorted motions and curved lines, as the reflected bodies and instruments appear to expand and contract, and the images of lines of sight follow the curvature of the mirror. Yet the outcomes of all measurements will, from the point of view of the people on the sphere, exactly agree with ours, and it will be apparent to them that their space is Euclidean. Our space, correspondingly, will appear to them as the image of a convex mirror. "And . . . if the men of the two worlds could converse together, then neither would be able to persuade the other that he had the true, and the other the distorted situation" (1870, 25/260). That is, they would measure the objects in their respective worlds by coincidences with their measuring instruments, which they must assume to be rigid, and so their verbal accounts of geometrical relations would agree in spite of the distorted appearance of each world to the inhabitants of the other. An objective decision would require not only that they meet, but also that they agree on what is a rigid body. Helmholtz goes on to suggest that, if we were willing to attribute the appropriate distortions to the bodies we commonly take to be rigid, we could "in a completely logical way" treat our own space as pseudospherical—"if we found it useful for any purpose" to do so (1870, 29/263).

To Helmholtz, the possibility of such a choice did not imply the conventionality of geometry, since it would require the revision of our laws of mechanics. Thus he was assuming that mechanical considerations must be decisive. Until we can see that there is a principled ground for that assumption, however, Helmholtz's view amounts to the reverse of Poincaré's conviction that considerations of mathematical simplicity would be decisive. And, as Poincaré implied, the seemingly absurd practice of adjusting mechanics ad hoc, in order to preserve our preferred geometry, is only the mirror image of adjusting our geometry to preserve our beliefs about mechanics. In either case a conventional choice would appear to have been made concerning which science to adopt as fundamental.[3] The objective facts are only the direct outcomes of our measurements, prior to their geometrical interpretation; in Helmholtz's example, they are only the coincidences of the measuring instruments with the objects measured. Thus Helmholtz's picture seems to point directly to the attitude espoused by Einstein (1916) and taken up by the logical empiricists: Geometrical measurements only determine "space-time coincidences," or "verifications of . . . meetings of the material points of our measuring instruments with other material points" (Einstein 1916, 117). This attitude explicitly asserts the objective equivalence of geometries that, like the two cases in Helmholtz's example, agree on verifiable coincidences. On such grounds did the logical empiricists include Helmholtz in the tradition of geometric conventionalism; even commentators who have rejected logical empiricism, especially its way of understanding theories of space and time, agree with this interpretation. Van Fraassen (1997), for example, understands Helmholtz's thought experiment as illustrating the relativity of geometrical descriptions; the existence of an isomorphism between two descriptions—in this case the mirror-image reflection—effectively dissolves the question which one is true.

Helmholtz's Analysis of the Concept of Space

A closer look shows, however, that Helmholtz's example is not intended to demonstrate the equivalence of different geometries, or the conventionality of the choice between them. His primary purpose, after all, is to show that we can imagine what a non-Euclidean world would be like—thus presupposing, more or less explicitly, that whether the world is Euclidean is a genuine matter of fact. The purpose of the thought experiment, then, is to illustrate the role that the rigid body plays in our conviction that space is Euclidean. The inhabitants of the mirror world will determine congruence by means of measuring instruments that they take to be rigid, since, with respect to each other and to the bodies of the inhabitants themselves, these instruments do not appear to change their dimensions. Therefore, though from our perspective the inhabitants and the mirrors seem to undergo systematic distortions, the inhabitants are having precisely the experiences—they are determining precisely the coincidences and congruences—that convince us that space is Euclidean. Helmholtz's remark that we could not dispute them on this point "as long as we introduce no mechanical considerations" (1870, 25/260) does not imply that a conventional choice needs to be made about which bodies are rigid; instead, it suggests that a comparison between the two worlds requires that both be brought under the same system of mechanical laws.

The philosophical question raised by the example, then, is not "How do we know whether our world is as we have supposed it to be, and not as the world in the spherical mirror?" It is, rather, "If rigid bodies behaved as if space were Euclidean—if all coincidences and congruences with our measuring instruments agreed with Euclid's postulates—what could it possibly mean to ask whether space is really Euclidean?" Exhibiting in this manner the role of rigid motion in our intuition of Euclidean space—as providing the very content of our claim to know that space is Euclidean—is what enables Helmholtz to construct an intuitive picture of a non-Euclidean world, and to exhibit the intuitive content of the claim that space is non-Euclidean. The inverse thought experiment, as it were, is derived from Beltrami's mapping of pseudospherical space onto a sphere in Euclidean space. In Beltrami's model, the pseudospherical surface is represented as the interior of a Euclidean disc, in which length contracts as a function of distance from the center; in pseudospherical space that function is essentially the metric, and the projections of these contracted lengths are the invariant lengths of the pseudospherical metric. In other words, in Beltrami's mapping, the image of a rigid body that moves infinitely far off in psuedo-spherical space is a body that contracts to a point as it moves away from the center of the disc. Now, Helmholtz imagines a physical sphere in whose interior Euclid's postulates hold, but some physical force causes bodies to contract in such a way that their pseudospherical projections remain rigid. He concludes: "Then observers whose own bodies regularly underwent this change would obtain results from geometrical measurements, made as they could make them, as if they themselves lived in pseudospherical space" (1870, 26/260).

This last inference is precisely the one notoriously used by Poincaré, and interpreted by the logical empiricists, to demonstrate the conventionality of geometry: If a Euclidean space with a peculiar force field, as a result of which systematic distortions are induced in bodies and measuring instruments, could thereby "mimic" a non-Euclidean space, then the two geometries are objectively equivalent, differing only insofar as they express the same situation in different languages. In the twentieth century, various philosophers (especially Reichenbach 1957, but also Putnam 1974 and Glymour 1977) have tried to exhibit methodological grounds on which one geometry, among several that are equivalent in Poincaré's sense, can be considered superior. Such discussions take for granted, in general, the conceptual equivalence of different geometries as argued by Poincaré and focus on the relative merits of the various hypotheses about distorting forces.

What Helmholtz intended to show by this example, however, is not that this imaginary world has two objectively equivalent descriptions. It is that in this world, the bodies we hypothesize to be distorted will actually satisfy, for the inhabitants, the concept of rigid body; the dispositions and displacements of these bodies will satisfy their concept of geometrical measurement. (This is analogous to Einstein's 1916 argument from the equivalence principle to space-time curvature: Our empirical conception of inertial frame is satisfied, locally, by any freely falling frame; thus it is satisfied by frames that, unlike Newtonian inertial frames, may be accelerated relative to one another.) These bodies would thus provide the unique and sufficient basis on which the inhabitants of this world could attribute any geometry at all to their space.

Thus, if the principle of free mobility is an analytic one, it is so in a different sense from the traditionally accepted ones. As Coffa points out, Kant's denial that geometrical principles are analytic makes sense only on his very restricted interpretation of analyticity, on which analytic judgments merely state that some predicate falls under a particular concept. Coffa himself offers a broader "contextual" interpretation, on which analytic judgments express the connection of a given concept to an entire system of judgments. On this interpretation, Helmholtz has recognized that "rigid body" is implicitly defined by the postulates of geometry; the principle of free mobility is therefore a principle constitutive of meaning for the framework of classical homogeneous geometry. In either of these senses of "analytic," Helmholtz's principle would be conventional, giving either an account of common usage or a mere stipulation. As employed by Helmholtz himself, however, the principle is analytic only in the sense of being analytic of our concept of measurement—that is, in the sense that it is the expression of a conceptual analysis of what is implicit in our theoretical and practical judgments about geometrical measurement. In this respect it precisely follows the pattern of Helmholtz's analysis of what it means to "imagine" a non-Euclidean world. For that was not merely a stipulation about how the word is to be understood, for the purpose of attaching some sense to the claim that non-Euclidean geometry can be imagined; it was an analysis of what is implicit in our claim that Euclidean geometry can be imagined, resulting in the recognition that in precisely the way that we visualize Euclidean space, we could, at least in principle, visualize any space of constant curvature; the possible abstract non-Euclidean geometries acquire empirical content in precisely the way that abstract Euclidean geometry does. By the same token, the claim that space is Euclidean, in circumstances where apparently rigid bodies behave as if the geometry were non-Euclidean, is for Helmholtz lacking in empirical content. For if the subject matter of geometry is the displacement of rigid bodies, then it would not be clear what the subject matter of that sort of claim could be. In the case of the spherical mirror, we could not even explain to the inhabitants of the distorted world that their geometry is non-Euclidean, unless we could convince them that their measuring instruments were not rigid. Analogously, if we were to find that our own rigid measuring instruments behaved as if space were non-Euclidean, the claim that space is nonetheless Euclidean would be essentially unintelligible.

CONCLUSION

Helmholtz's philosophy of geometry, from the point of view of twentieth-century philosophy of science, appears to be the first step in the dismantling of Kant's notion of pure intuition. Helmholtz showed that, to the extent that our knowledge of geometry involves intuition, it is a posteriori, founded on contingent empirical facts about rigid motion; then Poincaré and the logical empiricists showed that, to the extent to which that knowledge is pure, it is analytic, expressing not facts about the motions of bodies but only what is implicit in the concept of free mobility. In other words, either the synthetic a priori constitutive principles are empirical, as Helmholtz claimed, and therefore not properly constitutive in the Kantian sense, but only facts on which our ability to carry out geometrical measurement is contingent; or they are constitutive, but only insofar as they constitute particular concepts within a particular conceptual framework. Properly understood, however, Helmholtz's work does not show "pure intuition" to be an oxymoron. Nor does it occupy a confused middle position between Kant and conventionalism, reflecting Helmholtz's inability or unwillingness to follow through the conventionalist implications of his own ideas. Rather, it reveals the subtlety of the situation that Helmholtz was trying to understand.

Unlike Kant, he was prepared to acknowledge, even at the risk of apparent inconsistency, that our knowledge of fundamental geometrical principles is conceptual rather than empirical, or synthetic in Kant's sense; unlike Poincaré and the logical empiricists, he was not prepared to think of those principles as merely defining a formal framework, or of their empirical import as something fixed by arbitrary stipulation, because he had found them to be implicit in a systematic body of empirical knowledge and practice. In other words, what Kant called "pure intuition" derives its "intuitive" aspect from our practical engagement with physical contingencies, and its seemingly "pure" aspect derives from the concepts in virtue of which those practices become systematic for us.

What the twentieth-century assimilation of Helmholtz missed, by insisting on the stipulative character of the principle of free mobility, is precisely the process of conceptual analysis by which Helmholtz first arrived at it. The principle derives its empirical content, in spite of its apparently analytic character, from the fact that it defines the concept implicit in our sound empirical judgments about spatial measurement; its seemingly formal status reflects the fact that the conceptual analysis that produces it represents a kind of abstraction from those empirical judgments. Seen in this way, Helmholtz's work reveals an unexpected continuity with Kant's idea of pure intuition. For, in grounding our knowledge of geometry in pure intuition, Kant meant to show the dependence of our certainty on classical constructions—at least within the limitations of eighteenth-century logic—and also the independence of those constructions of any particular empirical act of construction. Pure intuition thus amounts to—in effect—the most general conditions on any geometrical representation, independent of the peculiarities of any particular drawing or mental image. What Kant thus provided was a conceptual analysis of what is implicit in any claim (in the eighteenth-century context) to formal knowledge of geometry, and a demonstration that, at least up to that time, such formal knowledge must presuppose the possibility of constructive proof—despite the claims of Platonistic philosophers to have some purely intellectual intuition of mathematics, or of the Leibnizeans to be able to reduce geometry to purely logical identities. What Helmholtz then provided, as the decisive critique of Kant's theory, was a further conceptual analysis, showing that Kant had not in fact identified the most general conditions on geometrical representation; the implicit basis of the Kantian conditions is the far more general principle of free mobility—not an empirical principle but one whose origin in empirical circumstances is evident. If Helmholtz's principle seems narrow and simplistic in the wake of twentieth-century physics, and in particular the special and general theories of relativity, it is not because the role he attributed to conceptual analysis must now be played by arbitrary stipulations. Rather, it is because the realm that such an analysis must comprehend goes far beyond the principles of ordinary spatial measurement, to include the dynamical principles governing the electromagnetic and gravitational fields. In that setting, it turns out, the concept constituted by our geometrical knowledge is not that of space, but of space-time.

Notes

1. See, for example, DiSalle 1993 for a discussion of the philosophical differences between nativism and Kantianism, and the relation of both to Helmholtz's view.

2. This is also emphasized by Carrier 1997. More circumspectly, Carrier identifies Dingler's constructivism as just one of three central tendencies whose roots may be found in Helmholtz, along with the (broadly empiricist) program of constructive axiomatics in general relativity (cf. Weyl 1921 and 1923; or Ehlers, Pirani, and Schild 1972), and conventionalism.

3. On Poincaré's view of the "hierarchy" of sciences, and the priority of geometry to physics, see Friedman 1996a and DiSalle 2002.

VIII

OPERATIONALIZING KANT: MANIFOLDS, MODELS, AND MATHEMATICS IN HELMHOLTZ'S THEORIES OF PERCEPTION

Timothy Lenoir

Although he considered himself a physicist, Hermann Helmholtz devoted more pages of his published work to physiological psychology and philosophical problems related to spatial perception than to any other subject. These writings constantly engage issues related to Kant's work on the perception of space, time, and the law of causality. Helmholtz's first published engagement with these issues came in his contribution to the 1855 centennial celebrating Kant's inaugural lecture in Königsberg, in a lecture entitled "On Human Perception." There Helmholtz outlined the difficulty he saw in some contemporary theories of vision that fashioned themselves as Kantian in adopting a so-called "nativist" position, which attributed spatial perception to the inborn neuroanatomical organization of the visual apparatus. He critiqued them as implicitly setting up two worlds, an objective physical world and a subjective world of intuition, somehow causally related to each other but existing as independent, parallel worlds relying on some unexplained preestablished harmony between perceptions and the real world as the basis for the objective reference of knowledge claims. Helmholtz, believing it was necessary to escape the subjectivity of idealist positions, asked: "But how is it that we escape from the world of the sensations of our own nervous system into the world of real things?" (1855a, 116).

In addressing this question over a number of years, Helmholtz consistently depicted himself as a Kantian—this in spite of the fact that he broke with Kant by defending the empirical origin of geometrical axioms. In many ways Helmholtz considered his work in epistemology not as contradicting Kant but as updating Kant's views in light of new developments in experimental physiology that challenged the nativist approach. Although disputed in Helmholtz's own day, Helmholtz thus considered himself to be more consistently Kantian than Kant had been himself. As I see it, Helmholtz pursued fairly consistently over many years a line of investigation centered around the questions, How do our mental representations of external objects get constructed? And how do those representations relate to the world of external objects (things in themselves)?

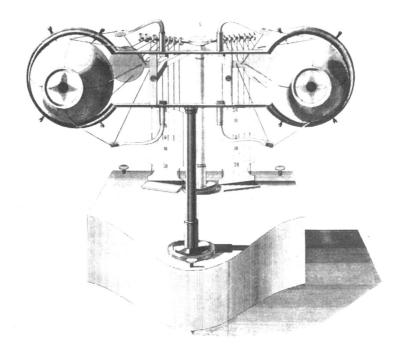

Figure 8.1
C. G. Theodor Reute, *Ophthalmotrope.* Source: Reute, C. G. Theodor. *Ein neues Ophthal-motrop.* Leipzig: Otto Wigand, 1857. See Hermann Helmholtz, *Handbuch der physiologis-chen Optik,* 2d ed. (Hamburg: Leopold Voss, 1896), 667.

His general approach to these questions was treated in the 1855 lecture as a sign theory of perception, which over the next decade became central both to Helmholtz's theory of the visual field and to his epistemological concerns. Helmholtz's theory of signs is not usually a subject of attention, but I regard it as a focal point from which to consider a number of deeply interacting and intertwined threads in Helmholtz's experimental and theoretical researches in physiological optics, physiological acoustics, psychophysics, and his later investigations into the relations between the foundations of mechanics and geometry. I have treated aspects of this network of investigations elsewhere (Lenoir 1992). Here I am primarily concerned with the interaction among Helmholtz's experimental researches in the theory of pigments, spectral colors, and color vision in the development of the sign theory. A further concern is the positive contribution to Helmholtz's theories of sensation and perception made by a variety of new media technologies, particularly electric, photographic, and telegraphic inscription devices. The material characteristics of these media served Helmholtz as analogues and as models of the sensory processes he was investigating. These devices not only assisted Helmholtz in understanding the operation of the eye and ear through measurement; more to the point, Helmholtz conceived of the nervous system as a telegraph—and not just for purposes of popular presentation. He viewed its appendages—sensory organs—as media apparatuses: The eye was a photometer and the ear a tuning fork interrupter with attached resonators. The output of these devices was encoded in the form of an n-dimensional manifold, a complex measure to which a sign, such as "red," "bluish-green," or "u" was attached. These materialities of communication were important not only because they enabled theoretical problems of vision and hearing to be translated, externalized, rendered concrete and manipulable in media technologies, but furthermore because in this exteriorized form analogies could be drawn between devices; linkages could be made between different processes and between various aspects of the same process. Crucial to Helmholtz's theorizing were analogies between sound and color perception. Indeed, I will argue that a crucial step in his development of the trichromatic receptor theory of color sensation came through analogies between the technologies of sound and of color production. The juxtaposition of media enabled by the materialities—the exteriorized forms—of communication was a driving force in the construction of theory.

Helmholtz was engaged in developing an empiricist theory of vision in which the visual field is constructed from experience and stabilized through repetition. Helmholtz adopted the radical stance that nothing is given in the act of perception. He denied even the empirically plausible claim of his mentor, Johannes Müller, that single vision in a binocular field is determined by neuroanatomical linkages between geometrically corresponding points in the two retinas and optical chiasma. For Helmholtz, all such associations are learned. Even more significant for his theory of representation, Helmholtz's radical empiricist stance led him to question and ultimately abandon all "copy" theories of representation. For instance, Helmholtz rejected straightaway the notion that a camera obscura–like image on the retina (even if it could be assumed that a sharp image forms on the retina) is a copy of an external reality, or that anything like a "copy" of the retinal image could be assumed to be conveyed via the nerves to the brain and reconstituted there as a virtual image. Even if there were somehow a copy of the image in the brain, in Helmholtz's view it would be totally inconsequential for a construction of the visual field, a sort of accidental by-product of brain anatomy irrelevant to the actual process of vision:

I maintain, therefore, that it cannot possibly make sense to speak about any truth of our perceptions other than practical truth. Our perceptions of things cannot be anything other than symbols, naturally given signs for things, which we have learned to use in order to control our motions and actions. When we have learned to read those signs in the proper manner, we are in a condition to use them to orient our actions such that they achieve their intended effect; that is to say, that new sensations arise in an expected manner. Any other comparison between perceptions and things exists neither in reality—on this point all schools of thought are in agreement—nor is any other sort of comparison conceivable and meaningful. This latter point is crucial for seeing how to get out of the labyrinth of contradictory opinions. To ask whether a perception which I have of a table, of its shape, solidity, color, weight, etc., is true and corresponds with the real thing or whether it is false and rests upon an illusion independently of the practical use to which I can put this perception has no more sense than to ask whether a certain sound is red, yellow, or blue. Perception and things perceived belong to two completely different worlds which admit of no more comparison to one another than colors and sounds or the letters of a book to the sound of the words which they signify. (1867, 443)

Further criticizing the notion that the truth of a perception depends on its being a copy of the object perceived, Helmholtz went on to question the sort of brain physiology required in order to guarantee some similarity between perceived object and its representation in the brain. What sort of similarity, for instance, should be imagined between the processes in the brain accompanying the perception of a table and the table itself? Should one assume that the shape of the table is traced by an electrical current in the brain? And when the percipient imagines that he is moving around the table, should a second person be electrically traced walking around the table in the brain? Even assuming such a fantasy could actually be the case, the problem of perception would not be resolved: "For the electrical copy of the table in the brain would now be a second object that would have to be perceived, but it would still not be a perception of the table" (1867, 443).

What was this symbolic language like? As the passage quoted indicates, and as Helmholtz elaborated elsewhere, the symbolic language must enable different modalities of sense to be linked within it. In addition, while the symbolic language does not provide copies of an external reality, Helmholtz argued that it is neither arbitrary nor independent of input from our sense organs. Principally the task of this symbolic language was to represent relationships between objects affecting one another and our sense organs. The structure of the relationships was the crucial aspect to be grasped in a representation. A copy of an object would in no way guarantee that the mind had grasped the relations between objects and sensations; and in fact a copy might be a very poor representation. A good representation is a symbol useful for organizing the practical activities in terms of which we interact with the external world through our senses:

> The perception of a single, individual table which I carry in me is correct and exact if I am able to deduce from it correctly and exactly which sensations I will have if I bring my eye and my hand into this and that position with respect to the table. What sort of similarity is supposed to exist between the perception and the object which I perceive by means of it, I am incapable of comprehending. The former is the mental symbol for the latter. The sort of symbol is not arbitrarily chosen by me, but rather is forced upon me by the nature of my sense organs and the nature of my mind. In this aspect the symbolic language [*Zeichensprache*] of our perceptions differs from the arbitrarily chosen phonetic and alphabetic symbols of our spoken and written language. A written expression is correct when the person who knows how to read it forms the appropriate perceptions in accordance with it; and the perception of a thing is correct for the person who knows how by means of it to predict which sensual impressions he will receive from the thing when he places himself in certain circumstances with respect to it. Moreover, of what sort these mental signs are is completely irrelevant as long as they form a sufficiently diverse and ordered system; just as it is irrelevant how words of a language sound as long as they are present in sufficient number and there are means for designating their grammatical relations to one another. (1867, 446)

In his series of lectures on the recent progress in the theory of vision held in Köln in 1868, Helmholtz refined this discussion of representation by drawing important distinctions between the notions of "sign" and "image." There he noted that in an image the representation "must be of the same kind" as the object represented. He went on to specify what he meant by "the same kind":

> A statue is an image of a man insofar as its form reproduces his; even if it is executed on a smaller scale, every dimension must be represented in proportion. A picture is an image, or representation of the original, first, because it represents the colors of the latter by similar colors, and secondly, because it represents a part of its relations in space (those which belong to perspective) by corresponding relations in space.
> The excitation of the nerves in the brain and the ideas in our consciousness can be considered images of processes in the external world insofar as the former parallel the latter, that is, insofar as they represent the similarity of objects by a similarity of signs and thus represent a lawful order by a lawful order. (1868a, 185–186)

In this discussion Helmholtz treated images as relations among signs in which a correspondence of structural relation obtains between representation and the things represented. Such structural correspondence is not given but is learned and constructed. Crucially for this theory of representation, while ideas or concepts are organized around structural correspondence with external affairs, sensations do not possess this property of structural correspondence with an external referent. They are signs or tokens for certain interactions between external bodies and our sense organs. The color "red" is the sign of the interaction between the retina and light of 6878 Angstrom wavelength. Indeed, the contribution from the side of the retina complicates things further, so that there is not a one-to-one correspondence of sign with external referent; for as we shall see, whether an object is seen as red or not depends on judgments concerning contrast and other factors. The properties attributed to light do not belong to it exclusively but to the interaction with the eye:

> Hence it is really meaningless to talk as if there were properties of light which belong to it absolutely, independent of all other objects, and which we may expect to find exhibited in the sensations of the human eye. The notion of such properties is a contradiction in itself. They cannot possibly exist, and therefore we cannot expect to find any correspondence of our sensations of color with qualities of light. (1868a, 187)

In Helmholtz's view, even at the level of sensation psychological factors involving experience and judgment are intermixed with physiological function and physical interaction with light.[1] Sensations considered as signs, Helmholtz tells us, do not possess the characteristics of constancy desirable in an ideal system of signs. At best, sensory signs have a relative constancy based on use (see 1868a, 192).

Essentially my claim is that Helmholtz's model of representation was that of an abstract system of relations among sense data. Like Bernhard Riemann, working independently at almost exactly the same time, Helmholtz treated the mental representation of sensations as n-dimensional manifolds. Different modalities of sense were characterized as manifolds obeying different metric relations. The sense data were organized into symbolic codes by a system of parameters due to the physical properties of each sense and adjusted by experience.[2] Helmholtz's ideas about the sign theory of representation were informed by a tradition of Kantian interpretation tracing its roots to the work of Johann Friedrich Herbart (1776–1841), particularly his *Introduction to Philosophy* (1813) and his *Psychology as a Science, newly founded upon Experience, Metaphysics, and Mathematics* (1824–1825). Herbart had been Kant's successor in Königsberg, and his collected works, issued in an excellent new six-volume edition in 1850–1851 by Gustav Hartenstein, had become part of a new "Back to Kant" movement that stressed a "realist" interpretation of Kant's transcendental aesthetic. Herbart's work was taken up and extended by a number of philosophers, mathematicians, and physicists in the early 1850s interested in developing a psychophysical approach to relating mental representations to the external world. Helmholtz, I argue, was deeply sympathetic to this work, and in many ways his own research agenda addressed issues that had been posed within this Herbartian revival.

Helmholtz wasn't the sort of guy to lounge around reading works of philosophy and metaphysics that didn't somehow pertain to his experimental research. If he had not been aware of the ideas on representation circulating in what I've called the Back-to-Kant Herbartian revival, he certainly became aware of them through his encounter with Hermann Grassmann, whose enigmatic *Doctrine of Extension* (*Ausdehnungslehre*) of 1844 was a prize-winning if still unappreciated work in that movement. Helmholtz learned about Grassmann in connection with his work on color mixture. The second aspect of my claim, and perhaps its more substantive part, is that Helmholtz's encounter with Grassmann stimulated his own ideas about representation through a lively dialogue with three lines of experimental investigation—color, sound, and electrotelegraphy—familiar to Helmholtz in the period 1850–1863, the period spanning his work on the speed of nerve transmission, physiological color mixing, and physiological acoustics, culminating with the publication of Part II of the *Physiological Optics*. For my purposes, the most innovative character of Helmholtz's work derived from the adaptation of a number of interrelated technical devices employed in telegraphy to the measurement of small intervals of time and the graphic recording of temporal events in sensory physiology. From as early as 1850 Helmholtz drew analogies between the electrical telegraph and the processes of perception. The telegraph began to serve as a generalized model for representing sensation and perception. In light of this telegraph analogy, Helmholtz, so I hypothesize, imagined the virtual image cast on the retina as dissolved into a set of electrical impulses, data to be represented by symbols as an "image" in the brain through a perceptual analogue of Morse Code (see Helmholtz 1850, 873). Throughout this period—between 1850 and 1855—Helmholtz was working intensively with the myograph and a variety of adaptations of electrical devices in the telegraph industry to measure the speed of nerve transmission and other features connected with nerve action and muscle contraction. Telegraphy was not only a useful model for representing and thinking about vision and hearing. Experiments involving those devices were also crucial in advancing his own program of sensory physiology.

This role of telegraphic devices and a variety of imaging devices became particularly important in the period between 1855 and 1860. During this period, reacting to a critique of his theory of spectral color mixing by Hermann Grassmann, Helmholtz retracted his earlier (1852) rejection of the Young trichromacy theory of physiological color mixtures. Interestingly, Helmholtz suggests that he arrived at this view via a comparative analysis with hearing (1868a, 181). I will pursue this suggestion in depth. Helmholtz pursued a similar research strategy of representing tone production and reception in terms of a variety of components of electrical telegraphic circuitry combined with several techniques for graphic display of wave motion, particularly sound waves. These devices were crucial in his investigation of combination tones, the analogue to forming color mixtures from primary colors. Helmholtz postulated retinal structures—three receptors sensitive primarily to wavelengths in the red, green, and violet ranges respectively—analogous to the arches of Corti in the ear, and then provided the path back to accepting the Young trichromatic theory previously rejected. Once again, the new media technologies were crucial in this transition; for Helmholtz drew upon processes of photonegative production in providing a physiological explanation of positive and negative afterimages crucial to refining the three-receptor hypothesis.

Rigid Bodies, Motion, and Causality: Back to Herbart

During the same months that Helmholtz was beginning to formulate the epistemological core of his program in physiological optics, the call to follow Herbart's lead in returning to Kant was moving into high gear. Throughout the late spring and summer of 1850 and into early 1851, Gustav Hartenstein published six volumes of Herbart's collected works (Herbart 1850–1851). If Helmholtz had not known about Herbart before, which is unlikely, the sensation aroused in Königsberg by the reappearance of the writings of the man who had been Kant's successor in Helmholtz's new home demanded attention, especially by a man who was beginning to fashion himself as the true intellectual heir to Kant. Even a cursory reading of Herbart would have alerted Helmholtz—as indeed it did Bernhard Riemann at about the same time—to the fact that here had been a kindred spirit; for Herbart had given profound consideration to the problems confronting Kant's theory of spatial intuition, and he had cast his own reformulation of the Kantian transcendental aesthetic in a form similar to the requirements of Helmholtz's own purposes.

Herbart wanted to reform elements of Kant's critical philosophy that had encouraged subjective idealism and emphasize the realist elements of

Kant's thought. A key part of this program was a fundamental revision of Kant's treatment of the relation between phenomena in space and time and the world of so-called "things in themselves." Herbart, like most philosophers of his generation, found Kant's treatment of that relationship enigmatic. For Kant, space and time are the forms of the faculty of human sensuous intuition in which things are given to us as appearances, while causality is a form of judgment for establishing links within these phenomena. In Kant's approach, therefore, causality only applies to the realm of possible experience, that is, to the realm of phenomena; it can never apply to the (noumenal) world of things in themselves.[3] On the one hand, while causality was supposed to apply only to relations within the realm of possible experience, Kant allowed that the things in themselves, which are not objects of possible experience for us, are yet somehow the *causes* of our sensations.[4]

Herbart's attempted solution to this conundrum was to rework the Kantian treatment of causality in order to allow a genuine causal connection between the objective, real world of things in themselves and the phenomenal world of our experience. His approach to this was to synthesize features of Kant's critical philosophy with the system of Leibniz and Wolff. In effect he attempted to retrace much of the same ground Kant himself had originally covered in his "precritical" period as he moved toward the construction of his early critical philosophy in the 1780s. As Michael Friedman has shown, from Kant's earliest publication, *Thoughts on the True Estimation of Living Forces,* in 1747, through his treatise *Physical Monadology* in 1756, and even through his *Inaugural Dissertation* of 1770, Kant attempted to refashion the Leibnizean-Wolffian tradition he inherited to harmonize metaphysics with advances in mathematics and Newtonian natural philosophy (Friedman 1992, 1–52). In the Leibnizean-Wolffian system, reality consists of an infinity of nonspatial, nontemporal, unextended simple substances or monads. These simple substances or monads do not interact with one another. The evolution of the states of each is completely determined by a purely internal principle of active force, and the appearance of interaction is explained by a preestablished harmony between the states of the diverse substances established originally at the creation by God. Since metaphysical reality is thus essentially nonrelational, neither space nor time is metaphysically real: Both are ideal phenomena representing the preestablished harmony—the mirroring of the universe from various points of view—between the really noninteracting simple substances.

In his early career Kant tried to revise aspects of this Leibnizean monadology by allowing that while (as in the monadology) reality consists of nonspatial, nontemporal, unextended simple substances, space, time, and motion are phenomena derivative from this underlying monadic realm. Kant broke with the Leibnizean monadology by allowing interaction between the monads in terms of active (Newtonian) force where one substance exerts an action on another substance, thereby changing the internal state of the second, rather than in terms of the Leibnizean notion of active force as an internal and hence nonrelational principle. Space for Kant is thus phenomenal and not ideal, as it was for the Leibnizean-Wolffian tradition. For Leibniz and Wolff space is ideal because relations between substances are ideal: Each substance mirrors the entire universe internally owing to its own inner principle, and space is an ideal representation of the underlying order of monads expressed in the preestablished harmony. Since each simple substance already expresses completely the order of the entire universe, nothing but the mere existence of substances is necessary to constitute phenomenal space. But Kant wanted to rid this system of the notion of preestablished harmony and conceptions leading to the ideality of space. For Kant, the committed Newtonian, relations of interaction between substances are in no way ideal. Thus, in his *Physical Monadology* Kant treated space as an external phenomenon derived from or constituted by the underlying nonspatial reality of simple substances (monads).

Kant encountered insurmountable obstacles in carrying this program through and, as Michael Friedman has shown, ultimately abandoned it (Friedman 1992, 25–27). The problems centered around Kant's initial conception of the nature of space as grounded in the law of interaction governing the external relations of nonspatial unextended monads. In particular, it was unclear how to derive properties of space, such as its three-dimensionality and infinite divisibility, from the underlying monadic realm. Friedman shows that although Kant did not frame his problem explicitly in terms of the limitations posed by the logical tools at his disposal—essentially those of monadic predicate logic—he nonetheless clearly stated the problem of representation encountered by a monadic predicate logic, and he took countermeasures in order to preserve the truth of Euclidean geometry (see Friedman 1992, 56–66). In particular, Kant knew that the syllogistic (monadic) logic he worked with prevented him from developing proofs that involved generating an infinity of objects, such as Euclidean demonstrations that require an infinity of points. Unable to capture the idea of an infinity of points through an iterative deductive logical procedure alone, Kant represented the idea of infinity intuitively through the iterative process of spatial construction with Euclidean procedures based on straightedge and compass: "I cannot represent to myself a line, however small, without drawing it in thought, that is gradually generating all its parts from a point. Only in this way can the intuition be obtained. . . . The mathematics of extension (geometry), together with its axioms, is based upon this successive synthesis of the productive imagination in the generation of figures" (B 203–204).[5]

Forced to abandon his original attempt to generate space from interactions among the elements of a nonspatial realm, Kant preserved both the particular character of spatial intuition and the generality of geometric proof by locating them in subjective human cognitive faculties—namely, a pure a priori faculty of representation, on the one hand, and on the other hand a faculty of understanding that provided rules, schemata, or general constructive procedures for the synthesis of objects in sensuous intuition. In this fashion Kant restricted the applicability of pure concepts of the understanding to the phenomenal world in space and time, and he excluded all access to an underlying monadic realm, whether that realm was conceived as constituting the phenomenal world or as in some sense the reality behind its appearance.

From a modern perspective, Kant's solution would appear too restrictive. And given a richer set of logical tools, it is possible to address these issues differently. Since the work of Frege, Pasch, Dedekind, Hilbert, and others who established modern polyadic logic with universal quantification, existential quantification, and quantifier-dependence based on a theory of order, it has been possible to overcome the problems Kant encountered and give an axiomatic characterization of a system of abstract relations from which the properties of Euclidean space can be derived logically. In contrast to Kant, this late nineteenth-century approach does not need to begin with intrinsically spatial elements. Nor does it need to construct its objects. Rather the elements of this abstract system of relations are completely undetermined, allowing for a distinction between pure and applied geometry, where pure geometry is analytic and based on concepts, while applied geometry is its model or spatial interpretation, a distinction Kant resisted.[6]

Herbart was no better off than Kant in having the logical resources to construct an axiomatic pure geometry, but he was keen on synthesizing certain elements of Leibniz's monadology with Kant's work on space and time. Herbart made two essential proposals. The first was to remove Kant's restriction on the applicability of the category of causality to the phenomenal realm alone. By treating causality as part of the real structure of the intelligible world (the world of things in themselves) rather than just a subjective category, Herbart saw the path toward removing the cleft between subjective and objective space. The notion of a rigid body was key to his solution of the problem.

Herbart's second proposal was to construct an abstract science of relations as a basis for various applied geometries, among which Euclidean geometry would be just one. For Herbart, intelligible space is an abstract science of consistent logical relations.

> Geometry assumes space as given; and it makes its constituents, lines and angles, through construction. But for the simple essences (and natural philosophy must be reduced to them in order to find the solid ground of the real) no space is given. It together with all its determinations must be produced. The standpoint of geometry is too low for metaphysics. Metaphysics must first make clear the possibility and the validity of geometry before she can make use of it. This transpires in the construction of intelligible space. (Herbart 1813, 310)

In Herbart's view the space assumed by the geometer as given is the space borrowed from the world of the senses, and that is Kant's world of spatial intuitions. The geometry of intelligible space is a higher form of geometry that does not borrow its concepts from our sensuous experience but rather constructs them on the basis of certain elementary notions (Herbart 1829, 165–166).[7] It starts from primitive notions such as "position," "between," "inside," "outside," and with rules for various logical relations. The concepts of "straight" and "rigid line" were the key concepts of the investigation.

Herbart's discussion of lines and planes exemplifies his approach and also relates directly to one of the central issues he confronted in relating psychological and objective space (1829, 187–188). I have pointed out above that Helmholtz regarded the key problem of a psychophysical account of vision to be the mechanism by which the input from the neurophysiological apparatus of vision is processed in the sensorium into the richly colored three-dimensional world of visual experience. This last set of steps had to be a psychological process and could not be a simple replication of the physiology, because somewhere along the way physiology had to give way to a system of meaning and intention capable of action. Helmholtz's sign theory was intended to address this problem. But the construction and operation of these signs required an abstract system capable of creating the schematisms that would translate into action on the physiological side of the psycho/physical divide. Herbart's approach to geometry provided the blueprint.

Herbart believed that an abstract science of relations would provide tools for solving two different but analogous problems. On the one hand, as we have seen, he saw it as relevant to solving the fundamental metaphysical problem of relating the nonspatial world of intelligible essences to the phenomenal world. But there was an analogous problem in empirical psychology of relating the world of externally caused sense data to the world of our three-dimensional visual experience. Dissenting from the views of some who thought that future physiological research would disclose that the nerves in the retina provide a one-to-one mapping of retinal data along neural pathways eventually terminating in the brain, Herbart argued that even if this nativist thesis should turn out to be true, a great divide persists between sense data and visual experience. Speaking directly to these concerns, Herbart had written that irrespective of the future advances physiology might make in establishing that images of external things are undistorted projections of real objects onto the retina, the entire "image" dissolves into an undifferentiated chaos as soon as the perceived object emerges in the sensorium: "The soul must now generate from the ground up the completely destroyed spatial relationships [within the retinal image]. And it has to do this without distorting its perceptions in the slightest. . . . But the perception of something spatial must have a certain similarity to the spatial thing itself, otherwise the perceived object resulting from this act of perception might be anything but something spatial" (Herbart 1829, 118).

The mistake of such views was to treat space as something real, and the phenomenal space of empirical psychology as "given" by the wiring of the biological substrate. Herbart, however, insisted that from the viewpoint of empirical psychology, space is not something real, a single container in which things are placed. Rather, it is a tool for symbolizing and representing the various modes of interaction with the world through our senses; and its measure is not given intrinsically, but rather "like all concepts of magnitude, must be considered merely as an aid to thought which has to be bent and shaped in accordance with the nature of the objects to which it is being applied, and never mistakenly conceived as delivering up their real predicates" (1821, 312).[8]

In his discussion of lines and planes, Herbart introduced the notion of lines as magnitudes with direction. He used this concept to define a straight line operationally as the shortest path between two points. To get to this result he began by operationally defining, in terms of compound directions, a perpendicular between a line and a point not on it. He then treated any line as the hypotenuse of a triangle; accordingly, such a line could be considered to be composed of compound directions by methods similar to constructing a parallelogram of motions.[9] An important consequence of this approach was that in principle there is no limit to the number of spatial dimensions in an abstract "philosophical" geometry. The same procedures deployed in constructing magnitudes of two or three dimensions could in principle be continued in the construction of spatial magnitudes of four, five, or any number of dimensions (Herbart 1829, 203–205). But such a construction, while logically possible, would be merely playing with concepts and rules of construction. "Intelligible space just like sensuous space," concluded Herbart, "can only have three dimensions" (1829, 204).

Herbart applied the same operational approach for generating other key concepts. Just as he treated straight lines as the path between two points requiring the least expenditure of motion, Herbart defined matter as a collection of points, lines, and surfaces having internal connections which are conserved (*Selbsterhaltung*) in all motions of the body (Herbart 1829, 212–218). Herbart dissented from definitions of matter as "collections of atoms," "cohesion," or relationships between repulsive and attractive forces giving rise to impenetrability—the sort of strategy found in Kant's *Metaphysical Foundations of Natural Science* (Herbart 1829, 248ff.). The way we go about determining a rigid body, Herbart argued, is by providing a consistent, constant rule for preserving the same systematic relationship between points, lines, and surfaces of an object as it undergoes all forms of motion, namely, translations and rotations (1829, 228–231, 249–250, and especially 252–253). Phenomena that do not satisfy this criterion are not rigid bodies or characterize (variable) relationships between more than one body.

Herbart applied these ideas on geometry as one among many applications of an abstract science of relations to the problem of constructing sensory space in his *Psychology as Science* (1824–1825). In order to provide a positive alternative to Kant's notion that space is given in intuition, Herbart, building upon the work of Locke and the associationists, aimed to show that psychological space could be constructed from simple sensations empirically. An essential requirement placed on such a construction was that it had to be capable of accounting for the apparent immediacy of visual experience. The task seemed self-contradictory; somehow visual space had to be empirically constructed in time and yet possess all the characteristics of being immediately given.[10] Herbart's attack on these problems stressed the following points: (1) the strength of associations between different perceptions or ideas should be distinguishable; (2) connections between perceptions often repeated in conjunction should be represented as flowing from one another with the strength of a mechanically determined necessity once a particular temporal threshold of association had been achieved; (3) once this degree of association between perceptions is attained, a' characteristic function is assigned to represent the path connecting them. The function would be retained in memory. When activated by the proper cue, the sequence of steps represented in the function would be run through extremely rapidly; and the more often the function was activated, the more rapidly the cascade of associations would be completed, approaching the limits of an immediate perception. After the appropriate learning period, such a mechanism serves as the foundation of our perception of visual space as immediately given (Herbart 1824/1825, 5:479–480).

The assumption that there are certain simple, pure sensations, such as principal colors, pure tones, and so on, provided the starting point for Herbart's psychophysics. But in order for these to be perceived, they required a contrasting sensation with some sensible difference between them. Already at the level of perception, according to Herbart, an active psychological process of comparison and measurement occurs, and it must be constantly reproduced. Thus constant exposure to the tone C without variation would result in its not being heard; staring at a patch of blue while keeping the eyes immobile would lead to its gradual disappearance (Herbart 1824/1825, 6:120–121). Perceptions in this scheme are treated as forces counterbalancing one another in varying degrees of intensity (1824/1825, 5:322). By describing sensations as forces, Herbart drew a direct analogy to the resistance or pressure felt in the sense of touch, and also to the intensity of the force experienced as pleasure or pain. He treated the sensation itself as the result of a force relation between an external agent and the physiological apparatus that interacts with it. The subject experiences this relationship as what Herbart termed an inhibition (*Hemmung*).

The treatment of perceptions as forces was essential to Herbart's goal of reducing them to the measure of his new psychophysics. The phenomenal world was to be constructed out of the dyamical and statical relations between elementary sensations and combinations of them in experience. At the center of this new theory of the faculty of sensibility was a general set of descriptive relations forming a mathematical calculus that he called sequential forms (*Reihenformen*), the objective of which was to express the relations between sensations in terms of their degree of intensity, quantity, and quality. Herbart's aim was to apply the formalism of this general calculus to each sensory modality. By representing them in terms of a common logical calculus it would be possible to combine sensations of different modalities into a common, unified experience. The same dynamical laws in more or less restricted form governed the construction of each modality of perception and their synthesis into higher ordered unities, which he referred to as "complexions." For example, a single color, such as blue, can vary continuously in intensity and saturation. Without working out the details quantitatively, Herbart argued that such a manifold of sensations could be represented in terms of a linear series of sensations of increasing intensity. Furthermore, according to Herbart, any color in the visual spectrum can be constructed from three primary colors, red, blue, and yellow (1824/1825, 5:323–324, also Herbart 1813, esp. 57, 489 and 489n.). Herbart's scheme was to represent each color in the visual field as a point on a color surface, with one of the primary colors, blue for instance, at the center of coordinates, and units of red and yellow as the abscissas and ordinates.

Herbart did not demonstrate how to carry out the construction of his color surfaces and tone lines in terms of the calculus of sequential forms. The objectives of his program and the rationale he offered for constructing it, however, offered an exciting blueprint for a constructivist theory of sensation. His rationale for arguing that each of the external senses is susceptible of a spatial representation was that each sense, such as hearing, color vision, visual acuity, or the tactile sense, is capable of comparative discrimination. Although themselves simple, colors, tones, and so on do not have their own absolute measures; there is no least perceptible unit of "blue," for instance, out of which all other shades of blue are constructed. Thus it is not possible to assign a quantity of blue to "royal blue" that matches some direct physical measurement. Nevertheless, it is possible to assign magnitudes to a manifold of sensations, to construct a psychometric, because of the fact that we make contrasts between sensations. We can discriminate whether one color is brighter or darker than a comparison color of the same hue. Between three pitches—P_1, P_2, P_3—we can say whether the interval (P_1, P_2) is greater than, equal to, or less than (P_1, P_3). The capability of making such contrasts between pairs of sensations in a manifold gives rise to the possibility of providing a psychometric coordination between numbers and sensations and of representing the manifold spatially.

Within the modality of color vision, for instance, Herbart suggested that a comparison between three arbitrary colors taken as primary could serve as the basis for constructing a metric in terms of which quantitative relationships between all other colors could be determined. Finally, and most interestingly in my view, when conjoined with certain other conditions (such as requiring that sensory magnitudes obey the associative, commutative, and distributive laws of arithmetic) the possibility of assigning a measure to contrasting sensations could be carried further and given a geometrical representation as color spaces, tone lines, and finally visual space. The general calculus of sequential forms, which Herbart conceptualized as differing degrees of restraints between perceptions, themselves treated as oriented magnitudes, meant that by following methods analogous to the combination of motions in the parallelogram of motions, discussed above in connection with Herbart's treatment of geometry, each external sense could be given a geometrical representation. Each "sensory-space" would have its own restrictions, of course, determined by the physiological conditions of its functioning. Thus the fact that, beginning from any one color as origin, any other color could be constructed as a mixture of two other colors meant that the spatial representation of colors would be a closed two-dimensional surface, such as a color triangle (*Farbendreieck*) (Herbart 1824/1825, 5:489). The tonal system could be similarly constructed from three variables—pitch, intensity, and timbre. Like the color system, by beginning at any particular tone, it would be possible to arrive at any other tone through a continuous series of changes in the three variables. But, as Herbart noted, there would be important differences. In the case of the color system, for example, it is possible to join the endpoints red and violet in a line via a continuous sequence of purples. This is not the case for the tone system, however. The lowest discernible tone, C_{-2}, and the highest discernible tone, D_8, are further separated from each other by any two other tones, and there is no continuous transition through different tones between them. Hence the system of tones would also have to be represented geometrically as a straight line.

Perhaps the most far-reaching suggestion in Herbart's work was the notion that the dynamical relations expressed in the sequential forms were a general set of operations performed by the mind in each of its activities of assembling sensations into perceptions and of relating perceptions in concepts—no matter what the sensory modality. An additional powerful idea in this work is the notion that space is not something given but is rather a tool for symbolizing our different sensory modes of interacting with the world and synthesizing them in a common experience:

> Sensory space, to be exact, is not originally a single space. Rather the eyes, and the sense of feeling or touch independently from one another initiate the production of space; afterward both are melted together [*verschmolzen*] and further developed. We cannot warn often enough against the prejudice that there exists only one space, namely phenomenal space. There exists no such thing as space; but there do exist motivations [*Veranlassungen*] for generating a system of perceptions by fusing them through a network [*Gewebe*] of laws of reproduction, whose perceived object is something spatial, namely for the perceiver. There are numerous motivations for undertaking such constructions, and they are not all equally successful; for many attempts to construct space remain incomplete and in the dark [e.g., in optical illusions]. (1824/1825, 5:489–490)

In elaborating upon this passage, Herbart explained that his objective was to link what he termed different manifolds of sense data:

Thus, in the case of the construction of binocular visual space, the different manifolds would include data from each retina, such as color and spatial localization relationships, data from the innervations associated with the muscle tensions on each of the six pairs of eye muscles for the corresponding positions of the visual axes, data on accommodation, proprioceptive data from the neck muscles, and so forth. In the sensorium, all of these different "spaces" would be reduced—or "fused" as Herbart expressed it—in terms of the generalized calculus of sequential forms into visual space. In order to accomplish this he developed numerous types of sequential forms which could form secondary series to primary series, side-sequences (*Seitenreihen*) that would take as their starting point a term within a sequence or series, and various other types of logarithmic series that were intended as structures for linking different series in a network. It is tempting to characterize Herbart's objective as analogous to seeking a master equation that would express the themes, variations, tempos, voices, each separately present but harmoniously interwoven in a fugue.

Herbart was seeking a reliable set of operations for establishing that sensory manifolds constituted from data transmitted from the retina to the sensorium correspond in an adequate fashion to the structure of the external object assumed to be producing those sense data. A key requirement, he argued, was that sensory manifolds must be bound and ordered, and the stability of objects in the world required that at every instant the order among elements in the manifold must be conserved in a constant structure of relations (1824/1825, 5:118–119). His proposal for generating a procedural rule was to follow the same course outlined in his treatment of the connection between subjective and intelligible space—namely, to determine whether the perceived object meets the mathematical criteria for a rigid body. Pursuing this strategy, Herbart sketched a rule for carrying out the determination: Any perception of a singular object must retain the same ordering of points from one instant to the next, no matter which point is taken as the starting point for determining the interconnections between all other points. If the object is a rigid body, then a system of determinations (equations) exists between all points taken pairwise such that, given one point, two others are determined; each of these two then determines two others, and so on. When the gaze is shifted from a point, a, on the surface—a corner, for instance—to another point, b, a new set of relation pairs is determined. If the object is a rigid body at rest, then some constant relation ought to exist between the two systems of relation pairs. The warrant of success in the construction was to be provided by an experiment that would establish the conservation of a constant structure of relations during the act of perception. According to Herbart, the eyes are constantly moving back and forth in small excursions over the surface of objects taken to be at rest, checking the constancy of these structural relationships: "When it wants to grasp an object the eye moves, not in a straight line but back and forth. In every forward motion a set [*Menge*] of reproduction rules is generated; in every return motion they are re-activated through the renewed sight of the object seen previously" (1824/1825, 5:126).

The final determination of whether a set of sense data corresponds to a single object, as well as the relation of that object to the viewer and to surrounding objects, is a determination requiring processing of various other forms of information. The information is not just given as fully constituted in the act of vision, as the nativist hypothesis argued; rather, it must be sought after and evaluated: "Behind this entire consideration lies a physiological presupposition; namely, that the eye moves in accordance with the impulse [*Antrieb*] given by the act of perception" (1824/1825, 5:132). The eye, therefore, is the tool employed by the sensorium to gather information and check the accuracy of the perceptions it constructs on the basis of the sense data it receives. In order to assemble these data in a self-consistent perception, the sensorium steers the eye in terms of several levels of conceptual input: "There are mainly four: enclosure of shape; the overlay of color against a background; the activity of the eye inside the contours of the field; and what is most important, the motion of the whole ensemble relative to a background" (1824/1825, 5:132). Consider this statement in light of Herbart's interest in the connections between intelligible and phenomenal space—that is, our relations to things in themselves on the one hand and our representation of those relations as phenomenal space on the other. The ultimate warrant for assuming a causally determined correspondence between our visual perceptions and the things they represent, the Archimedian lever that would guide the whole visual process, giving it unity and direction, would be the criteria provided by the laws of motion applied to rigid bodies.

This was the outline for a bold empiricist program. To transform it from a daring speculation to an empirically fruitful theory required a variety of experimental demonstrations that would make plausible the claim that visual perception is constructed and that various levels of conceptual information are employed in bringing it about. Herbart's work also suggested how to proceed with this part of the program: If the process of perception is fundamentally always concerned with shaping a perception and cross checking it for consistency in light of various regulative criteria, then each level involves a decision, and that decision can be in error. By exploring the various classes of illusion that occur in visual perception, it would be possible to elucidate the conceptual components involved in the construction of space.

To move beyond speculation, the program for empirical psychology envisioned by Herbart would require foundations grounded in experiment. It also required a more secure mathematical apparatus linking the laws of thought, the principles of mechanics, and the foundations of geometry. Several persons took up this project independently of one another in the 1850s. Moritz Drobisch, a former student and friend of Herbart, explicitly took up the task of providing more adequate mathematical foundations for Herbart's psychophysics (Drobisch 1850, iv–vii, xii). Bernhard Riemann was stimulated through his study of Herbart's works to pursue the alluring goal of examining the connection between the foundations of mechanics and geometry.[11] In many ways, Herbart's empirical psychology can also be seen as a blueprint for Helmholtz's own research.[12] For Helmholtz too found himself launched upon a broad-ranging project of revising Kant's transcendental aesthetic. At the center of that project was the establishment of an empiricist theory of the origins of both the geometrical axioms and visual space. These epistemological and physiological researches were themselves related streams of investigation in a network of reflections on the relationship between the principles of mechanics and geometry.

Mathematical Representations and Color Spaces

The development of a method for representing the various modalities of sense data within the sensorium was one of the fundamental problems confronting an empiricist theory of the origins of visual space. Herbart, as we have seen, had proposed treating each external sense as a particular type of spatial representation, such as a color space or a tone line. The idea of representing color mixtures and tonal relations spatially was not original with him, of course; it can be found in the works of Kepler, Newton, and J. H. Lambert, to name only a few. What was most interesting about Herbart's proposal was that it sought to treat these "spaces" as different applications of an abstract science of geometrical relations. Furthermore, as I have attempted to show, Herbart's conception of how one ought to proceed in constructing an abstract geometry consistent with psychological methods of construction was to focus on the concept of magnitude and to treat lines and points as directed magnitudes.

Herbart was not particularly concerned with mathematics per se but more especially with psychological processes involved in mathematical construction, and he cautioned that his speculations would not satisfy the mathematicians' requirements of rigor. A number of Herbart's colleagues, however, were interested precisely in the problem of constructing an abstract "higher mathematics," among them his illustrious Göttingen colleague Carl

Friedrich Gauß. Now, Gauß had made many powerful hints about how he might proceed to construct such a higher geometry,[13] and he had actually begun to lay the central groundwork for it in his theory of curvature, but he left the task suggestively unfinished.[14] Furthermore, these mathematical efforts were not motivated by the epistemological and psychological concerns that occupied Herbart, so there was no assurance that, if and when completed, the work of the mathematicians would meet the needs of empirical psychology or, for that matter, even support its primary objectives. A further crucial requirement of the empiricist program was that the mathematical structures used in attacking the construction of psychological space had also to be consistent with the physiological mechanisms involved. In my view, Helmholtz and Riemann brought these two strands of investigation together in a fruitful manner.

Gauß himself may not have been directly interested in this enterprise, but several of his friends and colleagues in Göttingen and Leipzig definitely were. In their view, Leibniz held the key to this new domain of higher mathematics, sketched tantalizingly in references to what he had called analysis in situ and the logic of characteristic. That Leibniz's monadology and mathematics were central to Herbart made Leibniz's views on this subject seem all the more relevant to their enterprise. Leibniz's brief but provocative comments on this subject became available for the first time through the publication of his correspondence with Christiaan Huygens (Ulyenbroek 1833). Leibniz claimed to have discovered a new system that was "entirely different from algebra and which has great advantages in representing to the mind, exactly and in a way faithful to its nature, even without figures, everything which depends on sense perception" (Crowe 1967, 3–4; Loemker 1956, 384–385). Whereas algebra expresses undetermined relationships between magnitudes and numbers, Lebniz's new logic of characteristic would also express situation, angles, and motion. The new system would combine the virtues of algebra and pictorial geometry, and with it one could mathematically treat mechanics as well as geometry. Leibniz noted that a chief value of his logic of characteristic would be to enable conclusions to be drawn by operations with its characters "which could not be expressed in figures, and still less in models, without multiplying these too greatly or without confusing them with too many points and lines in the course of the many futile attempts one is forced to make" (Crowe 1967, 3–4; Loemker 1956, 384–385). Leibniz's new calculus would provide methods of abridgment to lighten the burden of imagination in geometry and mechanics.

The history of attempts to reconstruct Leibniz's logic of characteristic is typically read as an episode in the history of mathematics and logic, particularly in connection with the history of vector analysis. But, for my purposes here, it is also interesting to read it in the context of problems occupying mathematically oriented physiological psychologists inspired by Herbart, particularly Helmholtz. I have argued above that in his empirical psychology Herbart sought a generalized formal calculus that could be applied to external sense data, rendering them as different forms of spatial representation. Moreover, I have argued that Herbart required—but never provided—methods for operationalizing the construction of sensory space without resorting to the Kantian theory that space is simply given in intuition. The key site for Herbart was the productive imagination and its connection to the "higher geometry" articulated as a generalized science of relations. For visual space, the notion of a rigid body and forms for quickly calculating identical mappings of sense data through translations and rotations were crucial. In this context Leibniz's logic of characteristic, by linking algebra, pictorial geometry, and mechanics would, I hypothesize, serve as an extremely suggestive stimulus to working out a concrete theory of how the mind interacts with the world through the construction of space.

Reconstructing Leibniz's logic of characteristic became a hot issue among several of Herbart's followers in the mid-1840s. Moritz Drobisch and August Ferdinand Möbius, both from Leipzig, took an interest in the problem in 1844. In early 1844 Drobisch proposed to the *Fürstlich Jablonskischen Scientific Society at Leipzig* the formulation of a prize question designed to spark interest in reconstructing Leibniz's analysis in situ.[15] The proposal was accepted and the prize question was published in the *Leipziger Zeitung* on 9 March 1844.

> There are a few fragments remaining of a geometrical characteristic which Leibniz invented (see *Christi. Hugenii aliorumque seculi XVII. virorum celebrium exercitationes mathematicae et philosophicae.* Ed. Uylenbroek. Hagae comitum 1833. fasc. II, p. 6), in which, without taking recourse to the magnitudes of lines and angles, the relative situation of position is immediately represented by means of simple symbols and determined through connections between them, and therefore differs completely from our algebraic and analytical geometry. The question is whether this calculus can be reconstructed and further developed, or whether a calculus similar to it can be provided, which by no means seems impossible. (Cf. *Göttinger gelehrte Anzeigen*, 1834, p. 1940)[16]

There was no winner in 1845, and the prize question was repeated in 1846, this time doubling the value of the prize from twenty-four to forty-eight gold ducats. Shortly after the original prize question appeared in 1844, Hermann Grassmann published his *Doctrine of Extension,* and he sent a copy of the work to Möbius, who had himself been working on a similar system of geometrical analysis, which he called the barycentric calculus. Möbius found the work interesting but not altogether felicitous in its presentation. The novelty of his methods, the peculiarity of his terminology, and the difficulties in his style of argumentation were persistent obstacles to the reception of Grassmann's work among mathematicians such as Möbius.[17] Nevertheless, when the prize question was reissued in 1845, Möbius wrote to Grassmann, encouraging him to submit an entry. Grassmann's 1847 treatise *Geometrical Analysis in Connection with the Geometrio Characteristic Invented by Leibniz* was awarded the prize by a committee consisting of Wilhelm Weber, Ernst Heinrich Weber, Gustav Fechner, Möbius, Drobisch, and others, and the prize was given to Grassmann on 1 July 1846 by Drobisch at the opening ceremonies of the newly founded Royal Scientific Society of Leipzig.

In spite of Grassmann's success, the full merits of his work were not appreciated. In private correspondence, in which they exchanged their evaluations of Grassmann's prize entry as well as the *Doctrine of Extension,* Möbius and Drobisch both resisted the conclusion that he had fully succeeded in establishing the sort of mathematics Leibniz had in mind, but they regarded his approach as an outstanding first attempt nonetheless. Möbius in particular complained about the abstract constructivist character of the work. Möbius wanted to see the calculus worked out and applied to problems with more content than Grassmann had yet attempted. But his main criticism concerned

> the foreign manner in which the author attempts to ground his methods of calculation. The foundation of geometry is intuition [*Anschauung*]; and even if analytical geometry has often been criticized for loosing sight of this, nonetheless, the formulae from which the calculator begins are founded in geometrical intuition and he remains conscious of the fact that the symbols with which he calculates signify lines, surfaces, or bodies. Things are quite different in the present [Grassmann's] work, where in accordance with a certain analogy to arithmetic, objects are treated as magnitudes, which in themselves are not magnitudes at all and which are totally incapable of being imagined. (Quoted in Engel 1911, 113–114)

Möbius was by no means clear about where he stood on the issue of intuitiveness in relation to higher geometry, however. In 1846, for example, he had received a letter from Ernst Friedrich Apelt, then professor of philosophy in Jena, and a strict Kantian of the old school, asking Möbius if he had read Grassmann's *Doctrine of Extension.* "It seems to me," Apelt wrote, "that a false philosophy of mathematics lies at the basis of it. Intuitiveness [*Anschaulichkeit*], the essential character of mathematical knowledge, appears to be banished in this approach. Such an abstract science of extensions as he attempts can only be derived from concepts. But the source of mathematical knowledge lies not in concepts; it is in intuition" (quoted in Engel 1911, 101). Möbius responded that until recently he had not studied Grassmann's book carefully, because, "as you have remarked yourself, the book rejects the essential character of mathematical knowledge, which is intuitiveness. In glancing through the work recently, however, I have been struck by many things, on the expansion of concepts, on generalizations, or whatever you want to call them, which could be extremely influential for mathematics itself and for the systematic presentation of its elements in particular. In this context belongs the addition and multiplication of lines, if one considers not just their length but also their directions" (quoted in Engel 1911, 101).

Grassmann did not discuss at length the relationship between his mathematics and the philosophical issues surrounding the discussion of Kant's transcendental aesthetic, but his contemporaries saw his work as taking sides in that debate. On which side he stood was made clear in the foreword and introduction to the *Doctrine of Extension*. Like Leibniz's references to the goals of the logic of characteristic, Grassmann noted that his aim was to construct a system of pure forms of thought (*Denkformen*) abstracted from the contents of any particular mathematical discipline, such as Euclidean geometry. Geometry and mechanics, he noted, were empirical disciplines based in an intuition of an objective reality, namely, physical space, time, and motion. The principles of mechanics and geometry could not be deduced from pure thought; nor could they be derived from experience. Rather they were based in what he called a basic intuition (*Grundanschauung*) in which the forms of pure thought were applied (*angewendet*) to what is given in intuition. Grassmann's goal was to develop forms of symbolic representation for creating mathematical concepts and theorems that embody the operations of the mind. He thought this would facilitate the development of empirical disciplines. Perhaps equally important, his new form of analysis would collapse the distinction between the analytic and synthetic treatments of geometry.

HELMHOLTZ'S FIRST ANALYSIS OF SPECTRAL COLORS AND HIS ENCOUNTER WITH GRASSMANN

Helmholtz's controversy with Hermann Grassmann on the proper representation of spectral colors was crucial to the development of Helmholtz's views on mental representations and their physiological correlates. Helmholtz first encountered Grassmann's work in 1852.[18] As part of the procedure connected with his appointment at Königsberg, Helmholtz chose as the subject for his

Habilitationschrift a critique of David Brewster's theory of color, which was based on the view that the spectral colors are mixtures of three elementary colors, red, yellow, and blue. Grassmann found Helmholtz's paper wanting in certain respects, but his primary interest lay in the opportunity it offered for overcoming the steadfast resistance to his ideas, partially outlined above, by illustrating once again the general applicability of the methods of the *Doctrine of Extension*. The exchange through papers in *Poggendorff's Annalen* between 1852 and 1855 not only led Helmholtz to revise his early ideas but brought him into direct contact with some exciting mathematical concepts for solving the problem of spatial representation in physiological optics.[19]

Of particular interest was Grassmann's development of the concept of a manifold, which consisted of systems of elements. Leaving the concept of "element" unspecified in content, Grassmann defined extensions as the collection of elements produced by continuous change of an original generating element. Different levels (*Stufen*) of elements and extensions were generated by specifying a different mode of change (*Änderungsweise*) for each level. An m-level system (*System m-ter Stufe*) of elements was defined in terms of m independent "modes of change." By interpreting elements as points and modes of change as directional coordinates, Grassmann defined Euclidean space as a geometrical interpretation of a three-level system. Thus he described points in a line as a one-level system; a two-level system would be interpretable as the elements of a plane, and so on (Grassmann 1844, 1.1:47–49). Interpreted geometrically, a line segment was a system of elements generated by motion in the same direction; parallel lines were two systems of elements with the same directional coordinates, and so forth. Having defined line segments as directed magnitudes, Grassmann interpreted addition geometrically as the resultant of joining the first and last elements in the collection of elements from which an extension is generated: "Since the same laws for conjoining the m original modes of change also obey the laws of addition and subtraction, we can summarize the results of the preceding discussion in the following extremely simple statement: If [ab] and [bc] represent arbitrary modes of change, then [ac] = [ab] + [bc]."

In a footnote to this passage, accompanied by an illustration of the parallelogram rule for addition, Grassmann wrote, "I cannot recommend strongly enough that one represent this result, which is one of the most difficult to grasp in this entire science, by means of the following geometrical construction" (1844, 1.1:52).

The parallelogram rule for combining forces thus served as the basis for the most fundamental conceptual operation in Grassmann's new science

of abstract forms. He obviously sensed that unless he could convince mathematicians and physicists of the power of this elementary aspect of the new
methods of representation he was developing, the entire mathematical edifice he was constructing would remain unappreciated. The critique of
Helmholtz's paper on color mixtures offered an occasion to illustrate this
point to an audience that had failed to take notice of the *Doctrine of Extension* for nearly a decade.

Helmholtz's papers on color mixtures featured the adaptation and
refinement of existing instruments and experimental practice that characterized all his early papers and aroused the admiration of his contemporaries
(Grassmann 1844, 1.1:12–13). In his first paper of 1852, the arrangement for
mixing pure spectral colors consisted of two slits forming a V in a black blind
(Helmholtz 1852). The slits were inclined by 45° to the horizontal and were
at right angles to one another (see figure 8.2). Different spectral colors passed
through each slit were combined at the point of intersection. To generate the
possible combinations of these two colors, a flintglass prism was placed vertically in front of the objective lens of a telescope, which was focused on the
intersection of the slits, both prism and telescope at a distance of twelve feet
from the V-slit blind. Helmholtz noted that a similar arrangement with a single vertical slit instead of a V-slit produces a rectangular spectrum in which the
different color bands and Frauenhofer lines run vertically parallel to one
another. The spectrum of an inclined slit is a parallelogram with two parallel
horizontal sides and two sides parallel to the slit. In this inclined situation, the
color bands and Frauenhofer lines run parallel in the direction of the slit. In
the case of Helmholtz's V-slit, the two spectra overlapped, with the two sets of
color bands and Frauenhofer lines running in the directions of the slits. When
viewed through the telescope, the area of overlap of the two spectra was a triangle, and within the triangle all the combination colors resulting from the
mixture were visible. Beyond the edges of the triangle in the remaining

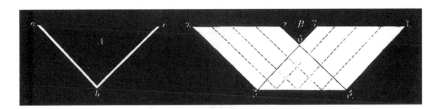

Figure 8.2
Hermann Helmholtz, V-Slit apparatus for mixtures of spectral colors. Source: Hermann
Helmholtz, *Handbuch der physiologischen Optik,* 2d ed. (Hamburg: Leopold Voss, 1896), 352.

portion of each parallelogram the spectral color admitted through each slit was visible. Helmholtz fixed cross-hair lines in his telescope, which he oriented at 45° so that they ran parallel to the Frauenhofer lines. This enabled him to estimate the relative distances from the dark lines in the spectra of the color bands entering the specific mixture. In this first approach to the problem, Helmholtz did not attempt to provide a quantitative determination of the wavelengths of his color mixtures. In order to compare the relative intensities of the two colors entering a particular mixture, Helmholtz noted that by rotating the prism about the axis of the telescope the surface area of the illuminated parallelogram changed, being greatest when the slit and prism were parallel. In that position the illuminated area was a rectangle. By rotating the prism relative to the slit, therefore, the same quantity of light would illuminate a larger or smaller surface area, and appear correspondingly less or more intense. In the original position of the V-slit, the intensities of the two spectral colors were equal. By rotating the prism, all combinations of relative intensities of the two spectra could be achieved.

Using this experimental design Helmholtz arrived at several remarkable conclusions. The first was that color mixtures formed from pigments or powders differ markedly from color mixtures formed from pure spectral colors. In contrast to the experience of painters for a thousand years, Helmholtz wrote, the mixture of blue and yellow spectra, for example, does not yield green but rather a greenish shade of white (1852, 15–16). The explanation, according to Helmholtz, is that the combination of spectral colors and the mixing of pigments rest on two different physical processes (1852, 17). In the case of colored pigments and powders, a portion of the light falling upon a colored body is reflected back as white light, while of the remaining light that penetrates the body one portion is irregularly absorbed, while another is reflected from the back surface and is taken by the observer as the color of the body. What transpires in the mixture of pigments is, Helmholtz concluded, the loss of different rays of colored light rather than the combination of the colors.

One of the aspects of the paper that attracted Grassmann's attention was the range of results achieved from mixing pairs of spectral colors. "The most striking of these results," Helmholtz wrote, "is that among the colors of the spectrum only two combine to produce white, being therefore complementary colors. These are yellow and indigo blue, two colors which were previously almost always thought of as producing green" (1852, 12–13). It was because previous investigators had used mixtures of pigments rather than basing their theories on the mixture of pure spectral colors that this incredible error had been propagated and reinforced, according to Helmholtz.

Another surprising result of Helmholtz's investigation was his rejection of Thomas Young's hypothesis that all colors of the spectrum can be generated from three primary colors (Helmholtz 1852, 21). Helmholtz concluded by contrast that the least number of colors out of which the entire spectrum could be generated was five. He arrived at this result by trying to construct a color circle. As the best method of construction he favored Newton's procedure of producing each simple color by combining it from the neighboring colors on either side, but he restricted Newton's approach even further by adding that the distance between the two combining colors should not be too great. Otherwise, he said, the resulting intermediate shades would not match those of the spectrum. Proceeding in this manner, Helmholtz concluded that the minimal list of colors required to imitate the spectrum would include red, yellow, green, blue, and violet.

Grassmann took issue with Helmholtz's claim that there is only one pair of complementary colors and set out to show that Newton's view was indeed correct: Every color has a complement with which it combines to produce white light. Grassmann's demonstration of this claim was remarkable in that he built his mathematical analysis directly upon certain structural features of the perception of color. He thus examined the purely phenomenal, mental side of color relations; here was exactly the analogue of the problem of generating the spatial components of visual experience from the phenomena. First Grassmann observed that any compound color can be imitated by mixing a homogeneous color of a particular intensity with white light of a particular intensity. This observation was the basis for his central assumption—namely, that the eye is capable of making three types of comparison. It distinguishes hue, the intensity of the color, and the intensity of the white light mixed with the homogeneous color producing the sensation, that is, the degree of saturation (Grassmann 1853, 161–162).

To establish the proposition concerning complementary colors, Grassmann assumed as well grounded in experience that (1) if one of the colors entering a mixture remains constant while the other undergoes a continuous change in intensity, saturation, or hue, then the resulting mixture also changes color continuously. As a final preparatory step, he observed that (2) by a continuous transition in wavelength it is possible to traverse the entire series of color tones in the spectral series. To this it was necessary to add that it is also possible to traverse a continuous series of purples from violet to red, thereby making it possible to complete an entire circuit of color perceptions. For his analysis Grassmann assumed that the transition from red to orange, yellow, green, blue, violet, purple, and back to red is positive, the reverse series negative. Three types of transition between two colors A and B were therefore possible: Namely, either the hue of the color A takes on all the hues in the positive transition series between A and B; or it takes on all the hues in the negative series; or the color undergoes one or more transitions through white light. With assumptions (1) and (2) Grassmann demonstrated the proposition that for every color there exists another homogeneous color with which it combines to produce white light.

The "proof" must have seemed strange to an experimentally oriented empiricist like Helmholtz, for it proceeded as a reductio ad absurdum. Instead of experimentally demonstrating that every color has its complementary, Grassmann proceeded in an abstract fashion by attempting to show that if this were not the case, our concept of continuity and our experience of the closure in the continuous transition of colors would be violated. Grassmann started with a given color of hue a, and he assumed that no other color could be mixed with it to form white. Next he chose an arbitrary color with color tone x and intensity y. He mixed the two colors, keeping the color tone, x, constant while continuously varying the intensity, y. According to assumption (1), the result of mixing the two colors will be a continuously changing color, and a point will be reached where the intensity of the color a will vanish relative to the intensity, y. Now, one way in which the mixture of the two colors could change would be for them to merge into white. But since it was assumed at the outset that the mixture of the two colors could not produce white, the only other alternative is that the color of the mixture will have to change continuously in terms of the other variable, x; so that a will gradually merge into x. The transition of hues can proceed either positively or negatively, depending on the hue x, but at some point a transition must occur in the series of mixtures from which point on x predominates. Since no transition between the two colors could take place through a band of white, it was required to establish the possibility of an abrupt transition between the two hues x and not-x (a). The proof strategy Grassmann pursued was to show that the assumption that the transition point was any color other than white led to a contradiction.

In order to show that his line of thinking was consistent with what was known experimentally to be the case, Grassmann proceeded to construct a color chart. He first showed that Newton's determinations of the beginning of each band of spectral color in terms of the mean determinations of the refractions and dispersion for his prism could be coordinated precisely with the Frauenhofer lines. To carry these considerations over into the construction of a color chart, Grassmann assumed in particular that Newton's red and violet fall together with the Frauenhofer **B** and **H** lines. This assumption determined that the colors would be represented as a circle. The ratios between Newton's color bands and the Frauenhofer lines permitted a placement of colors on the circle. For example, the transition line between red and orange lay between the **C** and **D** lines in the ratio of 7:6.

This placement compared well with Helmholtz's results. Helmholtz had found that mixtures of red with orange, yellow, and green produce the intermediate colors in the series; red mixed with green, for instance, gives a dull yellow, which by increased red goes over into orange and finally to red. Similarly, red mixed with violet, indigo, and blue gives all the intermediate colors. In particular, Grassmann declared, "Red mixed with blue produces a whitish violet, which goes over into rose and carmine as red begins to dominate the mixture. According to the theorem proved above the complementary color to red must lie between green and blue, therefore some tone of blue-green" (Grassmann 1853, 167–168).

Helmholtz had noted that red mixed with the blue-green tones produces a series of flesh-colored tones, but he had not described how it was possible for this transition to occur. Grassmann continued: "There is a gap here, therefore. Moreover, flesh tones are nothing other than red mixed with a great quantity of white, and no possible transition is imaginable other than that the red is diminished until it disappears relative to the white in the mixture, and then out of this white (or gray) the blue-green tones gradually emerge. In short, the usual transition through white occurs here" (Grassmann 1853, 168).

Grassmann concluded the paper with a discussion of the rule for combining colors. First he established a coordinate frame. He took an arbitrary homogeneous color **a** and its complement **a′**, placing them in a line at equal and opposite directions from each other. Next he chose a color **b**, which gave the same quantity of white when mixed with **a** and with **a′**. Its complement, **b′**, was chosen similarly, and the intensities of **b** and **b′** were chosen such that the intensity of the white produced from mixing them was equal to the intensity of the white produced from **a** and **a′**. Grassmann represented these conditions in terms of a right coordinate system with **aa′** bisected by **bb′**. In practical terms, if **a** is yellow, then **a′** is indigo, **b** a shade of green, and **b′** a purple. From mixtures of two of these four colors, any other colors could be constructed as the geometrical sum using his parallelogram rule:

> Once this has been done, one can find the hue and color intensity for any mixture of colors through construction. One needs only to determine the extensions which represent the hue and color intensity of the colors being mixed and then to sum these geometrically, that is to say, combine them like forces. The geometrical sum (the resultant of those forces) represents the hue and intensity of the mixture. It follows immediately from

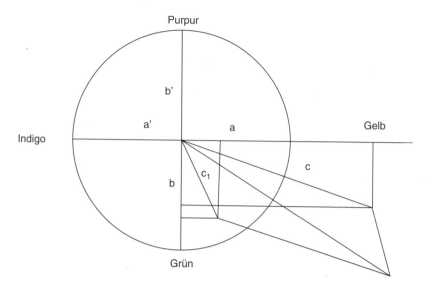

Figure 8.3
Grassmann's method for representing color mixtures. Source: Hermann Grassmann, "Zur Theorie der Farbenmischung," *Annalen der Physik and der Chemie* 89 (1853): 69–84, reprinted in *Hermann Grassmanns gesammelte mathematische und physikalische Werke*, ed. Friedrich Engel, 11:2 (Leipzig: Sächsische Akademie der Wissenschaften zu Leipzig, 1894–1911), 161–173.

this that the order in which one adds geometrically (combines the forces) is of no consequence for the result. (Grassmann 1853, 169–170)

To illustrate the procedure Grassmann constructed the arbitrary mixture of two colors, $c + c_1$, in the green-blue quadrant of his color circle (see figure 8.3). The color c could be represented as the diagonal of the parallelogram with sides βa and ϕb, and c_1 could be represented as the diagonal of the parallelogram with sides $\beta_1 a$ and $\phi_1 b$ Thus the mixture of the two colors could be represented as

$$(\beta a + \phi b) + (\beta_1 a + \phi_1 b) = (\beta + \beta_1) a + (\phi + \phi_1) b$$

This was the geometrical sum of the directed magnitudes representing the two colors entering into the mixture. In a footnote to this discussion, Grassmann indicated that a full exposition of the theory of geometrical sums was to be found in his *Doctrine of Extension* of 1844 as well as in Möbius's *Mechanics of the Heavens* (1843).

Grassmann's discussion of color mixtures led Helmholtz to rethink his own approach to the subject. From Helmholtz's point of view there were several interesting features about the paper, but at the same time it contained several loose ends. For one thing, Grassmann made it necessary to reexamine the treatment Helmholtz had given of complementary colors. The inadequacy of the instrumentation seemed to be the primary weak point in his work. Already in his first paper of 1852 Helmholtz had noted that a more refined instrumental arrangement for projecting the color mixtures onto a larger surface area and an improved method for measuring the distance of the color mixture from the nearest Frauenhofer line might lead to different results concerning the composition of the whitish hues (Helmholtz 1852, 13–14). But Grassmann's argument was totally inadequate. Among the community of measurement physicists Helmholtz respected, it was insufficient to establish a claim from some sort of abstract mathematical argument without also demonstrating the result empirically. The production of white light from complementary spectral colors had to be empirically demonstrated. Furthermore, Grassmann's reductio argument depended on assumptions concerning the discriminating power of the eye as a measuring device. These were extremely interesting physiological assumptions. But, having made them, Grassmann left the plane of physiological argumentation altogether. Now, the mathematical methods he applied as consistent with his assumptions concerning the different parameters of sensitivity governing the perception of color had led Grassmann to the result that four colors would be required to perform the parallelogram construction. Yet even in Grassmann's ingenious construction, one of the colors, purple, was itself a mixture of red and violet. Perhaps three colors would indeed suffice. By wedding himself to a circle as the method for representing color spaces instead of a triangle, Grassmann had introduced some unnecessary assumptions. Furthermore, the representation in terms of a triangle would correspond more adequately to the physiological character of the argument. This too needed to be checked empirically. In any case, the entire issue of Young's theory was reopened and the theory of color mixtures given a decisively physiological dimension as a result of the encounter with Grassmann.

Helmholtz's Reconsideration of Color Mixtures

Helmholtz took up these issues in a paper entitled "On the Composition of Spectral Colors," in *Poggendorff's Annalen der Physik and Chemie,* in 1855. The data and conclusions of that paper were repeated in much expanded form in sections 19–21 of the second part of the *Physiological Optics,* published in 1860, and my discussion will refer to the treatment in both texts.

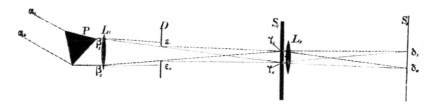

Figure 8.4

Helmholtz's experimental arrangement for color mixtures. Source: Hermann Helmholtz, *Handbuch der physiologischen Optik,* 2d ed. (Hamburg: Leopold Voss, 1896), 352.

In the new approach to color mixtures, Helmholtz refined each component of his apparatus for viewing and mixing spectral colors. He took his lead in this instance from the procedures developed by Foucault in a paper translated in *Poggendorffs Annalen* in 1853. First Helmholtz showed that in order to obtain the purest spectra it was necessary to use light from the sun or from the sky in the immediate vicinity of the sun. To this end he fixed a mirror on a heliostat in order to compensate for the sun's motion and to ensure light of constant purity (see figure 8.4). The light from the heliostat was reflected through a slit in a blind and then through a vertically positioned prism. Light emerging from the prism passed through an objective lens, then through another rectangular slit in a diaphragm, **D**, and projected onto a second diaphragm, S_1, placed at the focal length of the lens.

In place of the V–slit arrangement used in his first set of experiments, Helmholtz used two parallel Gravesande slits capable of refined adjustment, making it possible to align the slits on any two spectral colors (see figure 8.5). The size and position of the slit was controlled by two pairs of sliding plates mounted on tracks, one pair of which moved in direction **bb** inclined to the horizontal and a second, **fɸ**, adjustable in a horizontal direction. By adjusting the screws **d** and **Δ** the two slits could be given any distance relative to each other within the range of the spectrum projected onto the diaphragm. The spectrum itself was projected onto a silvered rectangle in the center of the diaphragm.

In his earlier experiments, Helmholtz controlled the quantity of light entering the mixture by rotating the prism in front of the V–slit, and hence changing the surface area of the parallelograms projected to the telescope. This simple arrangement had worked beautifully for obtaining most mixtures, but it led to difficulties in the whiter hues: first, because even when viewed through a telescope, the individual colors occupied a very small area, and second, because the difficulty of discriminating the hue was increased by the number of other colors in the field of vision. This had been the

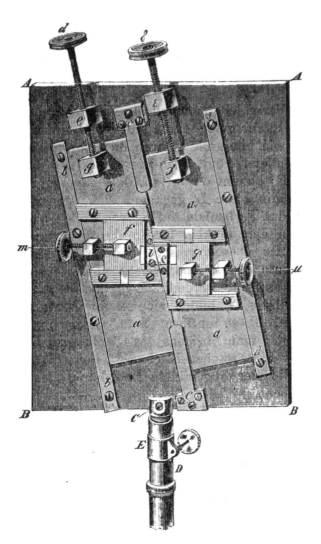

Figure 8.5
Helmholtz's adjustable slit diaphragm. Source: Hermann Helmholtz, *Handbuch der physi-ologischen Optik,* 2d ed. (Hamburg: Leopold Voss, 1896), 353.

source of the error in determining complementary colors. In the new apparatus the quantity of light was controlled directly by adjusting the width of the slit, by turning the screws **m** and **μ**. Hence both hue and intensity could be controlled more accurately in this arrangement.

After emerging from the slits, the two bands of color now passed through a second objective lens, L_2, of shorter focal length than L_1, and were projected onto a white paper as a uniformly colored rectangle. By covering the slits one at a time, the component colors of the mixture could be seen separately.

With this improved instrumentation Helmholtz determined anew the series of complementary spectral colors. He now was able to produce white from violet and greenish yellow, indigo blue and yellow, cyan blue and orange, greenish blue and red. He was, however, unable to produce white from mixing green with any other simple color, but only from mixing it with purple, "that is with at least two other colors, red and violet" (Helmholtz 1855b, 51, and 1867, 277).

In the next stage of his investigation Helmholtz determined the wavelengths of the complementary colors. He combined two spectral colors to produce white light and then removed the white screen S_2 in the previous setup. About six feet away he placed a telescope, and in front of its objective he positioned a glass containing a vertical thin wire-diffraction grating and a horizontal millimeter scale to determine wavelengths. Helmholtz graphed the results as shown in figure 8.6.

From the graph it was immediately obvious why his V-slit arrangement could not have been expected to reveal colors complementary to red and violet: The transitions between the blue-green color bands proceed extremely rapidly, being represented nearly as a vertical line in the graph, and these colors form extremely narrow bands difficult to detect in that arrangement. Indigo blue and yellow, on the other hand, had the advantage of being relatively wide bands of color.

These results also had certain consequences for the geometrical representation of the color table. Helmholtz praised Newton's idea of representing the colors in a plane by means of the center of gravity as one of the most ingenious of all his creative ideas (1855b, 64, and 1867, 288). But Newton himself had proposed the rule as a mere aid for summarizing the phenomena in a qualitative manner and had not defended its correctness as a quantitative explanation. Grassmann's contribution had been primarily to call attention to the mathematical assumptions underlying the center-of-gravity method, and Helmholtz was convinced by Grassmann's treatment that this was indeed the appropriate quantitative method to use. But he was

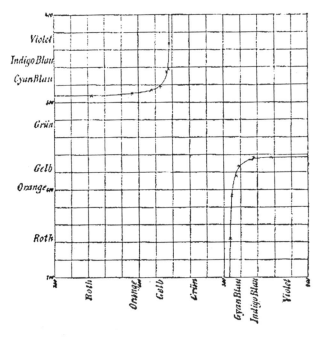

Figure 8.6
Graph relating complementary colors and wavelengths. Source: Hermann Helmholtz, *Handbuch der physiologischen Optik,* 2d ed. (Hamburg: Leopold Voss, 1896), 317.

not convinced that the color table should be represented as a circle. Grass-mann's analysis simply duplicated Newton's assumptions. But in fact, the center-of-gravity method for mixing colors was compatible with many geometrical representations.[20] In order to determine which of those representations fit the causal picture most closely, it was necessary to interpret the parameters in Grassmann's model in terms of empirical measurements. Subjected to this requirement, the choice of a circle turned out no longer adequately to fit the refined data Helmholtz had derived. In terms of the center-of-gravity method for determining color mixtures, white can be placed in the center of the line connecting the two colors only when the "weights" on the ends are equal. In his treatment of this problem, Grassmann had assumed that complementary colors have equal intensities, which implied that they should be placed on the circumference of a circle. This might seem like a natural assumption, since we tend to treat pure spectral colors as the most saturated colors, from which it would seem to follow that they ought to be placed on a scale at equal distances from white. But as

Helmholtz demonstrated in a related set of experiments, that assumption can easily be shown to be false. Different degrees of color saturation had to be assigned to the colors entering a mixture of white light (Helmholtz 1855b, 61, and 1867, 288–289).[21]

The diaphragm and slit apparatus Helmholtz constructed allowed him to derive an approximate measure of the relative quantities of light entering the mixture of white, and from these measurements to determine the shape of the color space. By adjusting the screws **m** and **μ** (see figure 8.5), the width of the slits and hence the quantity of light entering the mixture could be varied. After mixing white from a pair of complementary colors, Helmholtz measured the width of the slit of the brightest of the two colors with a micrometer. He then diminished the width of that slit until the two shadows cast by a wooden dowel appeared equally bright, and he measured the width of the slit once again. The ratio of the two widths of the slit was an approximation to the relative brightness of the two colors.

According to the physiological assumptions underlying Grassmann's construction of color space, quantities of colored light taken to be equal should be those quantities that, having a certain absolute intensity, *appear to the eye* as equally bright. In order to satisfy these requirements, the lengths of the lever-arms in Grassmann's construction had to be adjusted in accordance with the slit measurements discussed above. The result was a completely different shape for the color space. Instead of a circle or a circular lumen, it now turned out to be a hyperbola-like curve with violet, green, and red at the vertices (see figure 8.7).

Grassmann had stimulated Helmholtz to revise fundamentally his approach to the theory of subjective colors. From Grassmann Helmholtz had also learned the power of abstract, structural mathematical approaches to

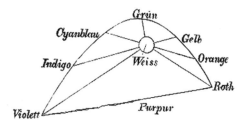

Figure 8.7
Helmholtz's approximate representation of color space for spectral colors. Source: Hermann Helmholtz, *Handbuch der physiologischen Optik,* 2d ed. (Hamburg: Leopold Voss, 1896), 332.

these problems. But while acknowledging that Helmholtz had profited from the interaction, we should also not overlook the object lesson Helmholtz was prepared to give Grassmann in relating abstract mathematical structures to the requirements of physics and physiology. A point Helmholtz would state explicitly in the next stage of his research in sensory physiology was that to be meaningful in reference to a physical problem, an abstract structure had to be embedded in measurements, and the internal logic of the abstract system had to be adapted to the requirements of the physical problem. That process of adaptation was achieved through a dialogue with the instruments. The result of the dialogue was a model of the physical system, in this case the physiological apparatus of the eye responsible for the production of color sensations. In his reference to Newton's approach, Helmholtz made clear that the proper mathematical description of a system was not one that served to represent usefully the phenomena, but rather one which corresponded to the causal properties of the physical system. Helmholtz's color space was such a model of the physiological apparatus employed by the eye in measuring color. For indeed the causal structure consistent with the model represented in the color space—a causal structure that immediately leapt from the representation—was the production of color sensations from the excitation of three receptors sensitive to violet, red, and green.

MUSIC TO THE EYE: PHYSIOLOGICAL ACOUSTICS, VISUALIZATION DEVICES, AND THE RECEPTOR HYPOTHESIS

It is tempting to assume that Helmholtz reversed his stand on Young's theory of color vision in the course of writing up the new experiments central to the 1855 paper. However, the original paper says nothing about the three-receptor theory. Indeed, when the paper was republished in 1883, Helmholtz added a footnote stating that his first recorded support of the Young hypothesis appeared in the second part of the *Physiological Optics,* published in 1862 (Helmholtz 1855b, 70). Nothing about the encounter with Grassmann would have forced a reversal of the position on the physiological bases of colors. The dispute with Grassmann had resulted in confirming Newton's center-of-gravity method for determining color mixtures, but led to the conclusion that the form in which the color chart should be represented geometrically was a shape approaching a conic section rather than a circle, as both Newton and Grassmann had assumed. Nothing concrete had been determined about the physiological causes of color mixtures. Helmholtz had shown that any three colors would suffice to generate the color chart, but those methods did not specify which three colors must be associated with the color receptors in the retina. Helmholtz had established, furthermore, that the center-of-gravity method for representing color combinations was indeed useful, but he wanted to go further and establish that its underlying principles embodied a more general calculating apparatus for the representation of sensations; that is, he wanted to establish it as a psychophysical principle as well. What led Helmholtz to change his mind about the Young hypothesis between 1855 and 1862? How did he arrive at red, green, and violet as the three primary physiological colors?

I want to suggest that work in physiological acoustics and a concerted effort to analogize the eye and ear provided the basis for Helmholtz's reevaluation of the Young hypothesis.[22] Strong support for this suggestion comes from Helmholtz's popular lectures on the recent progress in the theory of vision, which he delivered in Cologne in 1868. In that context Helmholtz explicitly pointed to the analogy with his work on the sensations of tone to convince his audience of the plausibility of the three-receptor hypothesis.[23] Moreover, the internal evidence of Helmholtz's papers in the period between 1855 and 1862 indicate that the receptor hypothesis was more strongly supported by his work in physiological acoustics before the line of investigation for establishing the receptor theory in physiological optics became clear. Indeed, the papers published in this period indicate that the juxtaposition of these two sensory modalities, the back-and-forth comparison of models in one domain with those in the other, guided Helmholtz toward the reversal of his earlier position on Young's trichromatic receptor hypothesis.

The analogy between physiological acoustics and color vision consisted in the assumption that just as in the ear a set of fundamental or primary tones is objectively based in the rods of Corti (discovered by Alfonso Corti in 1851), so in the eye a set of primary colors is based in specific nerve endings in the rods and cones. Neither assumption could be established in humans, although some evidence from comparative anatomy supported the analogy from the side of physiological acoustics. The analogy between eye and ear was a salient feature of Helmholtz's work on physiological acoustics. Comparisons between eye and ear are prominent in his first extended study in physiological acoustics in 1856 (1856a and 1856b),[24] and such comparisons abound in the first edition of Helmholtz's *Doctrine of the Sensations of Tone as Physiological Basis for a Theory of Music,* published in 1863. But such analogies were by no means new, and Helmholtz was certainly familiar with similar comparisons made by other authors, particularly August Seebeck, who had suggested an optical analogue to acoustical resonance of spectral colors with vibrating molecules in groups of nerves in the retina as the mechanism for the sensation of brightness.[25]

The objective Helmholtz set for his acoustical investigations was the solution of exactly the problem that a few months earlier had halted the further development of the Young three-receptor hypothesis: to determine that the mathematical form of the physical description of hearing had a material, physical basis in the physiology of the ear. Helmholtz wanted to show not just that Fourier analysis is a useful mathematical tool for representing the phenomena, but that the ear itself is a Fourier analyzer. Helmholtz's goal in the paper on combination tones was to establish that, like spectral colors, primary tones have an independent objective existence. He wanted furthermore to establish that combination tones have an objective existence independent of the ear, that they are not simply a psychological phenomenon:

> Is this means of analyzing forms of vibration which Fourier's theorem prescribes and renders possible, not merely a mathematical fiction, permissible for facilitating calculation, but not necessarily having any corresponding actual meaning in things themselves? . . . That . . . this [Fourier] analysis has a meaning in nature independently of theory, is rendered probable by the fact that the ear really effects the same analysis, and also by the circumstance already named, that this kind of analysis has been found so much more advantageous in mathematical investigations than any other. Those modes of regarding phenomena that correspond to the most intimate constitution of the matter under investigation are, of course, also always those which lead to the most suitable and evident theoretical treatment. (Helmholtz 1863a, 35–36)

Expressing a metaphysical assumption persistent in all his work, Helmholtz regarded the best mathematical model as one that mirrors the structure of both his measuring apparatus and the physical entities they are intended to represent. The reason Fourier analysis worked so well in capturing the phenomena was that the ear itself is a physical embodiment of a Fourier analyzer. And as we have seen, Helmholtz did not want simply to assume that the most elegant abstract mathematical model was best. Rather, his procedure was to arrive at the mathematical model through the give-and-take triangulation of experiment, theory, and model building.

Helmholtz found it necessary to approach the problem in this way because no adequate mathematical model existed. Work on the theory of vibrating strings provided a partial theoretical framework, but no complete mathematical model was in place to guide his efforts. As Helmholtz noted, the "mathematical analysis of the motions of sound is not nearly far enough advanced to determine with certainty what upper partials will be present and what intensity they will possess" (1863a, 55). A key feature of his work was in identifying the physical basis of tone quality, or timbre, the feature that distinguishes a violin from a flute or clarinet. A plausible assumption made by most physicists before Helmholtz was that tone quality was determined by the form of the sound wave, and that this corresponds to the form of water waves (1857, 85). Helmholtz, too, embraced this assumption, but found that "[u]nfortunately, the form of waves of sound, . . . can at present be assigned in only a few cases" (1857, 85). In order to make up for the defect in theory, he made use of a variety of devices for visualizing the form of a sound wave and mechanically analyzing it into its constituent primary tones.

Among the several experimental devices Helmholtz used were cleverly designed electromagnetic instruments for artificially producing and combining tones. Chief among these was the tuning-fork interrupter and tuning-fork resonator (see figures 8.8 and 8.9). In the tuning-fork interrupter a continuous direct current passing through a tuning fork was momentarily broken by electromagnets near the ends of the fork, which alternately attracted the ends of the fork, made contact, and transmitted current-pulses at the frequency of the fork. These currents were transmitted to a second tuning-fork apparatus. The fork in this apparatus was placed between an electromagnet activated by the incoming current-pulses. To get the best results, the lowest prime tones

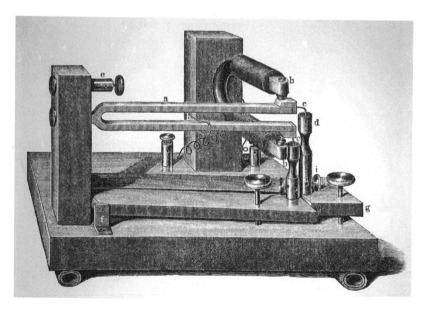

Figure 8.8
Helmholtz's tuning-fork interrupter. Source: Hermann Helmholtz, *Die Lehre von den Tonempfindungen als physiologische Grundlage für die Theorie der Musik* (Braunschweig: Vieweg, 1863).

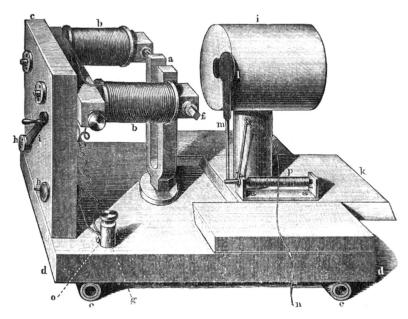

Figure 8.9
Helmholtz's tuning-fork resonator. Source: Hermann Helmholtz, *Die Lehre von den Tonempfindungen als physiologische Grundlage für die Theorie der Musik* (Braunschweig: Vieweg, 1863).

were required. This entailed using forks whose tones were barely audible. These tones were amplified by placing a resonator tuned to the proper frequency near the fork. By placing these resonator devices in connection, Helmholtz was able to combine numerous partial tones into tones indistinguishable from tones produced by musical instruments.

Helmholtz described several methods for making auditory vibrations visible. The first of these was what he termed the "graphic method": "To render the law of such motions more comprehensible to the eye than is possible by lengthy verbal descriptions" (Helmholtz 1863a, 20). He illustrated the graphic method with the phonautograph, which consisted of a tuning fork with a stylus on one prong of the fork. The vibrating fork produced a curve on paper blackened with lampblack and attached to a rotating drum, the same arrangement as in Ludwig's kymograph or in Helmholtz's own myograph. The most dramatic of these visualization devices was the so-called vibrational microscope, an instrument embodying methods described first by Lissajous for observing compounded vibrational motions. The microscope was constructed so that the objective lens was mounted in one of the prongs of a tuning fork.

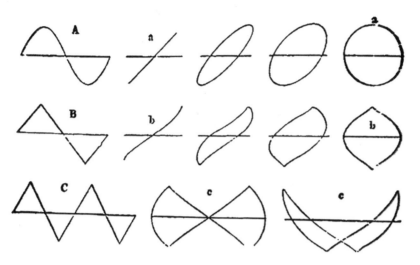

Figure 8.10
Lissajou's figures in the vibration microscope. Source: Hermann Helmholtz, *Die Lehre von den Tonempfindungen als physiologische Grundlage für die Theorie der Musik* (Braunschweig: Vieweg, 1863).

The eyepiece of the microscope was mounted on a plate so that the tube of the microscope was attached to the backing of the bracket holding the tuning fork. The prongs of the fork were set in vibration by two electromagnets, just as in the interrupter and resonator described above. When the tuning fork was set in motion, the object lens would vibrate vertically in a line. When the microscope was focused on a stationary grain of white starch and the forks set in motion, a white vertical line would be seen (see figure 8.10). By placing the grain of starch on a vertical string so that the grain was vibrating horizontally while the lens was moving vertically, the image viewed in the field of the microscope would be a line compounded of both motions inclined at 45°. As the phase of the resonator fork altered, the line visible in the field of the microscope shifted from a straight line inclined by 45° (when the two forks were in unison) through various oblique ellipses until the phase difference reached one-quarter of the period and then passed through a series of oblique ellipses to a straight line inclined 45° in the other direction from the vertical when the phase difference reached half a period (Helmholtz 1863a, 126).

The principal use Helmholtz made of these instruments was to manipulate phases of primary tones generated in his interrupter and resonator in order to demonstrate that phase differences in primary tones making up a compound tone—made visibly evident in the vibration microscope—had no effect on the perceived quality of the tone. According to Helmholtz, the quality of the tone was determined solely by the force of the impression on the ear, that is, on the amplitudes and primary tones entering the composition of the tone. As long as the relative intensities of the partial tones compounded into a musical tone remained the same, the tone would sound the same to the ear no matter how the alteration of phases of the partial tones affected the form of the wave. Helmholtz concluded from these experiments that the quality of the musical portion of a compound tone depends solely on the number and relative strength of its partial simple tones, and in no respect on their differences of phase (Helmholtz 1863a, 126).

These experiments, and the crucial role of the vibrating microscope in particular, were the basis of a direct comparison between the eye and ear. As the imaging device revealed, the eye is capable of detecting differences, even relatively minute differences, between wave forms. The ear is not. "The ear, on the other hand, does not distinguish every different form of vibration, but only such as when resolved into pendular vibrations, give different constituents" (Helmholtz 1863a, 128). With this conclusion, several of the differences as well as the fundamental similarities in the mechanism of color vision and hearing came into view. The reason the ear is able to distinguish the partial tones in a compound tone is that, among the 4,500 or so different nervous fibers in the arches of Corti, specific nerve fibers resonate in sympathetic vibration with the spectrum of primary tones composing musical tones. Simple tones of determinate pitch will be felt only by nerve fibers connected to the elastic bodies within the cochlear membrane that have a proper pitch corresponding to the various individual simple tones. The fact that amateurs, with minimal attention, are able to distinguish the partial tones in a compound tone, while trained musicians can distinguish differences of pitch amounting to half a vibration per second in a doubly accented octave, would thus be explained by the size of interval between the pitches of two fibers. Similarly, the fact that changes in pitch can take place continuously rather than in jumps found its explanation in the sympathetic vibration of arches with proper tones most nearly identical, while the elastic bodies in the membrane with more distantly separated proper tones were incapable of vibrating in resonance.

The physiological organization Helmholtz envisioned as the basis of tone sensation was modeled directly by the tuning fork and resonator apparatus. Indeed, the full set of resonators and connected tuning forks was a material model of the ear in reverse. The resonator apparatus was used to produce compound tones artificially out of the simple tones generated by the tuning-fork interrupter. But this transmitting device could also be imagined to run in reverse as a recording device. In this sense it was a material representation of the functioning ear, its resonators being the material analogues of the fibers in the Corti membrane and its tuning fork and acoustical interrupter being the device for translating, encoding, and telegraphing the component primary tones of the incoming sound wave analyzed by the resonators.

Moreover, the model also provided a resource for understanding the differences between the eye and ear, as well as a suggestion for further development in the theory of color vision. Helmholtz's apparatus enabled him to conclude that the sensation of different pitches would be a sensation in different nerve fibers. The sensation of tone quality would depend upon the power of a given compound tone to set in vibration not only those of Corti's arches that correspond to its prime tone, but also a series of other arches, and hence to excite sensation in several different groups of nerve fibers. Thus some groups of nerves would be simulated through shared resonance, whereas other nerves would remain silent. Helmholtz jumped directly from this conclusion to the analogy with Young's hypothesis concerning the eye:

> Just as the ear apprehends vibrations of different periodic time as tones of different pitch, so does the eye perceive luminiferous vibrations of different periodic time as different colors, the quickest giving violet and blue, the mean green and yellow, the slowest red. The laws of the mixture of colors led Thomas Young to the hypothesis that there were three kinds of nerve fibers in the eye, with different powers of sensation, for feeling red, for feeling green, and for feeling violet. In reality this assumption gives a very simple and perfectly consistent explanation of all the optical phenomena depending on color. And by this means the qualitative differences of the sensations of sight are reduced to differences in the nerves which receive the sensations. For the sensations of each individual fiber of the optic nerve there remains only the quantitative differences of greater or lesser irritation.
>
> The same result is obtained for hearing by the hypothesis to which the investigation of quality of tone has led us. The qualitative difference of pitch and quality of tone is reduced to a difference in the fibers of the nerve receptive to the sensation, and for each individual fiber of the nerve there remains only the quantitative difference in the amount of excitement. (1863a, 148)

The analysis of hearing seemed to suggest that just as the many musical tones capable of being distinguished is rooted in an organic Fourier analyzer that yields specific elemental sensations, so the eye could be conceived as generating color from a primary set of sensations rooted in specific nerve fibers. The delicate power of discrimination of which the ear is capable was explained by the large number of specific nerve fibers and related elastic resonating bodies in the Corti membrane. The inability of the eye similarly to resolve colors into elemental sensations would be explained accordingly as a result of the small number of different types of sensitive nerve fiber—three rather than roughly one thousand—and by the assumption that all three nerve types respond in different degrees to light stimulation. The assumption that fibers predominantly sensitive to red, green, or violet light nonetheless respond weakly to light of other wavelengths would explain the continuity of transitions in the sensation of color, as well as the inability of the attentive mind to analyze compound light into its elements. There is no music to the eye, because the eye has only three rather than the thousand "resonator" types of the Corti membrane.

YOUNG'S THREE-RECEPTOR THEORY REVISITED

In order to provide evidence for the three-receptor hypothesis suggested by his acoustical research, Helmholtz embarked upon an extensive series of investigations connected with afterimages. This research expanded the analysis of the 1855 paper on color mixtures and led to a further refinement of the physiological model of the color space by establishing that it is not the hyperbola-shaped figure arrived at in that paper (see figure 8.7), but rather an equilateral triangle with red, violet, and green as its vertices, and that these subjective colors are all more saturated than the spectral colors. Helmholtz introduced the three-receptor theory for the first time in the *Handbook of Physiological Optics,* where it formed the centerpiece of section 20, on color vision, and was elaborated throughout sections 21–23.

Helmholtz was led to reconsider the possible fundamental character of Young's theory by the fact that in constructing a color space he had to position complementary pairs of colors at different distances in order to accommodate the differences in degree of saturation of the colors forming the sensation of white (Helmholtz 1867, 369). This suggested that from a physiological point of view the spectral colors are not the most saturated or pure colors. This idea was further supported by his experiments on acoustical resonance, where resonators with similar proper tones would join in a compound tone. Young's hypothesis gave an immediate physiological ground for a similar expectation in the case of the eye. Young assumed that the physiological basis for color vision is the stimulation of three different types of nerve in the retina, and Helmholtz's experiments supported nerve types sensitive, respectively, to the red, green, and violet ranges of the spectrum as the appropriate choices. With these choices, one nerve type would be primarily sensitive to the lower frequencies, another to the middle frequencies, and a third to the upper frequencies.

To Young's hypothesis Helmholtz added several further auxiliary assumptions designed to provide a full physiological interpretation of the color-space model. He assumed, for instance, that while the stimulation of each nerve type produces a response chiefly in the range of its primary color, it also produces a weak response in the neighboring ranges (see figure 8.11). Thus the stimulation of the violet-sensitive nerve endings also generates responses in the red- and green-sensitive nerves. According to this condition, therefore, with light of medium intensity, the stimulation of the green-sensitive receptor predominates, initially producing the sensation of green but then shifting into yellow and red. Helmholtz also assumed that the strength of the nerve response or sensation in each nerve type is a function of the intensity of the light stimulating it. Reading the graph (figure 8.11) from right to left and considering the stimulation by white light, for instance, the violet-sensitive nerves are activated initially at low intensity; they rise more rapidly than the green- or red-sensitive receptors, reaching their maximum strength at lower intensities of light than the green- or red-sensitive nerves. Since the violet-sensitive nerves are activated at the lowest intensities, the sensation of violet will predominate at low intensity light. When stimulated by light of low intensity, for instance, the violet sensitive receptors generate sensations that begin with violet and then pass through blue over into green (Helmholtz 1867, 320).

Helmholtz's modification of the Young hypothesis brought with it the assumption that the sensation of one color, such as red, is always accompanied by the weak stimulation of nerve activity in the other ranges. This was particularly evident in Helmholtz's depiction of the middle ranges between yellow and blue. An immediate consequence of this assumption, therefore, was that the sensation of spectral green, for instance, is not the most saturated green possible, for it is always accompanied by stimulation of the red- and violet-sensitive nerves. If the activity of these nerves could be suppressed while at the same time leaving the capacity of the green-sensitive nerves undiminished, it would be possible to have the sensation of a saturated green lying well outside the boundaries of the color space constructed for

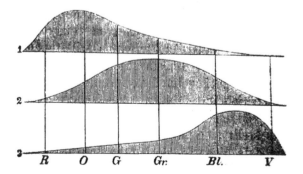

Figure 8.11

Helmholtz's representation of response ranges for three receptors. Source: Hermann Helmholtz, *Handbuch der physiologischen Optik,* 2d ed. (Hamburg: Leopold Voss, 1896), 346.

the spectral colors. The same would be true for red and violet. Against the sensation of these colors, the pure spectral colors would appear whitish. Here was a remarkable feature of the Helmholtz-Young hypothesis that was possible to test empirically. From the point of view of Grassmann's theory of color mixtures, any three colors that could combine to produce white would be satisfactory choices for producing the color space. From the perspective of the assumed physiological mechanisms of the Helmholtz-Young hypothesis, the most saturated colors should be those generated by the primary receptors. Experiments aimed at diminishing or eliminating the activity of one or two of the color-sensitive nerve endings in the retina would thus provide support for the distinction between three different nerve types, as well as dramatic evidence for the choice of red, green, and violet as the three primary physiological colors (Helmholtz 1867, 292–293).

Clinical investigations of color blindness provided support for the three-receptor hypothesis, but the most interesting experimental evidence for the theory resulted from the investigation of afterimages formed by fixing the eyes upon either a bright object against a dark background or a dark object against a bright background. Helmholtz distinguished between positive and negative afterimages. Like photographic images, positive afterimages were those in which bright areas in the object are also bright areas in the image. Negative afterimages reversed the relationship between bright and dark areas in the object, so that bright areas appeared dark in the afterimage. Colored afterimages follow similar patterns; only here the negative image is the complementary color. To form colored afterimages Helmholtz recommended fixing a colored object, such as a colored square or cross, and then observing the afterimage against either a very dark or against a white background of different brightness. The negative afterimage of a colored object is best achieved by fixating for a long period upon, for instance, a colored piece of paper viewed against a gray background. If the colored paper is suddenly removed, the afterimage appears as a negative in the color complementary to the color of the fixated object: The afterimage of a red object will appear blue-green, that of a yellow object will appear blue, that of a green object, red, and vice versa in each case.

The explanation of these phenomena followed directly from Thomas Young's assumption of three sensitive nerve types for the different colors. For, since colored light does not excite all of the nerves with equal strength, the different degrees of excitation must afterward be followed by different degrees of fatigue. If the eye has seen red, the red-sensitive nerves are strongly excited and fatigued, while the green- and violet-sensitive nerves are only weakly stimulated and only slightly fatigued. If, in this condition, white light falls on the eye, the green- and violet-sensitive nerves will be relatively more strongly affected than the red-sensitive. The impression of blue-green, the complementary color to red, will therefore predominate (Helmholtz 1867, 367). Continuing this line of experiment further, if the square section of the retina eye has been rendered insensitive to blue-green by fixating on a bright blue-green square against a red background, the square afterimage will be an even more saturated red than the background red itself. Helmholtz went on to devise a method for doing this experiment with pure spectral colors by repeating this experiment with the prism and color-slit apparatus he used for establishing complementary colors. From these experiments Helmholtz concluded that "*the most saturated objective colors which exist, namely the pure spectral colors, do not yet generate the sensation of the most saturated colors possible in the unfatigued eye,* but rather that we arrive at these colors only by rendering the eye insensitive to the complementary colors" (Helmholtz 1867, 370, emphasis in original).

With this result, Helmholtz had established the empirical basis for supporting his modified version of the Young hypothesis. The color-space of possible subjective colors was not the curve **RGV** worked out in terms of the construction of white from complementary colors (see figure 8.7). Rather, it was best represented as a triangle **VAR**, in which the spectral colors violet and red were displaced to V_1 and R_1 (see figure 8.12). In this scheme, therefore, the surface $V_1ICGrGR_1$ would contain the possible colors obtainable

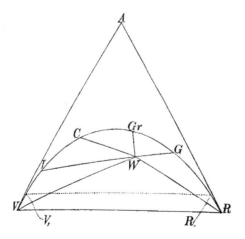

Figure 8.12
Helmholtz's color triangle for representing subjective color mixtures. Source: Hermann Helmholtz, *Handbuch der physiologischen Optik,* 2d ed. (Hamburg: Leopold Voss, 1896), 340.

by mixing objective colors, whereas **R**, **V**, and **A** represented the pure primary colors. The basis of these physiological colors was the tripartite physical constitution of the light-sensitive nerves and their specific nerve energies.

EYE MOVEMENTS AND THE PRINCIPLE OF EASIEST ORIENTATION

In discussing Helmholtz's work on color vision and physiological acoustics, I have emphasized his effort to construct physical models that incorporated measurements directly analogous to the neurophysiological processes of measurement being performed by the eye and ear. I have also pointed to the way in which Helmholtz's models were intended to generate measurements that could plausibly enter as parameters specifying the n-dimensional manifolds that he, like Herbart, believed the mind uses in building experiences of space and time out of sense data. In my view the most spectacular example of this strategy of relating models, instruments, and theory was Helmholtz's treatment of eye movements in connection with the construction of the three-dimensional visual field.

Helmholtz's work in this area built on the researches of Adolf Fick and Wilhelm Wundt. Fick, a student of Helmholtz's friends Carl Ludwig and Emil du Bois-Reymond, was the first person to attempt an exact physical treatment of eye movements based on Helmholtz's work on the conservation of force. Fick's proposal was to treat eye movements as a problem in statics in which the moment of rotation of the eye at rest is the sum of the moments of rotation produced by the six muscles holding it in equilibrium, and that in moving from one position to another the sum of muscle contractions is made with the least total expenditure of force (Fick 1854, 120). Unfortunately, Fick could not provide a general solution to this problem, and was forced to proceed on a case-by-case basis.

Wilhelm Wundt, Helmholtz's assistant in Heidelberg at the time, attempted to improve on Fick's model in 1862. Wundt's approach was to develop his theory directly on a physical model of the eye known as the ophthalmotrope, first introduced by Reute in 1847, with a much improved version in 1857 (see figure 8.13).[26] The idea was to construct the device in such a manner that springs would completely replicate the activity of the eye muscles. Wundt emphasized the need to develop an equation that incorporated a measure of the forces actually exerted by contracting muscles at any given point, rather than a general rule capable of being fulfilled in an infinite number of ways, such as Fick's least-expenditure-of-force approach had done. It was crucial to take the contractile and elastic forces of muscles into account, the opposition between agonist and antagonist muscles in a

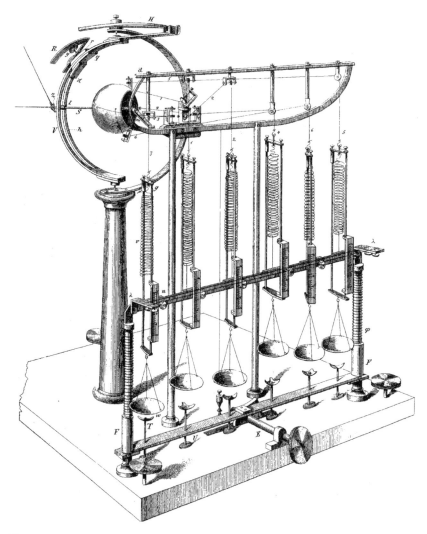

Figure 8.13
Wundt's ophthalmotrope. Source: Wilhelm Wundt, "Beschreibung eines künstlichen Augenmuskelsystems zur Untersuchung der Bewegungsgesetze des menschlichen Auges im gesunden und kranken Zustanden: Part I." *Archiv für Ophthalmologie* 8 (1862): 88–114.

given movement, as well as the resistance to motion offered by external obstacles, such as the optic nerve, fat, and so on. Wundt called this the principle of least resistance. Wundt's principle was stated as follows: In every voluntarily determined position of the visual axis, the eye always adopts that position in which the opposition to its muscles is least (Wundt 1862b).

Moreover, Wundt was keen on incorporating a psychological component into his treatment of eye movements. As Wundt saw it, the problem of eye movements is not just a problem in mechanics but also part of the intentional mental act of achieving visual orientation. Wundt argued that whatever solution works for the eye as a mechanical instrument must be compatible with the psychological regulative mechanism the mind uses in operating and guiding the eye. As we shall see, Helmholtz thought the psychological component postulated by Wundt fell short.

Drawing on work done by the Weber brothers on muscles while he was still a student in the lab of du Bois-Reymond, Wundt had established experimentally that when a muscle shortens or lengthens by a magnitude **e**, the resistance it experiences in each small part of the path **de** is proportional to **e.de**. The form of the equation describing the muscle contraction was, in short, similar to that used in describing the motion of springs, thus justifying the use of the ophthalmotrope as an appropriate physical model of the eye. The earlier work had also established that the resistance is directly proportional to the diameter of the muscle and inversely proportional to the length of the muscle. Where **q** is the diameter, **l** the length of the muscle and **μ** the coefficent of muscle elasticity, Wundt obtained:

(1) $dw = \mu \dfrac{q}{l} ede$

Integration of this equation gives the total resistance **w** experienced by the muscle contracting through distance e:

(2) $w = \mu \dfrac{q}{l} \displaystyle\int_0^e ede$

If the resistances, lengths, cross-sectional diameters, and displacements are similarly represented for each of the six muscles, we then have:

(3) $W_1 = \mu \dfrac{q_1}{l_1} \dfrac{e_1^2}{2}, \quad W_2 = \mu \dfrac{q_2}{l_2} \dfrac{e_2^2}{2},$

Wundt supposed the eye to be a system in which all of the muscles acting like elastic springs bring the visual axis to a particular position according to the condition that the sum of the resistances is the smallest possible for that position; i.e., $W_1 + W_2 + W_3 + \ldots + W_6$ is a minimum. Neglecting the constant $\mu/2$ Wundt expressed the condition as:

(4) $\quad \dfrac{q_1}{l_1} e_1^2 + \dfrac{q_2}{l_2} e_2^2 + \ldots + \dfrac{q_6}{l_6} e_6^2 = \text{minimum}$

(5) $\quad \sum\limits_{i=1}^{i=6} \dfrac{q_i}{l_i} e_i^2 = \text{minimum}$

In my view the most interesting aspect of Wundt's analysis for Helmholtz came in his discussion of an alternative interpretation of the minimum equation and its significance as a regulative principle for guiding the visual axis. Wundt noted that equation (4), expressing the condition of equilibrium for the eye, agrees in form with the general equation expressing the least squares of the observational error for six measurements, each weighted differently:

> If we were to suppose $e_1, e_2, e_3, \ldots e_6$ the observational errors for any phenomenon, $q_1/l_1, q_2/l_2, \ldots q_6/l_6$, those magnitudes which are assigned as weights of observation, our formula would be nothing other than the fundamental equation of the method of least squares, which asserts that the sum of the products of the squares of the observational errors multiplied by the weighted observations must be a minimum. Therefore, whenever we move the visual axis into a new position the eye proceeds just like a mathematician when he compensates for errors according to the rules of the probability calculus. The individual muscles behave like the individual observations, the lengthening and shortening which they experience in the transition to the new position behave like the unavoidable errors in observation, and the coefficients of resistance of the muscles behave exactly like the observational weightings.
>
> Indeed, it is easy to see that it would have been possible for us to have grounded the principle in a completely different way than we have pursued. We could have deduced it directly from the fundamental postulates of the probability calculus in the same way that the method of least squares is grounded. (Wundt 1862a, 58–59)

Wundt's source for this idea was undoubtedly a paper published in *Crelle's Journal for Pure and Applied Mathematics (für reine und angewandte Mathematik)* by Carl Friedrich Gauß in 1829, entitled "On a New General Fundamental Law of Mechanics." The general law referred to was the principle of least constraint, which stated that each mass in any system of masses connected to one another through any sort of external constraints moves at every instant in the closest agreement possible with free movement, or under the movement each would follow under the least possible constraint, the measure of the constraint the entire system experiences at each instance being considered as the sum of the products of the masses and the squared deviation from the free path. Gauß had concluded this brief but powerful paper as follows: "It is very remarkable that free motions, if they cannot exist with the necessary conditions, are modified by nature in exactly the same manner that the mathematician calculating according to the method of least squares compensates experiences which are related to one another through the necessary dependence of connected magnitudes. This analogy could be followed further in still other directions, but this does not accord with my present purpose" (1829, 28).

Wundt's conclusion relating the probability calculus to his own principle of least resistance in eye movements was almost a literal quotation from Gauß's conclusion to the paper on the principle of least constraint. The comparison was completely appropriate; for the eye and the six attached muscles operate as a system of constrained motion closely analogous to the situation described by Gauß in deriving his law.

By calling the eye a mathematician operating in terms of the probability calculus, Wundt signaled his intent to incorporate a psychological component in vision. By emphasizing least resistance as the psychological regulative principle guiding the eye, however, Wundt had reduced the problem of vision to one of mechanics. Although Wundt claimed to invoke a psychological process, that process seemed more directly related to resistance the eye encounters in moving rather than to any conscious acts related to vision. As Helmholtz would point out, it was unclear, for example, how the principle of least resistance could illuminate how single vision comes about. Moreover, Wundt seemed to imply in the passage quoted above that it would be possible to start from the probability calculus and arrive at constraints on eye movements, but he did not provide this derivation. The derivation of the conditions of eye movement from a psychophysical principle was Helmholtz's goal in his paper published in *von Graefe's Archiv* the following year (1863b).

Helmholtz introduced his treatment of eye movements by stating that the conclusions Wundt and Fick had reached regarding the role of least constraint were probably correct. But, he noted, "even if the principle . . . should prove to be completely applicable, it would not follow that an optical principle is not actually decisive" (1863b, 160). By "optical" Helmholtz had in mind some principle directly and primarily concerned with visual perception but one that in being executed would have the same effect in coordinating muscle movements as Wundt's principle of least resistance. According to Helmholtz, the primary goal in vision is to avoid double images by fusing the images in both eyes into a single visual object. In order to bring this about, the eyes must be properly oriented. Given that all sorts of variation in the strengths of the eye muscles would be compatible with Wundt's principle of least resistance, it seemed more reasonable to consider the purposes of achieving singular vision as primary in the selection among the myriad muscle actions compatible with Wundt's law. Indeed, Helmholtz could cite numerous experimental examples where voluntary exertion produces the position of the eye best suited to vision. The principle actively guiding this process Helmholtz called the principle of easiest orientation. At first glance this might seem like a semantic variation on Fick's and Wundt's proposals, but, as we shall see, Helmholtz took Wundt's idea that the eye is a mathematician as much more than a metaphor.

Crucial to Helmholtz's development of the principle of easiest orientation were the laws of Donders and especially of Listing. The eye has three primary degrees of freedom. It performs abductions and adductions around a central vertical axis through the eye; it performs elevations and depressions around a central horizontal axis; and, finally, it can perform torsions, cyclorotations, or wheel-like rotations around the visual axis itself, the anterior-posterior axis of the eye (see figure 8.14). Helmholtz thought the work of achieving single vision would be greatly simplified if, in order to maintain the orientation of the eye as it moves from one position to another, the mind did not have to attend to elevations, abductions, and cyclorotations, but rather that cyclorotations would be excluded, thus leaving only two variables to be attended to. This would be accomplished if the cyclorotations made by the eye were completely dependent upon and determined by abductions and elevations. The laws of Donders and Listing, were they to be confirmed, would guarantee this dependence of eye movements on two variables alone and at the same time exclude cyclorotations.

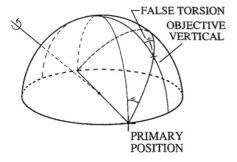

Figure 8.14

Eye movements without cyclorotations. Source: Based on Hermann Helmholtz, *Handbuch der physiologischen Optik,* 2d ed. (Hamburg: Leopold Voss, 1896), 622.

The law of Donders states that there is one and only one orientation for each position of the line of fixation (the visual axis). Helmholtz pointed out that this fact alone would greatly facilitate orientation, but it would not ensure optimal conditions for a stable visual field during eye movements, for it would not guarantee a smooth transition from one configuration of points to an adjacent or remote configuration of points. This problem would be resolved, however, if in moving from one position to an adjacent position, the rotation of the line of fixation takes place around a single axis. Listing's law states that every movement of the eye from the primary position to any other position takes place by rotation around an axis perpendicular to the meridian plane that passes through the final position of the line of fixation. It can easily be shown that each of these axes of rotation lies in an equatorial plane perpendicular to the visual axis. This is Listing's Plane (see figure 8.15).

Helmholtz's principle of easiest orientation is a rule that restricts eye movements as closely as possible to axes in Listing's Plane. In order to discuss motions limited to rotational axes in Listing's Plane, Helmholtz introduced what he termed the *atropic* or nonrotated line. The atropic line was a line perpendicular to the plane of the rotational axes of the eye, that is, the XZ plane in figure 8.15, or Listing's Plane. In the primary position the atropic line coincides with the visual axis. In treating the motion of the eye from a primary to secondary position Helmholtz showed that it is possible to resolve the motion into two rotations by the parallelogram of motions. Thus a sec-

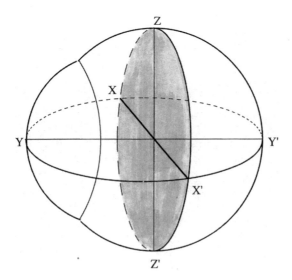

Figure 8.15
Principal planes for eye orientation: Listing's Plane.

ondary position could be reached by first executing an elevation, α, around the horizontal axis XX′ in the XY plane, and then an abduction ω around the vertical axis, ZZ′. Alternatively, the motion could be represented as a single rotation around an axis in the equatorial XZ plane (Listing's Plane) intermediate between the horizontal and vertical axes and perpendicular to the plane containing the primary and secondary point. The direction and magnitude of that rotation could be determined by the parallelogram of motions according to familiar rules of projections in terms of direction cosines—our vector product. In terms of the atropic line, this meant that in general, for motions occuring in Listing's Plane, any motion of the visual axis could be decomposed into rotations around axes perpendicular to the atropic line. If, however, the motion of the visual axis was generated around an axis not in Listing's Plane, Helmholtz showed that it would always be possible to decompose that displacement into a rotation around an axis perpendicular to the atropic line (i.e., an axis in Listing's Plane) and a rotation, ρ, around the atropic line itself. The principle of easiest orientation demanded that the components of rotational motion having the atropic line as axis must be zero.

Helmholtz confirmed this approach experimentally by mapping the trajectory of the visual axis on a wall-grid placed opposite his experimental subjects. He projected the expected paths according to Listing's law in selected angles of 10° (represented in figure 8.16). Those lines consistent with motion around axes in Listing's Plane, Helmholtz went on to show, are the right projections onto the wall of great circles passing through the point

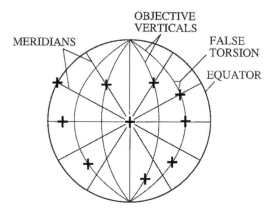

Figure 8.16
Projection of afterimages when the eye rotates around axes in Listing's Plane. Source: Based on Hermann Helmholtz, *Handbuch der physiologischen Optik,* 2d ed. (Hamburg: Leopold Voss, 1896), 622.

Y′ in figure 8.15, which he called the occipital point. For reasons that will become apparent in their connection with visual orientation, Helmholtz designated these "direction circles." These are, in a sense, the eye's own internal coordinate system. Also represented in figure 8.16 are the objective verticals (that is, lines perpendicular to the "horizon line" on the wall-grid) as projected from the occipital point. Except for the primary vertical, which is also a meridian passing through the occipital point, the projections of these lines onto the wall are not straight lines but arcs, as indicated in the diagram. In practical terms, the implication of the grid construction was that if a cross is fixated by the eye in the primary position—indicated by the center of figure 8.16—the afterimage of the cross will retain the same orientation with respect to the primary position.

Helmholtz used this grid for experiments to investigate whether in fact the visual axis can move freely from one point to any other in the visual field—and not simply from a primary to a secondary position—without producing a cyclorotation. He showed that, for the most general case, if eye movements are restricted to a very small section of the visual field, the principle is satisfied. This results from the fact that for infinitesimally small movements, the sine of the angle of rotation is nearly identical with the rotational arc, so that the axes of rotation all lie virtually in the same plane. For larger displacements, however, this was not generally the case. The upshot was that for a continuous movement of the visual axis over a large angle of the visual field, the succession of axes for each of the infinitesimally small rotations making it up could not all lie in the same plane. Hence cyclorotations are unavoidable. Helmholtz did show, however, that some paths are torsion free, namely, along the direction circles. A cross moving along one of these lines always remains upright (see figure 8.17). Motion

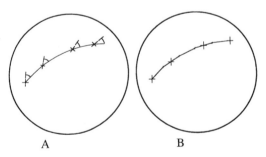

A   B

Figure 8.17
Saccadic motion of eye movements around axes not located in Listing's Plane. According to Helmholtz, the eye breaks up the motion into partial motions approximating motions in Listing's Plane.

along the direction circles can be executed around axes in Listing's Plane, so that cyclorotation will not occur. Thus the direction circles are crucial to visual orientation. In any given part of the field the path between two points not connected by a meridian can be approximated as a succession of small paths along neighboring direction circles (see figure 8.17). Any deviation from a directional circle would be regarded as an error. The task in visual orientation as the visual axis moves from point to point is to keep the sum of these errors a minimum, and this can be accomplished by the process of successive approximations along direction circles: "If we consider, therefore, every rotation of the eye around the atropic line as an error, we can reduce the principle of easiest orientation to the following demand: The law of eye movements must be so determined that the sum of the squares of the errors for the total of all the possible infinitesimally small movements of the eye taken together shall be a minimum" (Helmholtz 1863b, 169). Helmholtz went on in the next paragraph to reclaim Wundt's metaphor, only now recast as a psychophysical principle: "The squares of the errors have to be understood here in the same sense as the well-known principle of least squares in the probability calculus" (Helmholtz 1863b, 169).[27]

Wundt probably felt he had been upstaged by Helmholtz, his mentor and lab director, whose own work on the principle of easiest orientation was so clearly dependent on Wundt's work on the ophthalmotrope and discussion of least constraint. But Helmholtz had superseded Wundt's attempt to construct a psychophysical account by showing that the principles the mind employs in making judgments in pursuit of the fundamental aim of visual orientation—single vision—have as their physical consequence motion according to Listing's and Donders's laws, both of which are the optimal conditions for eye movements in a mechanical system operating under least constraint. The psychological and physical aspects were intimately linked in Helmholtz's solution, like two sides of the same coin.

This work on eye movements was a crucial part of Helmholtz's empiricist theory of vision and provided a key element of his sign theory of perception, with which I began. In Helmholtz's empiricist theory, visual sensations involve psychological acts of judgment. The collection of points we judge to be a stable object is generated by stimulation of the retina. But that judgment is based on confirmation of a sensory hypothesis: To see a collection of points in a visual image as a stationary object, the points in the collection must retain their identical configuration as a group with respect to one another when the eye makes successive passes over the supposed object; or, in the case of a moving object, the collection of points must retain its character as a connected group as it moves across the visual field. In effect, the mind is performing a series of experiments with the eye, testing the hypothesis of an object, and the outcome is judged a success if the squares of the errors after several passes or in different parts of the field are an acceptable minimum. The comparison base with respect to which we make this judgment of "fit" is the distribution of departures from Listing's law in different parts of the visual field as indicated by the cross experiments discussed above. Listing's law is crucial, for in a mechanical system such as the eye, it is the simplest method for approximating a distortion-free mapping of connected points on a sphere as the sphere adopts different successive orientations. As infants we may not have made such controlled experiments with horizontal and vertical lines, but we undoubtedly did some sort of eye exercises with the similar outcome of our becoming intimately familiar with the direction lines of our visual system, those distortion-free lines in the visual field that obey Listing's law perfectly. This measuring device embedded in the eye is in fact sensory knowledge, part of the system of local signs we carry around with us at all times, active but below the level of consciousness, like the violin player's skill, ready to be activated whenever we view external objects.

Conclusion

In his 1878 lecture "The Facts of Perception," Helmholtz pointed to the frequent references in his previous work to the agreement between the findings of contemporary sensory physiology, including his own physiological researches, and the teachings of Kant. While cautioning that he would not "swear in *verba magistri* to all his more minor points," Helmholtz went on to note:

> I believe that the most fundamental advance of recent times must be judged to be the analysis of the concept of intuition into the elementary processes of thought. Kant failed to carry out this analysis; this is one reason why he considered the axioms of geometry to be transcendental propositions. It has been the physiological investigations of sense perception which have led us to elementary processes of understanding inexpressible in words which remained unknown and inaccessible to philosophers as long as they inquired only into knowledge expressed in language. (Helmholtz 1878, 244; compare Kahl 1971, 390–391)

Kant and his followers had assumed that the contents of intuition are given immediately in sensation (Helmholtz 1878, 379–380). As I have attempted to show in this essay, a central organizing theme of all of

Helmholtz's physiological research was the persistent pursuit of the question, What is given in intuition? And, as we have seen, central to Helmholtz's empiricist theory of spatial perception was the rejection of the claim that anything is given immediately in perception. Perception is a process constantly involving active measurement, hypothesis, and learning. It may transpire incredibly quickly, so as to give the appearance of immediacy; but a clever physiologist will be able to trick the senses and tease apart the stages of construction. Nevertheless, Helmholtz agreed that Kant's theory of the apriori forms of intuition was a clear and important point, but these forms had to be completely general, empty of content and free of all restrictions, in order to accommodate any content whatsoever that could enter a particular form of perception (Helmholtz 1878, 408). Where Helmholtz parted company with Kant and his followers was in their assertion of a pure geometry that has its foundations, including its axioms, in a transcendental intuition having nothing to do with physical bodies and their movement. In this geometry it is possible, according to Kant and his followers, to have a transcendental intuition of the congruence and equality of geometrical figures unmediated by experience. Helmholtz emphatically rejected Kant's argument for a pure geometry given in intuition, and in doing so he criticized Kant for not being critical enough in the *Critique*: What Kant should have claimed was that the form of space can be transcendental without its axioms being so.

What did Helmholtz have in mind by this notion of space as transcendental, in partial agreement with Kant? Helmholtz's theory of space as transcendental was set out in other sections of the essay on the facts of perception and in his papers on the origin and meaning of the geometrical axioms. In those papers Helmholtz makes it clear that what he had in mind was the general, abstract notion of space explored by the new field of "metamathematics." Unavailable in Kant's own day, the new metamathematics explored the logical consistency and logical foundations of spatial forms as a pure analytical science—Helmholtz characterized it sporadically as "scientific geometry" and "algebraic geometry"—careful to exclude any unwarranted inferences drawn from "constructive intuition" inadvertently entangling everyday experience with the necessities of thought. The new metamathematics, Helmholtz wrote, had been prepared by the work of Gauß, Lobachevsky, Beltrami, and others, but especially by Bernhard Riemann (Helmholtz 1870, 253)

The key to the new approach Helmholtz characterized as "scientific geometry" is the observation that all spatial relations are measurable; even the axioms make reference to magnitudes, such as a straight line being the *shortest* distance between two points. Riemann, in Helmholtz's view, had called attention to the possibility of treating geometry as a particular application of the more general analytical treatment of relations among measurable quantities, an approach "which obviates entirely the danger of habitual perceptions being taken for necessities of thought" (Helmholtz 1870, 253). Thus, following Riemann, a system in which one individual, a point, for instance, can be determined by *n* measurements is called an *n-fold* extended aggregate, or a *n* aggregate of *n* dimensions. Casting his own research in these terms, Helmholtz explained,

> Thus the space in which we live is a threefold, a surface is a twofold, and a line is a simple extended aggregate of points. Time also is an aggregate of one dimension. The system of colors is an aggregate of three dimensions, inasmuch as each color, according to the investigations of Thomas Young and Clerk Maxwell, may be represented as a mixture of three primary colors in definite quantities. In the same way, we may consider the system of tones as an aggregate of two dimensions, if we distinguish only pitch and intensity and leave out of account differences of timbre. (Helmholtz 1870, 253–254)

Thus space as considered by geometry is just one application of a general aggregate of three dimensions, and as Riemann showed, by adding the special conditions of free mobility of solid bodies without change of form in translations and rotations, and postulating the special value of zero as the measure of curvature, we arrive at Euclidean space as we know it.

The relevance of the work on *n*-fold aggregates extended well beyond questions of geometry for Helmholtz. I have argued that for Helmholtz, just as for Herbart before him, the work on *n*-fold aggregates was relevant to solving a fundamental problem of psychophysics—namely, how to construct the mental machinery for relationships that are necessary, valid, prior to all experience, and at the same time related to the world of outer sense. This was the problem of the relation of ideas and the schematisms of the categories to the forms of perception Kant had bequeathed his followers. Helmholtz was initially led to the power of *n*-fold aggregates in treating these problems through his effort to sort out the relationship of the mind and the sensory apparatus of the eye in color perception and three-dimensional vision.

> While Riemann entered upon this new field from the side of the most general and fundamental questions of analytic geometry, I myself arrived at similar conclusions, partly through seeking to represent in space the system of colors (involving the comparison of one threefold extended aggregate with another) and partly through inquiries into the origin of

our ocular measure for distances in the field of vision. Riemann starts by assuming the above-mentioned algebraic expression, which represents in the most general form the distance between two infinitely close points, and deduces therefrom the conditions of mobility of rigid figures. I, on the other hand, starting from the observed fact that the movement of rigid figures is possible in our space, with the degree of freedom that we know, deduce the necessity of the algebraic expression taken by Riemann as an axiom. (Helmholtz 1870, 255–256)

From the perspective of Helmholtz's Kant-inspired empiricist theory of perception, the organs of sense played a role exactly analogous to the specifications of rigid bodies, constancy under rotations and translations, and curvature provided in Riemann's geometry. The sensory organs are measuring devices that specify a given n-fold aggregate as color, tone, or geometrical surface. As Helmholtz commented, "Here our own body with its organs is the instrument we carry about in space. Now it is the hand, now the leg that serves for a compass, while the eye turning in all directions is our theodolite for measuing arcs and angles in the visual field" (Helmholtz 1870, 259). Helmholtz does not name Grassmann or Herbart in commenting on his path to these ideas. But it is difficult not to see them in the picture, perhaps, as I have argued, even at the center of the picture.

I have suggested that, in understanding how Helmholtz arrived at these results, we not view him as relentlessly pursuing a Kantian research program. Rather, I have proposed that we see him as attempting to unravel the complex problems of color vision and hearing with the tools of experimental physiology and physics. In this process the dialogue among his instruments and physical models of sensory apparatus provided the primary impetus, and in the course of working out mechanisms for sensory perception he was led both to dissent from the Kantian orthodoxy and to operationalize many of the elements of outer sense Kant had taken to be given immediately in the act of perception. In working out these details Helmholtz's selection of telegraphic apparatus was particularly crucial. The telegraph was certainly more than a useful metaphor. Telegraphic apparatus and its modification in various ways to achieve the ends of experiment had a significance beyond the fact that these devices were readily available and familiar objects of investigation for Helmholtz. Telegraphic devices were not only important as means for constructing models and experiments; telegraphy embodied a system of signification central to Helmholtz's views about mental representations and their relationship to the world, the sign theory of perception. I have suggested that the adaptation of Grassmann's approach to the specific requirements of physiological modeling was crucial. Grassmann's proposal of a

formalism that operated in terms of three quantifiable measures was important and timely because it was a system, based on the notion of *n*-dimensional
manifolds (or *n*-fold aggregates, in Riemann's terms), for constructing representations that operated similarly to the way messages were encoded in the
telegraph. Viewed in this light, the telegraphic system with which Helmholtz
was familiar in his daily experience and upon which he and his friends
worked intensively was important in suggesting a direction for the development of his own ideas about representation:

> Nerves have been often and not unsuitably compared to telegraph wires.
> Such a wire conducts one kind of electric current and no other; it may
> be stronger, it may be weaker, it may move in either direction; it has no
> other qualitative differences. Nevertheless, according to the different kinds
> of apparatus with which we provide its terminations, we can send tele-
> graphic dispatches, ring bells, explode mines, decompose water, move
> magnets, magnetize iron, develop light, and so on. So with the nerves. The
> condition of excitement which can be produced in them, and is con-
> ducted by them, is, so far as it can be recognized in isolated fibers of a
> nerve, everywhere the same, but when it is brought to various parts of the
> brain, or the body, it produces motion, secretions of glands, increase and
> decrease of the quantity of blood, of redness and of warmth of individual
> organs, and also sensations of light, of hearing, and so forth. Supposing
> that every qualitatively different action is produced in an organ of a dif-
> ferent kind, to which also separate fibers of nerve must proceed, then the
> actual process of irritation in individual nerves may always be precisely
> the same, just as the electrical current in the telegraph wires remains one
> and the same notwithstanding the various kinds of effects which it pro-
> duces at its extremities. On the other hand, if we assume that the same
> fiber of a nerve is capable of conducting different kinds of sensation, we
> should have to assume that it admits of various kinds of processes of irri-
> tation, and this we have been hitherto unable to establish.
>
> In this respect then the view here proposed, like Young's hypothe-
> sis for the difference of colors, has still a wider signification for the phys-
> iology of the nerves in general. (Helmholtz 1863a, 149)

Telegraphic apparatus was central to Helmholtz's work in physiological
acoustics. He relied on the representation of the ear as a tuning-fork resonator.
The juxtaposition and comparison of differences between media, between
hearing and vision, was a positive resource for revisiting Young's hypothesis.
Indeed, as the passage quoted above implies, the analogy between the Young
hypothesis for color vision and Helmholtz's model for hearing, and the assim-
ilability of both to the telegraph as apparently generalizable for sensory physiology, seemed to provide further convincing support for the approach. In the

years 1858–1860, when Helmholtz returned to the elaboration of the Young hypothesis through work on afterimages and color blindness, culminating in his general presentation in the second part of the *Physiological Optics* (1860), the search for a material embodiment of his models once again served as a resource for advancing theory (Helmholtz 1858 and 1859).

In his earlier work on subjective color Helmholtz had used the center-of-gravity method as a tool for constructing a color chart. As noted above, in response to Grassmann Helmholtz argued that a graphic representation that embodied relationships based on actual measurements of the brightness of the complementary colors that form the most intense white light should have the shape of a hyperbola. But Helmholtz decided to go even further and map the space corresponding to the combination of complementary spectral colors the eye actually sees.[28] The crucial resources enabling Helmholtz (and independently Maxwell) to generate such a color chart were studies of color blindness. In these studies the methods employed in constructing the graphical representation for combinations of spectral colors were turned to graphing the colors perceived by color-blind individuals, persons presumably suffering from the absence or malfunction of one of the three receptors. A color chart could be constructed for persons suffering the loss of the red receptors, for example, completely from mixtures of yellow and blue (Helmholtz 1859, 348). Helmholtz, as I have shown, took this line of work a step further in his experiments on afterimages by demonstrating that colors more saturated than spectral red, green, or violet can be generated by fatiguing one or two of the hypothesized three receptors through intense exposure to spectral light of the complementary color. For Helmholtz, the construction of the color chart as a graphical representation embodied the very principles used by the eye in encoding signals perceived as color in the brain. In his studies on tone sensation Helmholtz had constructed a mechanical simulacrum for advancing his theory. In the final stages of his work on color vision the graphic trace itself became both the material embodiment of theory and the source of its improvement. Through his work on physiological color and acoustics, Helmholtz became convinced of the power of *n*-fold aggregates as a general abstract calculus the mind uses in processing sense data. That idea was even more powerfully confirmed through Helmholtz's work on eye movements, where the mapping of connected groups of local retinal signs by means of rotations around axes in Listing's Plane served as the physiological analogue in optical space played by the assumption of rigid bodies in Riemann's geometry. When understood as the abstract space of *n*-dimensional manifolds emptied of all content, Kant's notion of space as the transcendental form of intuition was a powerful stimulus both to scientific geometry and to psychophysical research.

Notes

1. This point is a source of confusion for some treatments of Helmholtz's theory of sensation. The theory of color mixture is presented in Book II of the *Physiological Optics,* in which the physiological components of vision are treated. We discover in Book III, devoted to psychological factors in vision, however, that pure sensations of color unmixed with phenomena of contrast and compensatory judgments concerning expected states of affairs can occur only through careful experimental arrangements and with trained observers.

2. Helmholtz concluded his lecture on the progress in vision research with the following statement concerning the correspondence between perceptions and the external world: "Beyond these limits—for example, in the region of qualities—we can prove conclusively that in some instances there is no correspondence at all between sensations and their objects. Only the relations of time, space, and equality and those which are derived from them (number, size, and regularity of coexistence and of sequence)—mathematical relations, in short—are common to the outer and the inner worlds. Here we may indeed anticipate a complete correspondence between our conceptions and the objects which excite them" (1868, 222).

3. See Herbart 1813, 258: "Kant lost sight of the true concept of causality. His concept of causality, as the rule for a temporal succession, belonged completely to the world of phenomena. And so must it be, if the succession of events is supposed to proceed directly out of the causal relation. But it need not happen this way if any particular occurrence, such as the origin of perceptions within us or free action, is to be deduced from an intelligible cause [i.e., a thing in itself in Kant's sense]. The concept of causality cannot be restricted to phenomena. It remains indispensable for real—and especially for mental—passivity and activity [*Leiden und Thun*]."

4. Kant himself explicitly acknowledged this problem in his precritical philosophy in a letter to Marcus Herz of 21 February 1772: "In my dissertation I had contented myself with expressing the nature of intellectual representations merely negatively: namely, that they were not modifications of the soul by means of the object. But how then a representation is possible that could otherwise relate to an object without being in any way affected by it I passed over in silence. I had said: the sensible representations represent the things as they appear, the intellectual as they are. But through what are these things then given to us if they are not [so given] through the mode in which we are affected" (quoted in Friedman 1992, 35n58). Herbart and his contemporaries felt that this problem still remained even in Kant's writings of the critical period.

5. Elsewhere in the *Critique of Pure Reason,* Kant discussed this application of the understanding to sensibility as follows: "We can think no line without drawing it in thought, no circle without describing it. We can absolutely not represent the three dimensions of space without setting three lines at right angles to one another from the same point. And even time we cannot represent without attending in the drawing of a straight line (which is to be the outer figurative representation of time) merely to the action of synthesis of the manifold, by which we successively determine inner sense—and thereby to the succession of this determination in it" (B154–155).

6. Thus Kant wrote: "Space is represented as an infinite given magnitude. A general concept of space (which is common to both a foot and an ell alike) can determine nothing in regard to magnitude. Were there no limitlessness in the progression of intuition, no concept of relations could by itself, supply a principle of their infinitude" (A25).

7. This text was written by Herbart in 1828 for his lectures in Göttingen.

8. In a similar fashion a few pages earlier (306), Herbart characterized the space in which real objects are placed as "the symbol of the possible community of things standing in causal relationship."

9. Herbart establishes the straight line as the shortest path between two points by showing that by mentally tracing any other path between AC, such as AβC, we would have to make at least one superfluous motion. He assumes that AC is the hypotenuse of the triangle AβC; for if AC is the hypotenuse composed of directions Aβ, βC, and if each of these directions can be similarly conceived as compounded of other directions; then to get from A to C he imagines a perpendicular βf to AC, so that Aβ = Aφ + φβ, βC = βφ + φC. And Aφ + φC = Aφ + φβ + φβ + φC. But the path along the hypotenuse, AC = Aφ + φC. Hence, the path via β involves two extra motions, φβ and βφ.

10. Helmholtz's mentor, Johannes Müller, writing in the 1830s, rejected a similar constructivist approach set forth by Steinbuch. Herbart noted that the problem was even more complex than might at first glance be imagined; for it was necessary that "thousands or millions of perceptions which are simultaneously in consciousness have to mutually inhibit one another and become stabilized. . . . Instead of two or three tones . . . we have to consider a multitude of simple sensations which are infinitely near to one another and flow together [in the same unified experience]" (Herbart 1824/1825, 5:349–350).

11. See Riemann 1854, 255; also see the draft of the uncompleted 1876, especially p. 490, where the goals of the program are related to Herbart.

12. Helmholtz wrote that Herbart had initiated the task of providing the philosophical foundations for revising Kant and rejecting the nativist thesis in sensory physiology. See Helmholtz 1867, 496, 595.

13. Perhaps the most direct statement made by Gauß on this subject came in an 1831 review of a work on ternary quadratic foms by Ludwig August Seeber. In criticizing the prolixity of Seeber's methods, Gauß mentioned that one could arrive at some extremely powerful results that go well beyond problems in higher arithmetic by giving ternary quadratic forms a geometrical representation. Through this system points, lines, surfaces, and solids are represented in spaces constructed from parallelograms and parallelopipedal solids, and all the operations using determinants provide the means for solving problems of geometrical constructions. Gauß concluded his brief illustration of the methods with the statement that "[i]t is at least possible to recognize what a rich field these researches could open up which would not only be of interest for high theory but they would also lead to convenient as well as general treatments of all the relations in which for example crystal forms can be used" (Gauß 1831, 194–196).

14. Two persons who took up Gauß's suggestive ideas and attempted to develop them further were C. G. Jacobi and G. Lejeune Dirichlet. Both men published papers in 1848–1849 on quadratic forms and the powerful results that could be obtained from treating them as the structure for constructing a geometrical system of representation. See Dirichlet 1849 and Jacobi 1848.

15. This episode is recounted from the archival sources of the *Jablonskischen Gesellschaft der Wissenschaften* by Friedrich Engel in his biography of Hermann Grassmann (Engel 1911, 110).

16. *Leipziger Zeitung,* 9 March 1844, 877. The reference to the *Göttinger Anzeigen* concerned a review of the Uylenbroek edition and a discussion of Leibniz's characteristic.

17. The most thorough treatment of Grassmann's work and the problems of its reception is Crowe 1967, 54–96. The relationship between Leibniz's analysis in situ and Grassmann's work is also discussed in Couturat 1901. Also see Heath 1917.

18. For outstanding treatments of Helmholtz's developing theory of color vision, see Turner 1994, especially 95–113, and Kremer 1993. My account here differs from Turner's and Kremer's in emphasizing the controversy with Grassmann as a source for stimulating Helmholtz's interest in the theory of manifolds and in applying these ideas to geometry and the construction of sensory space. In my view, Herbart's empirical psychology provided a blueprint for much of Helmholtz's approach to physiological psychology, while the encounter with Grassmann stimulated Helmholtz to pursue concrete strategies for implementing it. For earlier accounts of Helmholtz's color theory, see McKendrick 1899, Koenigsberger 1902–1903, Millington 1941, Boring 1942, and Sherman 1981.

19. In their commentary on the epistemological writings of Helmholtz, Schlick and Hertz create the impression that Helmholtz, like everyone else at the time, did not appreciate the work of Grassmann. This must be qualified. The methods of the *Doctrine of Extension* may not have been applied, but they could not have gone unnoticed, at least not by Helmholtz. Grassmann made such direct reference to the basic concepts of the methods in his paper on color mixtures that it would have been impossible for anyone who read that paper not to have been acquainted with the general strategy of Grassmann's abstract approach. See the note of Schlick and Hertz in Helmholtz 1977, 112n27.

20. In his discussion of Grassmann's center-of-gravity method in the *Physiological Optics,* Helmholtz noted that, according to the conditions of the method, one begins with three quantities of three colors, none of which is a mixture of the others, and assigns them positions on the color plane such that they do not all lie in a straight line. The position of two of the colors on the plane must always remain arbitrary, so that four conditions remain that determine the form of the curve of the color space. One might demand, for instance—as Grassmann had done in his construction of the color table— that five primary colors should be placed equidistant from white. In this case the perimeter of the color space would be a circle, except for the portion connecting red and violet, which would have to be a straight line as in figure 8.9 (see Helmholtz 1867, 287).

21. If a thin object, such as a pencil or a wooden dowel, were placed in front of the screen upon which he projected his color mixtures, shadows were cast in each of the two colors entering the mixture. If the field of white were composed from violet and yellow-green, the violet shadow came out dark but sharply defined, whereas the yellow-green shadow appeared weak, distinguishable almost only by its color and not at all by its brightness. In a mixture of red and green-blue, the red was very weak in comparison; whereas in a mixture of orange and cyan blue, both shadows appeared equally bright.

22. For an excellent analysis of Helmholtz's physiological acoustics and its epistemological relevance, see Vogel 1993.

23. "I have myself subsequently found a similar hypothesis very convenient and well suited to explain in a most simple manner certain peculiarities which have been observed in the perception of musical tones, peculiarities as enigmatic as those we have been considering in the eye. In the cochlea of the internal ear, the ends of the nerve fibers, which lie spread out regularly side by side, are provided with minute elastic appendages (the rods of Corti) arranged like the keys and hammers of a piano. My hypothesis is that each of these separate nerve fibers is constructed so as to be sensitive to a definite tone, to which its elastic fiber vibrates in perfect consonance. This is not the place to describe the special characteristics of our sensations of musical tones which led me to frame this hypothesis. Its analogy with Young's theory of colors is obvious, and it explains the origin of overtones, the perception of the quality of sounds, the difference between consonance and dissonance, the formation of the musical scale, and other acoustic phenomena by as simple a principle as that of Young" (Helmholtz 1868a, 181).

24. "In analogy to the primary colors of the spectrum we intend to call such tones simple tones in contrast to the compound tones of musical instruments, which are actually accords with a dominant fundamental tone" (Helmholtz 1856b, 257).

25. See Seebeck 1844, which was a response to Melloni's proposal to treat the sensation of color as a resonance phenomenon analogous to acoustical resonance (Melloni 1842).

26. On the ophthalmotrope and its improvement, see Reute 1845 and 1857. Helmholtz discussed the work of both Reute and Wundt in his *Handbook of Physiological Optics* (1867, part III, section 27, p. 526).

27. In an accompanying appendix to the paper, Helmholtz went on to give a mathematical derivation of this proposition. He developed an expression for a small rotation around the atropic line as a function of the angles of abduction α, elevation θ, and cyclorotation ω, in the primary coordinate axes of the eye, postulating as a condition for motion that ω be a single-valued function of α and θ. He showed that when α, θ, and ω all undergo a small displacement, the resultant rotation ρ around the atropic line constructed in terms of the composition of small rotations can be represented as:

$$\rho = d\omega + \cos\alpha\theta = \frac{d\omega}{d\alpha}\,d\alpha + \left(\frac{d\omega}{d\theta} + \cos d\alpha\right)d\theta.$$

The sums of the squares of these magnitudes for all infinitely small displacements, **ds** of the atropic line over the entire visual field were supposed to be made a minimum.

Carrying out this complex calculation over the next five pages of the article confirmed the result that a minimum value for this equation would indeed produce a rotation obeying Listing's law.

28. Paul D. Sherman has observed that it is surprising Helmholtz did not immediately perform the calculations necessary to produce such an "actual" color chart. James Clerk Maxwell was the first person to undertake the construction of a color chart representation based on quantitative data (see Sherman 1981, 114–115 and 153–183).

IX

"The Fact of Science" and Critique of Knowledge: Exact Science as Problem and Resource in Marburg Neo-Kantianism

Alan Richardson

A puzzle, rarely remarked upon and almost never specifically addressed, in contemporary analytic philosophy is the existence of two expert communities concerned with knowledge. There are epistemologists and there are philosophers of science. Of course, there are common concerns between these two groups and also some common members of them. There is, however, no compelling general account of the relations between the projects of the two groups or of how we, in historical fact, arrived at this division of philosophical labor.

Simple accounts do not work. Some may be tempted to think of philosophy of science as that branch of epistemology that concerns itself with scientific knowledge. This, however, partially understates and partially misstates the place of science in philosophy of science. Not all philosophy of science seems peculiarly concerned with the knowledge status of science. The motivations for examination of the reduction of psychology to physics, for example, do not seem to be exclusively epistemological. Still less do quite specific concerns with, say, the interpretation of quantum nonlocality or the ontological status of biological species seem exactly epistemological. Philosophy of science seems not to be exclusively a branch of epistemology.

There is another, perhaps deeper, reason to resist the view that philosophy of science is a branch of epistemology. We should, if we are at all concerned with historical truth, avoid the suggestion that the canonical issues of epistemology have been arrived at independently of work in philosophy of science and that the latter work simply applies those pregiven issues to science. A quite contrary case can be made for some central concerns of contemporary epistemology. Contemporary discussions of a priori knowledge, I submit, would have neither their current form nor their current prominence but for detailed and significant work in the methodology of physics and mathematics in the early twentieth century.[1] Similarly, certain general approaches within epistemology, such as social epistemology, seem to owe much of their origin and plausibility to prior arguments in post-Kuhnian philosophy of science regarding the relations of scientific knowledge and scientific community.

The question of the relations of epistemology and philosophy of science in the twentieth century is historically interesting and complicated; there have been prominent calls for a discipline of philosophy of science that explicitly presented philosophy of science not as a branch of epistemology but as a replacement of it. This is an argument we may associate with logical empiricism, especially with Rudolf Carnap, in the 1930s. Throughout the mid-1930s, Carnap repeatedly expressed a view that "the logic of science" or "the logical syntax of scientific language" was the scientifically acceptable successor to epistemology. This was the topic of his 1936 paper for the Paris Conference on Unity of Science entitled "From Epistemology to the Logic of Science," and his views on this matter had by then already been expressed in his 1934 book *The Logical Syntax of Language,* his 1934 pamphlet "The Task of the Logic of Science," and elsewhere.

In his Paris address Carnap invited his audience to see the transition to the logic of science as the third stage in the development of scientific philosophy. The first stage was the overcoming of metaphysics, whereas the second stage marked "the transition from speculative philosophy to epistemology" (Carnap 1936, 36). The third step overcomes epistemology as well, but with an important difference from the way metaphysics had been overcome: "Epistemology is not, as were metaphysics and a priorism before, completely repudiated, but rather purified and decomposed into its constituent parts" (Carnap 1936, 36). The constituent parts, which theretofore had been confusedly mixed together in epistemology, according to Carnap, were logical and psychological. The logic of science clarifies philosophically appropriate questions about knowledge, assigning science as their topic and logic as their province and method. Psychological questions about mental processes or mechanisms of belief formation are placed squarely outside philosophy and in the domain of empirical psychology.

This is not the place to consider in any detail the proposals that Carnap put forward in the 1930s. Carnap's calls, however, point us in an interesting direction. Recent work in the history of logical empiricism, and of Carnap in particular, has uncovered historical and thematic connections between Carnap's work and the work of the last great generation of neo-Kantians—Ernst Cassirer, Bruno Bauch, Jonas Cohn, and others. These philosophers, in turn, connect us to a previous era in which debates revolved around exactly the two issues at stake for Carnap in the move to the logic of science: first, the relations between epistemology, logic, and psychology, and, second, science as the appropriate topic of epistemology. These were key issues in the development of epistemology in Germany in the 1860s through 1880s. Among the leading participants in those debates were the founders of the school of scientific neo-Kantians centered in Marburg, whose work was carried forward by Cassirer, Bauch, and Cohn. Here I will concentrate on the key figure in the formation of the Marburg School, Hermann Cohen, making particular use of his 1883 *Principle of the Infinitesimal Method and Its History*.[2]

KANT ON TOPIC AND METHOD OF TRANSCENDENTAL CRITIQUE

Before considering the work of Cohen, we should note why neo-Kantians would be especially likely to have views on the appropriate topic of epistemology and on the relation of the method of epistemology to logic and to psychology. The reason is, of course, that this thicket of issues is a Kantian inheritance. Kant's transcendental philosophy was taken to speak to all sides of both issues. It bequeathed to German epistemology of the late nineteenth century both its central status and its most vexed problems.

Consider, first, the question of science as the appropriate topic of epistemology. There is a natural reading of Kant's *Critique of Pure Reason* that finds Kantian authority for tying epistemology directly to the achievements and methods of the exact sciences.[3] The characteristic question of the first *Critique* is "How is synthetic a priori knowledge possible?" This question serves the purposes of the critique of pure reason by underscoring the fact that metaphysics *as a science* would have to deliver synthetic a priori knowledge. Indeed, Kant explicitly, in both the *Prolegomena* and the preface to the second edition of the first *Critique*—and at least implicitly in several places in the *Critique* itself—appeals to the actuality of synthetic a priori knowledge in physics and mathematics to ground the significance of his fundamental question. It is the existence of synthetic a priori knowledge in the exact sciences that at once lends poignancy to the transcendental question and embarrasses metaphysics, which unlike mathematics and physics has not found "the secure path of a science" (B xv).

This reading of Kant's *Critique* takes seriously his Newtonianism and, ultimately, the "special metaphysics of nature" found in his *Metaphysical Foundations of Natural Science,* which completes the transcendental grounding of Newtonian physics. This reading not only has good textual support in the *Critique* and throughout Kant's literary corpus, it also connects his philosophy centrally to the greatest achievement of modern natural philosophy, the development of Newtonian physics.

There is nothing in Kant's first *Critique* that fails to generate serious interpretative disagreement, however, and this Newtonian reading of Kant has rivals. Followers of Kant who wish to sever the relations of Kant's project to natural science in general or to Newtonian science in particular point to other but equally characteristic Kantian rhetoric. In the Transcendental Deduction and the Schematism, Kant's search for the conditions of the possibility of synthetic a priori knowledge finds characteristic expression as a search for the a priori conditions of the possibility of experience. Here is a typical passage expressing this aspect of transcendental philosophy:

> In this way synthetic a priori judgments are possible if we relate formal conditions of a priori intuition, the synthesis of the imagination, and its necessary unity in a transcendental apperception to a possible cognition of experience in general and say: The conditions of the *possibility of experience* are at the same time conditions of the *possibility of the objects of experience,* and on this account have objective validity in a synthetic judgment a priori. (B 197)

The search for the conditions of the possibility of experience or of the objects of experience might in itself suggest that the transcendental philosophy issues in no special scientific content—experience is available also to those who are not (even implicitly) Newtonians. In any case, it could be argued that neither the categories of the understanding, nor the transcendental unity of apperception, nor the principles that follow from the transcendental synthesis of the imagination in themselves import any peculiarly Newtonian conditions on experience.

I have no desire to argue directly for or against either side of this interpretative issue. I do want the issue itself, however, to be reasonably clear. Ultimately, it is a question of the extent to which the Kantian notion of experience is impregnated with a specifically Newtonian theoretical flavor. Is the experience that is grounded by the conditions of the intuition, the understanding, and their a priori synthesis experience of a specifically Newtonian world, or do the a priori conditions ground an experience with a less specific physical and mathematical structure?

The second issue concerns the legitimate theoretical resources of transcendental philosophy itself. The Copernican revolution in philosophy that Kant sought to inaugurate advised the philosopher "that we can cognize of things a priori only what we ourselves have put into them" (B xviii). That is, the conditions of objective experience are located in and explained as functions of the mind. Kant pursues transcendental philosophy within the language of faculty psychology, appealing to the faculties of sensibility, understanding, imagination, and reason. In this sense, transcendental philosophy can appropriately be understood as transcendental psychology.

Notwithstanding this appeal to psychological language, the discipline that occupies the bulk of the first *Critique* and comprehends both the transcendental analytic and the transcendental dialectic is, according to Kant, appropriately called "transcendental logic." Transcendental logic is the philosophical discipline that reveals the a priori conditions of the understanding and shows how they provide the conditions of a priori synthesis of the manifold of intuition, thereby grounding the possibility of objective thought.

Moreover, transcendental logic discovers the a priori conditions of the understanding—the categories—not by rummaging around in the mind to see how concepts are in fact put together but through a much more exact procedure. Kant looks to the formal conditions of judgment as revealed by formal logic (what Kant calls "pure general logic") for the key to the discovery of the categories. That is, there is a specific synthetic moment that is relevant to epistemology: Knowledge is expressed in judgment, not in any and every relation of ideas. Thus logic as an account of judgment is the appropriate key to a properly epistemological discussion.

The interpretative issue is this: Should we emphasize the generally psychological language of Kant's explanation of the transcendental conditions of a priori knowledge and, through such emphasis, seek the key to transcendental philosophy in an a priori phenomenology or an a posteriori psychology of mind? Or should we, rather, stress the way in which the transcendental conditions of a priori knowledge are, according to Kant, formal conditions of objectivity revealed through attention to the formal conditions of judgment? The first way of interpreting Kant opens up a path to a continuing connection with psychology by stressing the (transcendental) subjectivity of conditions of objective knowledge. The second way of interpreting Kant invites an intimate connection between epistemology and logic, since the formal conditions of objectivity are expressed in the formal conditions of judgment, a matter appropriate to formal logic.

There is no logical link between the two issues we are considering— exact science or some less theoretically informed notion of experience as the topic of transcendental philosophy, and logic or psychology as its method— but there does historically seem to have been a connection between the two. Those who stress transcendental logic are drawn to finding the a priori conditions of objective knowledge in the contents of theoretically explicit knowledge in exact science. Those who stress transcendental psychology frequently see in synthesis a temporal activity of the mind and seek the conditions of objectivity in the processes of a mind in interaction with the world. Thus there is a historical division between those who stress transcendental logic and interpret Kant as providing transcendental grounding

to exact science, and those who stress transcendental psychology and interpret Kant as grounding experience in a less scientifically structured sense.
Of course, the latter group sometimes imports a highly scientific and theoretical psychology or highly recondite phenomenology, but typically does
not use that psychology or phenomenology in the service of grounding a
highly scientific and theoretically informed experience.

The issues involve, therefore, subtle questions of science as topic and as
resource for epistemology. Kant's transcendental project seems both to take
for granted a specific and robustly positive epistemological status for exact
science and to undertake a weighty philosophical burden in transcendentally
grounding that science. That is, mathematics and mathematical physics issue
in synthetic a priori knowledge, a fact that serves both to curb the preten-
sions of metaphysics and to institute an important philosophical burden on
transcendental logic, which must explain how synthetic a priori knowledge
is possible. In this way, science serves both as a resource in the fight against
metaphysics and its skeptical antithesis and as problem for transcendental philosophy. More precisely, and to anticipate some language we shall see in use
among the Marburgers, the fact of science explodes skepticism and humbles
metaphysics, while the philosophical account of scientific objectivity becomes the highest speculative burden of transcendental philosophy.

Moreover, the uneasy place of transcendental logic between formal
logic and psychology raises another problem of science for Kantian philosophy. After all, most epistemologists undertake sooner or later to explain the
success of exact science. There is a question of circularity or reflexivity that
arises when the epistemologist uses the resources of science in giving such
an explanation, however. The more transcendental logic becomes a pure or
empirical psychology, the more it looks to be grounding the possibility of
science in general in the offerings of a particular, and rather problematic, science. On the other hand, the more transcendental logic is tied to formal
logic, especially if intuition becomes unattractive as a separate source of
knowledge, the more transcendental philosophy seems itself to become a
formal discipline that can appeal neither to acts of the mind nor to facts of
nature in grounding the possibility of knowledge.[4]

COHEN ON EXPERIENCE AND THE ANTICIPATIONS OF PERCEPTION

Let us leave Kant and fast forward to neo-Kantianism. In the 1870s, a movement, divisible into several schools, in German philosophy arose that sought
to go "back to Kant" on questions of the proper methods and topics of
philosophy. The Marburg School was an important element of this more

general movement. Its two most systematically minded early figures were Hermann Cohen and Paul Natorp, although Friedrich Lange, who was called from Switzerland to Marburg just a few years before his death, was also important in founding the school.

At a general level, neo-Kantianism had two goals in theoretical philosophy.[5] The first was to establish epistemology as the central philosophical discipline. In this they understood themselves to be following the spirit of Kant's transcendental critique. Several neo-Kantians, including Cohen, preferred to make this explicit by eschewing the *theory* of knowledge (*Erkenntnislehre* or *Erkenntnistheorie*) in favor of the *critique* of knowledge (*Erkenntniskritik*). The second goal, for which the first served as a means, was to establish a scientifically respectable philosophy in an era in which empirical science seemed to be going its own way without need for philosophical guidance or, more hubristically yet, seemed to be trying to co-opt the objects of philosophical knowledge as its own. As other schools, such as positivism, did, the neo-Kantians saw the excesses of systematic philosophy in the German Idealist tradition as an important reason why philosophy in the late nineteenth century labored under a cloud of epistemic suspicion while, as Nietzsche wrote in the same time period, "science [was] flourishing . . . and [had] her good conscience written all over her face" (Nietzsche 1886, 123).[6]

In their effort to ground the critique of knowledge in a return to Kant, Cohen and Natorp inherited the interpretive difficulties discussed above and thus had to find their own Kant to whom to return. Moreover, their work went forward in a scientific context somewhat altered from Kant's own. Although relativity theory was still in the future in the 1880s, the rise of non-Euclidean and nonmetrical geometries had already occurred. Moreover, what is now typically called "the rigorization of analysis" was in full swing— a fact of particular relevance, one might think, for Cohen's 1883 book on the infinitesimal. Within physics, several important conceptual changes had occurred—the rise of field theories, the rise of statistical theories, and the enunciation of new conservation principles, for example. Further afield from the direct concerns of Marburg neo-Kantianism, we find Darwinism on the increase in philosophy in the 1880s, although arguably more so in English-language philosophy than in Germany. Central to any account of neo-Kantianism and scientific philosophy in the nineteenth century, furthermore, is the development of the humanities or *Geisteswissenschaften* and the questions of the methodological and ontological relations of these new sciences to the natural sciences and philosophy itself.

Here I want to concentrate most attention on the developments in mathematics and physics in relation to our first interpretive issue regarding

Kant: Is the experience that the critique of knowledge investigates the richly structured experience of the world of the exact sciences? If so, how is that world to be conceived by the neo-Kantians, given the changing circumstances of the sciences themselves? In other words, what precisely is the fact of science that the critic of knowledge brings to our attention and seeks philosophically to explain? We will, however, see that this question is not wholly to be disentangled from the question of logic or psychology as the proper method of the critique of knowledge.

Hermann Cohen was admirably clear about some aspects of our problem. His attitude toward the question of the exact sciences and experience is suggested already in the title of his first great work, his 1871 *Kant's Theory of Experience*. In that work he presented a reading of Kant that begins precisely from the claim that Kant's great merit was to place into question the notion of experience itself. Cohen reminded us that the first sentence of the introduction to the second edition of the first *Critique* begins, "There is no doubt whatever that all our cognition begins with experience," but ends (several lines later) by suggesting that "the activity of our understanding . . . work[s] up the raw material of sensible impressions into a cognition of objects that is called experience" (B 1). From this Kant concludes that "although all our cognition commences *with* experience, . . . it does not on that account all arise *from* experience." This conclusion, wrote Cohen, "presents experience as a puzzle. The solution to this puzzle is the content of Kantian philosophy. *Kant discovered a new concept of experience*" (1871, 3).[7] Throughout the rest of this book and the rest of his systematic work, Cohen presented this new concept of experience in close relation to the exact sciences. Cohen's Kant was in speculative matters primarily a radical student of Isaac Newton who completed Newton's drive to systematic scientific completeness by grounding Newton's "mathematical principles of natural science" in the a priori principles of transcendental philosophy.[8]

Perhaps Cohen's most sustained and systematic attempt to present Kantian philosophy in historical connection to mathematical natural science is his 1883 book on the principle of the infinitesimal method, which he subtitled "a chapter on the founding of critique of knowledge" (*ein Kapitel zur Grundlegung der Erkenntniskritik*). This monograph is in essence an attempt to explicate Kant's general project by placing a small section of the first *Critique*—the "Anticipations of Perception"—in historical context. The Anticipations present the principle of the understanding associated with the categories of quality; this principle is enunciated in the second edition of the *Critique* as "[i]n all appearances the real, which is an object of the sensation, has intensive magnitude, i.e., a degree" (B 207).

Now, it is not our business here to launch into a long digression on the Anticipations as they appear in Kant. For our consideration of Cohen's book, we need only recall the general place of the Anticipations in the project of the first *Critique*. The Anticipations present the principle of the pure understanding that is the application of the categories of quality—reality, negation, and limitation—to the spatio-temporal manifold of pure intuition. The principle provides the transcendental guarantee that the spatio-temporal world contains intensive magnitudes. Kant explains intensive magnitudes as those magnitudes that "can only be represented as a unity, and in which multiplicity can only be represented through approximation to negation = 0" (B 210). Kant's examples in the *Critique* suggest what he is after: a notion of magnitude that is connected with continuity and for which a precise determination of value can only occur through processes of limitation. The connection with the calculus is thus clear. His example of "the moment of gravity" (B 210) also suggests that mass is a—perhaps the—important case of intensive magnitude for Kant.

The connection between the Anticipations and natural philosophy is yet clearer when we note that the section of the *Metaphysical Foundations of Natural Science* associated with the Anticipations provides, according to Kant, the first principles of the pure physical doctrine of dynamics. Dynamics differs from phoronomy for Kant in that the latter considers matter simply as the movable in space, while the former concerns itself with matter as filling space. The distinction between empty and full space—between space itself and space in which there is a real physical object—reveals itself in resistance to motion. Thus dynamics does, whereas phoronomy does not, deal with force: the repulsive and attractive forces that explain the way matter fills space, that is, that explain the real distinction between matter in space and space itself.

So, especially for the scientifically minded Kant interpreter, at least the following is at stake in the Anticipations: The principle is meant to guarantee the applicability of the methods of the calculus in the spatio-temporal world of physics. Calculus is needed to comprehend instantaneous velocity and instantaneous acceleration and to connect the latter to force—and ultimately to mass. In Kantian pure physics the real in space is grounded in matter that fills space in virtue of being constituted as a moving force. Returning to the Anticipations, we can say that the principle is a synthetic a priori principle of objective experience because it guarantees that intensive magnitudes can be found in nature and intensive magnitudes are our sole means of distinguishing between an object (*realitas phaenomenon*) and the lack of an object (*nihil privativum*) in space (B 347–349).

Bearing in mind these subtle and complicated conceptual connections helps us understand Cohen's methods and his central claim in the 1883 book on infinitesimals. For Cohen there are two key features of the Anticipations. First, the systematic significance of the principle of the Anticipations in Kantian philosophy is as an a priori principle that guarantees our cognitive ability to determine the real. I shall follow him in calling it a principle of realization, although the mentalist connotations of that term in contemporary English are exactly wrong.[9] Second, the principle serves as an illustration of the importance of Kantian philosophy as a whole—Kant is the ultimately scientific systematist who indicated the key philosophical task of completing the system of principles of the possibility of knowledge, a system that was begun in the Newtonian achievement in natural philosophy.

Cohen's account of the "principle of the infinitesimal method" was meant to show us that Kant's work was the moment of coming into full self-consciousness of the scientific drive for a principle that governed the real applicability in experience of the infinitesimal and the calculus. Cohen presented his case through a historical narrative that claimed that all the great founders of mathematical physics sought the principles of proper application of the calculus in the physical realm. In this historical account Newton and Leibniz play, not surprisingly, key roles, but Cohen presents the development from pre-Newtonian efforts to understand continuity and infinite divisibility of space and instantaneous velocity and acceleration in Galileo Galilei, Isaac Barrow, Johannes Kepler, and others, right through eighteenth-century developments in mathematical physics and the calculus in work by Leonhard Euler, Sadi Carnot, Christian Wolff, Johann Heinrich Lambert, and others, and into post-Kantian developments in Ernst Fischer, Hegel, Jakob Friedrich Fries, Johann Friedrich Herbart, and others. There is neither world enough nor time to present or evaluate Cohen's historical account. Let us, then, attend simply to its argumentative point and its way of motivating the project of the critique of knowledge. I shall summarize the conceptual framing of the narrative and then return to the critique of knowledge and the fact of science.

Cohen's philosophical framing of the history of the infinitesimal method is roughly this: The development of a mathematically expressed science of nature in the early modern period required that sense be made of notions such as instantaneous velocity and instantaneous acceleration. Mor over, techniques of mathematically determining such quantities were developed and applied in the natural sciences. Yet doubts arose about the conceptual or metaphysical sense that one could assign to these quantities. The mathematical techniques seemed to require quantities—the differentials or infinitesimals—that were all indefinitely small but that stood in determinate, finite, and distinct relations to one another. This led to two unsatisfactory ways of attempting to ground the concept of the infinitesimal. Those of an intuitionist metaphysical temper sought to ground the infinitesimal in a direct awareness of the spatial order. Those of an intellectualist metaphysical temper sought to ground the infinitesimal in a proper conceptual understanding of infinite divisibility or continuity.

Why were these philosophical attempts at comprehending the infinitesimal insufficient? Cohen's distinctive Kantianism finds expression just here. The philosophical alternatives, for Cohen, illuminate neither the use and proper development of the mathematical techniques of the calculus nor the employment of the calculus in physics. Indeed, pushed in certain directions, the intellectualist alternative was sometimes a source of skepticism about the knowledge value of the calculus—a skepticism that is gainsaid by the use of the calculus in our best theories of the physical world. Cohen seeks to illuminate such employment of the calculus—he seeks to show how and why the calculus makes objective knowledge possible in physics.

Cohen's alternative proposal—one he sees conflated with metaphysical concerns in the work of the early moderns—stems from Kant's transcendental philosophy, then, and splits the difference between wholly intuitive and wholly intellectualist alternatives. Continuity is neither a concept that derives from a prior spatial intuition nor a concept that arises (or exhibits itself as confused or impossible) wholly independently of intuition and that thus might be used to criticize the new mathematics. Rather, continuity is appropriately understood as a fundamental law of the understanding that at once underpins our general ability to form species-concepts and that produces the determinate spatio-temporal order when applied to intuition. That is, the law of continuity that underpins the calculus and the mathematical sciences of nature expresses within the realm of nature the fundamental intellectual act of generation (*Erzeugung*). The infinitesimal is neither abstracted from a pregiven spatio-temporal trajectory nor capable of being criticized from a purely intellectualist standpoint. The differential expresses the lawful generation of a trajectory through space and time in the first place, and in this way serves as a precondition of the possibility of determinate physical relations among real objects in space. The philosophical lessons that Cohen drew from this are concisely collected in these two passages:

> The infinitely small participates in the tangent as the *generative* moment, and, thus, a positive, creative meaning comes into limitation. . . . And this new basic law, which is to be distinguished from the negative limit

principle, at the same time secures the connection of the *geometrical* motive of the concept of the differential with the motive that lies in *arithmetic* and *analytic* geometry and, especially, with the *mechanical* motive. . . .

The new principle is the principle of *continuity*. . . . Continuity refers to a *general characteristic of consciousness,* similar to identity. It is, thus, a *special* expression of the *general law of unity of consciousness.* (Cohen 1883, 34–35)[10]

This introduction of consciousness in Cohen's work expresses a commitment to transcendental idealism, but Cohen was quite concerned that this not be understood as any form of psychologism or ontological idealism. Cohen made a strict distinction between his views and a conceptualism he associated with Hegel. The differences he stressed between his view and Hegel's are two. First, for Cohen, pure concepts are explicated in and through the basic principles of the possibility of experience in which they occur, whereas, he argued, for Hegel, principles are meant to be founded in concepts. Second, and related to this, Hegel's ambition to rise through natural science to metaphysics is exactly what the critique of knowledge abjures. This is one role for "the fact of science" in Cohen's philosophy: Exact science is not merely apparent knowledge to overcome philosophically. These themes come together for Cohen:

Quality, thought as reality, is at the same time an example of the difference between so-called conceptual philosophy and critique of knowledge. One must start from principles, not concepts. Science, however, exhibits these principles.

If we are not to make the great controversy over method into a verbal controversy, what matters is that the fact of mathematical natural science is appreciated. It in no way suffices for this appreciation nor aids a clear orientation to designate natural science as a "mere point of passage." The difference between a fact, whose intellectual presuppositions are to be discovered, and a "point of passage" (to what?) marks the difference between transcendental and metaphysical methods. (1883, 119–120)[11]

The "fact of science" also provides Cohen with his most efficient rejection of psychologism—a rejection that underscores the way in which "consciousness" must be understood in its positive use. He writes:

I take . . . cognition not as a type or means of consciousness, but rather as a fact that has been accomplished in science and proceeds toward its accomplishment on given foundations; thus, the investigation is not based on what is after all a subjective fact [*Tatsache*], but rather on a state of affairs that is objectively given (however much it is to be enlarged) and grounded in principles, not on the process and apparatus of cognizing [*Erkennen*] but on their result, science. . . . The investigation of cognition [*Erkenntnis*], the testing of its validity and sources of justification, orients itself toward the fact [*Tatbestand*] of science. (1883, 5–6)[12]

Here we see the characteristic tasks of Cohen's neo-Kantianism. It understands Kant to have placed the explanation of the possibility of objective knowledge at the center of epistemology, to have insisted that this question proceeds in full awareness that objective knowledge has been achieved in the exact sciences, and, thus, to have oriented the epistemologist to the proper topic of epistemological research, namely, the question regarding the formal structure of mind as expressed in the edifice of exact scientific knowledge.

NEO-KANTIANISM AND PHILOSOPHY OF SCIENCE

The fact that Cohen's critique of knowledge orients itself toward the fact of science does not mean that Cohen uncritically accepted all scientific claims or found everything in science equally epistemologically relevant. As we have noted, Cohen found both philosophers and scientists to mix appropriate philosophical questions together with ill-defined and misdirected philosophical questions; philosophers had, on Cohen's view, no special epistemological incompetence. What is more interesting, given our contemporary understanding of the importance of nineteenth-century developments in the foundations of analysis, is what Cohen thinks is epistemologically irrelevant. He begins his discussion of Cournot with this extraordinary claim: "The internal mathematical ground [of the infinitesimal, AR], except where it can be met with in the boundary method in general, lies outside our competence and our concern. For properly mathematical explications, which seek to establish the concept of the differential as a definition, belong to the general theory of the function and can be tested, just as it can be achieved, only within that theory" (1883, 121).[13] This passage could be taken to exhibit the irrelevance of neo-Kantian concerns to any concerns that we as philosophers now have or should have regarding the foundations of analysis. Indeed, Rolf George does just that (1997, 7). George feels the need to do little more than display this passage in order to establish it as the "astonishing ignorance" (*erstaunliche Unkenntnis*) of his essay's title.

There is, however, a more interesting response to the passage. Cohen's lack of interest in the theory of the function is indicative of a deep commitment in neo-Kantian epistemology, namely, the epistemological centrality of the question of the applicability of mathematics in science. Cohen's attitude prefigures Cassirer's attitude toward Russell's logicism. In Cassirer's view, Russell's logicism could not provide any epistemologically interesting foundation to mathematics because it entirely missed the problem, let alone the solution (Cassirer 1907). The problem for the critique of knowledge is the role of mathematics in providing the a priori conditions of the possibility of empirical knowledge in general. For Cassirer and for Cohen before him, no philosophy of mathematics that hives off a pure edifice of mathematical knowledge and abstracts from the role of mathematical structure in the exact sciences of nature has properly understood the role of mathematics in knowledge.

We began our remarks by noting the contemporary division of epistemology and philosophy of science and looking at Carnap's interesting remarks on the transition he saw in the 1930s from epistemology to the logic of science. Our brief discussion of Cohen suggests that Carnap's account of this transition and the one he saw earlier from speculative philosophy to epistemology is historically misleading. The epistemology on offer from the 1860s in Germany was not so much a single confused project as it was a variety of projects divided along important dimensions. There were projects, such as Cohen's, that presage Carnap's later moves: Critique of knowledge in Cohen's hands has mathematical natural science as its topic and its method is closely aligned with logic and explicitly distinguished from psychology.

Indeed, Cassirer's own version of neo-Kantianism, which received its canonical formulation in his 1910 *Substance and Function,* is perhaps the preeminent example of this type of neo-Kantian project. In 1910 Cassirer clearly places the explanation of the possibility of exact natural science at the center of epistemology and offers a version of "logical idealism" that sharply distinguishes that question from any question of the empirical psychology of thought. Cassirer, unlike Cohen, moreover, had by 1910 taken on board the significance of developments in formal logic and no longer tied his own logical idealism to the resources of Aristotelian logic; in particular, the whole structure of quality, quantity, relation, and modality used by Kant and by Cohen in their respective transcendental logics has, in the work of Cassirer, disappeared in favor of a logic that is based in a primitive notion of the function as the lawful coordination of objects to one another. Indeed, Cassirer's work shows quite eloquently how changes in logic and mathematics did not merely change the proper topics of neo-Kantian epistemological investigation, but also altered the philosophical resources the neo-Kantians employed in their transcendental philosophy.[14]

There are important differences, then, among the young Carnap, Cassirer, and Cohen, including their respective conceptions of and commitment to formal logic as the proper resource for philosophical understanding; Carnap's own employment of formal logic as the essence of philosophy had moved him far from orthodox neo-Kantianism by the 1930s. My point is that if we take Carnap's calls in the 1930s for a philosophy of science to raise canonical issues for the relations of epistemology and philosophy of science, then we have to recognize that those issues were already being explored by the founders of epistemology in Germany fifty to seventy years earlier. Indeed, those issues arose in large measure in working out tensions in Kant's own positions. If this is right, then we owe contemporary philosophy of science as much to the work of Immanuel Kant as to any other philosopher or scientist.

NOTES

1. The most thorough arguments for this point of view have been given by Michael Friedman over the past twenty or so years. See, especially, chapter 1 of Friedman 1983 and the essays collected in Friedman 1999.

2. On Carnap's transition from epistemology to philosophy of science, see Richardson 1996; 1997a, chap. 9. On the general relations of logical empiricism to neo-Kantianism, see Friedman 1999 and Richardson 1997a. Due to space limitations, I have not pursued interesting connections between the work of Cohen and early work by his Marburg colleague, Paul Natorp. Natorp's 1887 key essay "On the Objective and Subjective Grounding of Knowledge" (*Über objektive und subjektive Begründung der Erkenntis*) provides a particularly interesting expression of views that parallel and complement Cohen's.

3. The greatest contemporary expression of this reading of Kant is Friedman 1993.

4. This set of issues is examined in detail in Natorp 1887/1981. Put in the form of a slogan, the issue troubling Natorp is that while grounding the objective in the objective might seem empty, grounding the objective in the subjective is dumb. Thus Natorp seeks to explicate a notion of objective grounding that is not philosophically circular or empty.

5. The practical philosophy of the neo-Kantians is neither wholly distinct from nor less important than their theoretical philosophy, but for the sake of coherence and brevity I shall not deal with it here.

6. For a preliminary account of the history of nineteenth- and twentieth-century philosophy that emphasizes the role of scientific philosophy, see Richardson 1997b.

7. "In diesem Satz wird die Erfahrung als ein Räthsel aufgegeben. Die Auflösung dieses Räthsels ist der Inhalt der Kantischen Philosophie. *Kant hat einen neuen Begriff der Erfahrung entdeckt.*"

8. Cohen's 1904 address on the one-hundredth anniversary of Kant's death presents this view succinctly.

9. That is, the principle is not one of how objects come to our attention. I reject the "principle of reification" for similar reasons: Reification is typically taken to be a metaphysical, not a transcendental, principle.

10. "Das Unendlichkleine betätigt sich in der Tangente als das *erzeugende* Moment, und somit kommt positive, schöpferische Bedeutung in die *Begrenzung*. . . . Und dieses neue, vom dem negativen Grenzprinzip zu unterscheidende *Grundgesetz* vermittelt zugleich die Verbindung des *geometrischen* Motivs des Differentialbegriffs mit dem in der *Arithmetik* und der *analytischen* Geometrie gelegenen, besonders aber mit dem *mechanischen*.

"Dieses neue Prinzip ist das Prinzip der *Kontinuität*. . . . Die Kontinuität bezeichnet einen *allgemeinen Charakter des Bewußtseins*, ähnlich wie die *Identität*. Sie ist daher ein *Spezial* ausdruck des *allgemeinen Gesetzes der Einheit des Bewußtseins*."

11. "*Die Qualität als Realität* gedacht ist zugleich ein Beispiel für den Unterschied von sogenannter Begriffsphilosophie und Erkenntniskritik. Nicht vom *Begriffe*, sondern vom *Grundsatze* muß ausgegangen werden. Diesen jedoch legt die *Wissenschaft* dar.

"Wenn also der große *Streit um die Methode* nicht zu einem Wortstreit werden soll, so kommt es darauf an, das *Faktum der mathematischen Naturwissenschaft* anzuerkennen. Zu dieser Anerkennung genügt jedoch keineswegs, noch verhilft es zur klareren Orientierung, die Naturwissenschaften als '*bloßen Durchgangspunkt*' zu bezeichnen. In dem Unterschiede von Faktum, dessen gedankliche Voraussetzungen zu entdecken sind, und 'Durchgangspunkt' (zu welchem anderen?) liegt der Unterschied der *transzendentalen* und der *metaphysischen Methode*."

12. "Nehme ich . . . die Erkenntnis nicht als eine Art und Weise des Bewußtseins, sondern als ein *Faktum*, welches in der *Wissenschaft* sich vollzogen fortfährt, so bezieht sich die Untersuchung nicht mehr auf eine immerhin subjektive Tatsache, sondern auf einen wie sehr auch sich vermehrenden, so doch objektiv gegebenen *und in Prinzipien gegründeten* Tatbestand, nicht auf den Vorgang und Apparat des Erkennens, sondern auf das Ergebnis desselben, die Wissenschaft. . . . Auf den Tatbestand der Wissenschaft richtet sich die Untersuchung der Erkenntnis, die Prüfung ihres Geltungswertes und ihrer Rechtsquellen."

13. "Die intern-mathematische Begründung liegt, außer sofern sie in der *Grenzmethode* im Allgemeinen getroffen werden sein, *außerhalb unserer Kompetenz und unseres Anliegens*. Denn eigentlich mathematische Ausführungen, die den Differentialbegriff als eine *Definition* aufstellen, gehören der allgemeinen Funktionentheorie an und können nur innerhalb derselben so geprüft wie geleistet werden."

14. More information on this aspect of Marburg neo-Kantianism and, especially, on Cassirer is provided in Richardson 1997a, chap. 5; Friedman 1999, chap. 6, and, especially, 2000a.

X

Kantianism and Realism: Alois Riehl (and Moritz Schlick)

Michael Heidelberger

The recent efforts to identify and analyze the neo-Kantian roots of logical positivism have primarily concentrated on the impact of the Marburg School on Carnap and Schlick. Among the Marburgers, attention has focused primarily on Ernst Cassirer. From among the contemporaries of the neopositivists he is commonly regarded as their leading representative. In view of the different currents of the many-faceted neo-Kantian movement, this seems quite a reasonable approach at first sight. This impression is even reinforced if one takes into account Cassirer's intense occupation with the philosophy of mathematics. A second look, however, reveals a further possibility for a neo-Kantian root of neopositivist thought that is in opposition to the Marburg School—the so-called "critical realism" of Alois Riehl. I think the following appraisal of Riehl's work in 1895 by a fellow realist can—at least prima facie—render this plausible:

> Only in Alois Riehl's great work *Philosophy of Criticism and Its Significance for Positive Science* has Kant's criticism found a form which makes it practical for the contemporary necessities of thought and appealing to the different problems of science. This sort of criticism presents itself with the greatest resoluteness as positivism, because it sees the only real source of knowledge in the individual scientific disciplines and because it rejects any fabrication of philosophical systems [*Systemdichtung*]. In the same vein, it dismisses as illusory every belief in the possibility of knowledge that is not rigidly scientific. It determines the task of philosophy in nothing else but science [*Wissenschaft*] and critique of knowledge [*Kritik der Erkenntnis*]. (Jodl 1895, 374)

In the following, I will investigate Riehl's special version of neo-Kantianism and compare it to the views of the early Moritz Schlick. The first part of this essay provides some information on Riehl's life, work, and influence on twentieth-century philosophy. In the second part, Riehl's special blend of neo-Kantianism with realism and positivism is investigated. I will show that many of Riehl's tenets agree in a striking manner with later logical positivism. It even seems that Riehl's Kantian doctrines are much closer to logical positivism than to most of the views of his neo-Kantian fellows. In the third part, I will deal with Riehl's criticism of Helmholtz's epistemology, which played an essential role in shaping his realist outlook. The fourth part demonstrates that Schlick's early philosophy (i.e., before he took his phenomenalist turn under the influence of Wittgenstein around 1925) can be seen as an attempt to adjust Riehl's heritage to the problem situation of science of Schlick's own time.

This modernizing of Riehl's program in the hands of Schlick admittedly afforded some remodeling and constructive rebuilding, but on the whole Schlick did not have to introduce very much new into Riehl's critical realism in order to come to grips with Einstein's general relativity theory. This modifies the claim of Michael Friedman that the decisive break with Helmholtz's conception of space occurred with and was necessitated by the general theory of relativity and that Schlick was the one to take this crucial step (Friedman 1997, 21–22, 42, and passim). It is true that "Schlick's conception constitutes a radical transformation or transmutation of the former theory of Helmholtz" (Friedman 1997, 21), yet this rupture is not so much Schlick's own achievement but has already taken place in the work of Riehl in 1879 in the context of a debate that was partly neo-Kantian, partly sensory physiological and psychological, and partly concerned with the foundations of mathematics. All Schlick had to do was take away from Riehl any notion of a unifying role of concepts that goes beyond the internal unity of a formal mathematical system. But this he could easily do because this notion had already become an idle wheel in Riehl's work that did not accomplish any real work anymore, anyway.

The story told in this essay also supplements Barry Gower's recent claim that "structural realism is a theme that enables us to link Schlick's prepositivism with Cassirer's neo-Kantianism" (Gower 2000, 101). In order to establish such a link one must go back to the major realist among the neo-Kantians, Alois Riehl.

RIEHL IN CONTEXT

Born in Bozen (today Bolzano, in South Tyrol, Italy, then Austria) in 1844, Alois Riehl[1] took part in the Herbartian movement, which was the dominant philosophy in Austria at the time (see Röd 1986). After his *Habilitation* in 1870, he was a professor in Graz from 1873 until he moved to the University of Freiburg im Breisgau in 1882. His views on morality and dogma brought him into conflict with the local archbishop and he converted to

Protestantism. The problems he encountered as a result made him accept a call to the University of Kiel in 1895. Soon afterward, in 1898, he changed again and went to the University of Halle. His career finally reached its summit with a call to the University of Berlin in 1905, where he taught until around 1921 as the successor of Wilhelm Dilthey.[2] Riehl died in Neubabelsberg, a suburb of Berlin, in 1924.

His main work, *Der philosophische Kriticismus und seine Bedeutung für die positive Wissenschaft,* appeared in two editions—the first volume of the first edition in 1876 and the second volume in two parts, in 1879 and 1887 respectively (cited hereafter as PK1, PK2, and PK3). The heavily reworked and partly posthumous second edition came out in three volumes in 1908, 1925, and 1926. It should be noted that the last installment of the first edition bears the subtitle *Zur Wissenschaftstheorie und Metaphysik.* This is one of the first usages of the term *Wissenschaftstheorie,* which is the standard expression for "philosophy of science" in German down to the present day (compare also Köhnke 1991, 245). This volume is also the only one that has been translated into English (Riehl 1894).

In the first volume (PK1), Riehl portrays the development of Kant's critical philosophy and its prehistory. He sees its roots not in Descartes but especially in Locke and Hume. The first installment of the second volume (PK2) deals with epistemology. Its first part is concerned with the "sensory basis of experience" and includes a chapter on the "origin and meaning of the ideas of time and space," which takes up almost 40 percent of the volume and with which we will mainly be occupied. Its second part comprises a discussion of the "logical principles of scientific experience," including an account of the principle of identity, causality, substance, force, and magnitude. The second installment of the second volume (PK3, Riehl 1894) is something of a mixed bag of problems. Its first part is entitled "problems of general philosophy of science" and deals mainly with the concept of "philosophy as the science and critique of knowledge." Its second part, on "metaphysical problems," criticizes idealistic treatments of the problem of the external world, deals with the mind-body problem, the problem of determinism and free will, the concept of the infinite in cosmology, and the notions of necessity and finality in nature.

For many years Riehl was also editor or co-editor of the *Vierteljahrs-schrift für wissenschaftliche Philosophie*—a journal that promoted Riehl's and others' conception of "scientific philosophy" (see Richardson 1997b for the history of this concept) and documents the strong alliance between neo-Kantianism and positivism at the time (Köhnke 1991, 252–262). Riehl is also to be noted for his book on Friedrich Nietzsche, which was the first monograph ever written on this philosopher and saw its seventh edition in 1920 (Riehl 1897). Like many other "German mandarins" (compare Ringer 1983), Riehl delivered an address during World War I in which he enthusiastically greeted the war and which is otherwise alien to the spirit of his work (Riehl 1915, 313–325; Kusch 1995, 214–215).

Riehl was one of the few German philosophers of his time who took note of the development of logic in England (Riehl 1877). As the first volume of his main work (PK1) amply testifies, he was also very well informed about English empiricism. Although Riehl published comparatively little, his teaching seems to have been successful, wide ranging, and influential (see Rickert 1924/1925, 178). An impression of this is best conveyed by his published lectures on the *Philosophy of the Present Age* (*Zur Einführung in die Philosophie der Gegenwart*), which are very well written and preserve some of the liveliness of the spoken word (Riehl 1903). These lectures saw at least six editions by 1921.

Riehl enjoyed some international renown, because in 1913 he received an honorary doctorate from Princeton University, together with the French philosopher Émile Boutroux and others, on the occasion of the opening of the graduate school (Siegel 1932, 16; Riehl 1913). His wartime aversion against England notwithstanding, Riehl was invited to the University of London in 1923, but he could not accept the invitation because of his old age (Siegel 1932, 17). Through the Japanese philosopher Kunaki Genyoku, who had studied with Riehl in 1907–1908, Riehl gained some renown in Japan (see Gülberg 2003). The lectures in the *Philosophy of the Present Age* have been translated into Japanese.

Riehl's work is full of (explicit and implicit) references to Helmholtz, and he tried to come to terms with Helmholtz's thought in two articles (Riehl 1904 and 1921). In fact, Riehl was regarded as being very close to the philosophical views of Helmholtz. This becomes evident from a report of the *philosophische Fakultät* of Berlin University to the state ministry that concerned three possible candidates for the professorship vacated by Dilthey in 1905: "No one stands in a closer relationship to Helmholtz's philosophical work among the contemporaries of Helmholtz and the thinkers that followed him than Riehl" (Erman and Planck 1905, 11r).[3] The report stressed Riehl's capacity to fill a "painfully felt" gap at the University in the subject of "general theory of science" (*Naturwissenschaft*) and "history of the inductive sciences." "Vienna," the report continued, "has in Mach its separate ordinary professor for this field. Through his previous studies, Riehl is prepared to satisfy these needs, for he has also repeatedly held lectures on the fundamental concepts of science" (Erman and Planck 1905, 13v).

Maybe some members of the commission that produced this report were still remembering that Helmholtz himself had brought up Riehl on an earlier occasion in the summer of 1893, when another post for philosophy was to be filled. As the protocol of this session reads: "Hr. v. Helmholtz referred to Riehl in Freiburg whose book on *Criticism* [i.e., PK1–PK3] has made an excellent impression on him" (Protokoll 1893, 103v). This must have had much weight because, as is well known, Helmholtz rarely showed any patience with philosophers.

One should remember that the sciences at Berlin University still belonged to the *philosophische Fakultät* before they formed their own *natur-wissenschaftlich-mathematische Fakultät* in 1936. Accordingly, the report of 1905 was signed not only by an Egyptologist, the dean Adolf Erman, but also by a physicist who acted as vice dean, none other than Max Planck.[4] Planck, of course, was a deep admirer of Helmholtz and was also Schlick's doctoral adviser. His mature philosophical outlook—manifested so famously in his devastating 1908 critique of Ernst Mach's sensationalism—has a lot in common with Riehl's kind of critical realism (Planck 1909).

For many of the dissertations written under the psychologist Carl Stumpf in Berlin, Riehl acted as second reader, as in the case of Kurt Lewin, Wolfgang Köhler, and Kurt Koffka. In the case of the psychologist Adhémar Gelb (who wrote on Gestalt qualities in 1910), he acted as the main supervisor. Mitchell Ash reports that many of the later Gestalt psychologists followed Riehl's lectures (Ash 1995). Among his students were such diverse figures as Oswald Spengler (who wrote about Heraclitus's philosophy in 1904 when Riehl was still in Halle), the art critic Carl Einstein, and the philosopher Richard Hönigswald (author of a trenchant critique of Ernst Mach's philosophical outlook in a realist vein in 1903; see Hönigswald 1903). Riehl had a profound influence on the philologist Werner Jaeger, on the founder of the most important tradition in formal logic in Germany, Heinrich Scholz, and on the pedagogue Eduard Spranger. All three were part of a philosophical discussion circle that regularly met in Riehl's house (Neumeyer 2001).

In his student days in Berlin between 1911 and 1913, Hans Reichenbach also attended some of Riehl's lectures, as his dissertation testifies (Reichenbach 1915, 228). Perhaps Reichenbach's lifelong rejection of positivism and his embrace both of realism and of anticognitivism in ethics, as well as his distinction between the context of discovery and the context of justification, were influenced by this experience.

Moritz Schlick received his doctorate in physics from the University of Berlin in 1904, a year before Riehl took his chair there in 1905. They must nevertheless have met soon thereafter in Berlin, as becomes evident from a postcard in Schlick's *Nachlass*. Riehl invited the twenty-nine-year-old Schlick, in March 1910, to discuss some treatises Schlick had sent him. Another postcard, of 1921, documents that Schlick was still in contact with Riehl in Berlin at that time (Riehl 1910–1921). Both the *General Theory of Knowledge (Allgemeine Erkenntnislehre)*, a copy of which Schlick had sent to Riehl, and the notes to the Helmholtz centenary edition (Helmholtz 1921), as well as other material and excerpts in Schlick's *Nachlass,* reveal that Schlick was quite familiar with Riehl's work.[5] In 1921, when Schlick applied to a professorship at the University of Erlangen, he wrote in a letter to Paul Hensel of the philosophy department of that university that in addition to Planck, Einstein, and others, Riehl could also speak in his behalf (Schlick 1921).

From the physics context, Ilse Schneider (later Rosenthal-Schneider), another of Riehl's students, should be mentioned. Schneider wrote a dissertation with Riehl on the space-time problem in Kant and Einstein for which Albert Einstein's colleague Max von Laue acted as second reader (Schneider 1921). This brought Schneider into very close contact with Einstein, Planck, and von Laue, who subsequently discussed her work with her. The contacts continued after her emigration to Australia with her husband in 1938; in Einstein's case they lasted until his death (Rosenthal-Schneider 1980, esp. chap. 6). Schneider came to teach philosophy of science at the University of Sydney and wrote an article for the Schilpp volume on Einstein to which Einstein responded (Schilpp 1949).

Not only Schlick himself but also his Vienna assistant Herbert Feigl were strongly influenced by Riehl. In an autobiographical manuscript Feigl wrote that "perhaps the most decisive turn [away from physics and physical chemistry to philosophy] came with his [i.e., Feigl's] reading (in 1920) of Alois Riehl's *Philosophy of the Present Age* [*Philosophie der Gegenwart*] and Moritz Schlick's *General Theory of Knowledge*" (Feigl 1966, 3). The realism imbibed through this reading proved tenacious, because during the days of the Vienna Circle he "opposed Carnap's phenomenalism (as formulated in *The Logical Construction of the World* [*Der Logische Aufbau der Welt*]) and defended the sort of critical realism which Külpe and Riehl had so beautifully formulated and which Schlick had convincingly presented in his earlier work" (Feigl 1966, 8).[6] In the same manuscript Feigl admits that he "was temporarily overwhelmed by Carnap's and Wittgenstein's criticisms, but he had his triumphant vindication when Carnap later turned to his version of physicalism in which his earlier phenomenalism was completely repudiated" (Feigl 1966, 9).

In a draft for his contribution to the Schilpp volume on Carnap, Feigl gave a "brief sketch of the historical background" of physicalism in which he dealt with "three main currents of philosophical thought," which, in a revised and reformulated form, he saw eventually culminate in contemporary physicalism. Besides the positivism of Mach, Avenarius, and James and the "psychophysically-monistic forms of realism, notably of Russell and Schlick," he identified as decisive above all "Neokantianism, especially in the 'philosophical monism' of Alois Riehl" (Feigl 1963b, 1; compare Feigl 1963a, 254, 261). (Unfortunately, Feigl does not make clear the difference he sees here between Riehl and Schlick.) And in 1974, in the preface to the English edition of Schlick's *General Theory of Knowledge,* Feigl still testified to the influence in this book of the "sadly neglected" Riehl on Schlick's critique of the "philosophies of immanence" and on his solution of the mind-body problem (Feigl and Blumberg 1974, xxi, xxiii).[7]

To end this short and incomplete sketch of Riehl's impact on the development of philosophy in the twentieth century, one should not forget his anecdotal involvement with (and influence on?) the development of modern architecture.[8] In 1907 Riehl entrusted the construction of his villa in the suburbs of Berlin to the still unaccomplished twenty-one-year-old student of architecture Ludwig Mies van der Rohe. They must have developed an immediate mutual sympathy, because Riehl reached his decision to entrust Mies with this task very soon after their first meeting (Neumeyer 2001). Mies in return appears to have gotten some inspiration from Riehl's writings, especially from his *Philosophy of the Present Age* and from his book on Nietzsche (Neumeyer 1991, 36–50, 57–61). It is likely that in Riehl's circle Mies came to meet some of his later clients in Berlin's high society. Riehl had also a strong leaning toward the philosophy of art; he supervised some dissertations in this direction or again acted as second reader with art historians (especially H. Wölfflin and V. Goldschmidt).

SURVEY OF RIEHL'S APPROACH

If Riehl's work is dealt with at all in the historiography of philosophy, he is usually accorded just a few words after treatment of the Marburg and the South-West German Schools of neo-Kantianism. He is usually described as a "realist" interpreter of Kant's philosophy. A closer reading of his work, however, reveals that the Kantian element is comparatively limited and that his treatment of Kant is very idiosyncratic and inventive. Riehl's philosophy is a peculiar and at times highly original blend of English empiricism (especially in the spirit of John Locke), contemporary sensory physiology with a heavy Darwinian bias, and German positivism. The Kantian element comes in with antipsychologism and with "logic," which, according to Riehl, deals with the function of the unity of reason as provided by judgment.

Riehl uses the term "criticism" (*Kritizismus*) to characterize his own philosophy. It is therefore adequate to start with a clarification of this concept. By criticism the neo-Kantians in general mean a philosophical program beyond dogmatism and skepticism that is critical of metaphysics and subscribes to certain a priori and/or antinaturalist principles. The goal is to examine the conditions and presuppositions, the range and the limits, of knowledge, and to separate in experience the a priori from the empirical. The original method to be employed in executing this task is to be transcendental in a more or less Kantian sense. For Riehl, however, the state of philosophy and especially the necessities caused by the rapid development of science demand an expansion of Kant's original program and an examination of all preconditions of the faculty of knowledge. "Critical epistemology," Riehl wrote in an unmistakably Helmholtzian manner, "examines the sources of our knowledge and it determines the degree of its justification" (Riehl 1907, 88). "There remains for philosophy in the narrower sense of the word no other problems than such as are treated by the critical science of knowledge" (Riehl 1894, 18).

In investigating the sources of knowledge, Riehl also made use of the results of sensory physiology and psychology. Questions of justification, however, he wanted to restrict to considerations concerning the objective order of epistemological concepts (Riehl 1907, 89–90). He advocated a strict and sharp separation between psychological questions that deal with the genesis and development of representations, and epistemological questions that investigate the objective preconditions of knowledge and its meaning.

Riehl saw critical philosophy in a very close relation to science or *Wissenschaft:* "The true system of knowledge . . . is the whole of the sciences themselves" (Riehl 1883, 230). Or in another place he wrote: "True philosophy follows science; in constant connection with science, it is ever obtaining a clearer and more complete understanding of science" (Riehl 1894, 17). This attitude is one of the reasons why Riehl was also regarded as a member of the German positivist movement and a follower of Eugen Dühring (Köhnke 1991, 243–246; Vaihinger 1893, 732). Riehl had indeed taken over his view of the prehistory of Kantianism in Locke and Hume from Dühring, and he rejected metaphysics as Dühring did: "We regard the conviction that metaphysical systems are impossible as one of the most important results of the general theory of science" (Riehl 1894, 118). Another element of positivism in his view is his complete philosophical reliance on sensations, although this is supplemented by his realism.

Criticism, for Riehl, represented only one side of philosophy, its scientific or *wissenschaftliche* dimension, which is coextensive with epistemology and rejects metaphysics. Scientific philosophy does not and cannot aim at building a philosophical system or a worldview. There is, however, also a legitimate role for an unscientific or *nichtwissenschaftliche* philosophy that is concerned not with knowledge but with the "art of leading a good life" (*Kunst der Lebensführung* or *Geistesführung*), or with the "ideals of action" (*Ideale des Handelns*) (Riehl 1883). Philosophy is thus "twofold": "it includes the general theory of science, and the theory of practical wisdom" (Riehl 1894, 23 and the whole of chap. 1).

Riehl's special brand of "criticism" manifests itself in two guises that make up its peculiarity: "critical realism" and "critical monism." To be a realist meant to acknowledge that the process of cognition is related to something outside of consciousness that is not wholly constituted by cognitive categories alone. This still allows for two different versions of realism. There is the naive claim of common sense to have direct access to reality, which was commonly attributed to the materialists. And then there is the critical version, which refused to take perception at face value as the "crude" materialists did and which brought the role of the perceiving subject into play in the manner of Kantian philosophy. Riehl seems to have been alone among the neo-Kantians in clinging to Kant's concept of the thing in itself, in the sense of a "cause of the appearances" (B 344, 522). Both the Marburg and the South-West German Schools of neo-Kantianism were united in their conviction that Kant's philosophy has to be cleansed from the thing in itself in the sense of an ontologically independent basis of being. It does not come as a surprise, therefore, that Riehl's special brand of realism was regarded as opposed not only to phenomenalism (and in the end also to narrow positivism like that of Ernst Laas; compare PK3, 141, 154), but also to idealism and all idealist interpretations of Kantianism prevalent in his time (see Rickert 1924/1925, 169, 177). Riehl sometimes expresses this by saying that he is opposed to "subjectivism." Moritz Schlick later came to call these opponents of the thing in itself, as was common during his time, as advocates of the "thought of immanence" or "philosophy of immanence" (Schlick 1925, §25).

Accordingly, realism for Riehl means to accept that there is something behind our sensations, that all appearances given in experience are tied and related to objects beyond our direct acquaintance. These objects have to be taken as transcending our direct experience, i.e., as different from and independent of our consciousness. "Perceptions are the phenomena of things themselves existing" (Riehl 1894, 162, PK3, 171). Although we cannot infer things in themselves by reason, we can know of them through our sensation of phenomena. "Sensory knowledge is the knowledge of the relations of things through the relations of the sensations of things" (quoted in Case 1911, 237). This is clearly a version of structural realism.

Riehl produces two main arguments in support of his brand of realism. The first tries to show that idealism does not succeed in contesting the appearances' right to direct access to external objects and that therefore realism is the more plausible position of the two. In contrast to realism, all idealistic positions must have recourse, Riehl maintains, to convoluted and implausible constructions in trying to make sense of the relations between appearances. "Idealism," he argues,

> cannot explain the only reality it recognizes [i.e., the appearances]. The content of consciousness remains for it an aggregate of facts, inconceivable because disconnected. Every step toward the explanation of these facts undeniably leads beyond the content of consciousness as immediately given. We can be sure of no law in perception, unless the conditions of perception lie outside of consciousness. Idealism is not compatible with the principle of causation, which underlies all knowledge. (Riehl 1894, 152–153, PK3, 161)

Riehl thus comes to question how John Stuart Mill's view of matter as the permanent possibility of sensations can possibly explain the simultaneous existence of one and the same complex of sensations in different knowing subjects, either as actual sensations or as possibilities, and how two people can ever agree on one and the same object. He also asks how on earth the practical manipulation of a complex of sensations can have the same effect for two different perceiving subjects (PK2, 21).

Accordingly, he sees the greatest challenge for idealism in our readiness to coordinate one and the same object with different groups of sensations from different sense modalities at different times, as well as in our promptness to coordinate our own sensations to those of other persons. In fact, Riehl values the role of intersubjective agreement among perceiving subjects so highly that he even tries to give a "social proof," as he says, of the existence of the external world (Riehl 1894, 163, PK3, 173). The object, he writes, "is the common cause, the foundation of all perceptions, whether actually or only potentially given; it is the rule from which the perceptions as seen from the subject's perspective can be developed with an intuitively appealing consequence" (Riehl 1903, 54).

When he articulates the second main argument in favor of realism, Riehl reads almost like an analytic philosopher. He tries to show that the very concepts of perception and appearance, and with them also the concepts of sensation and representation (*Vorstellung*), already presuppose a relation to real objects beyond the directly given. "I think I am able to show," he writes, "that already the sensation viewed by itself has objective meaning insofar as every sensation is . . . *localized,* i.e., that it includes an indication of something that is not a sensation. Its relation to an object does not first originate in pure, spontaneously produced concepts" (PK2, 32). Every sensation in itself therefore already includes an act of judgment.

According to Riehl, a sensation is never the result of a single stimulus but always of a difference of stimuli, and it is formed by a unity of quality and feeling (*Gefühl*) (PK2, part I, chap. 1). The force of the latter depends on the intensity of the sensation and on the strength of the stimuli and of the psychical activity of grasping them. In its strongest manifestation, feeling comes as pleasure or displeasure. Every sensation thus consists of a reciprocal and correlative action of objective and subjective elements. Riehl sees the doctrine of the pure subjectivity of sensations as a remnant of Cartesian dualism that must be rejected. In addition, he claims that Darwinism also suggests an objective ingredient in every sensation, since a sensation that is not directed toward movement and external activity of an organism would be useless and would therefore not have evolved (PK3, 60, Riehl 1894, 62). Riehl has no qualms about generally combining "Darwinism and transcendental philosophy" with each other (PK3, part I, chap. 4).

Riehl goes even so far as to claim that consciousness of the self and of the object outside the self arises out of the apperception of the subjective as well as objective sides of sensation. No side can be understood without simultaneously grasping the other one. Riehl therefore sees reason to radically transform the Cartesian *cogito* into a "sentio, ergo sum et est": "To talk about the *Ego* without thinking its opposite, the *non-Ego,* does not make any sense, since the Ego exists only through this opposition. . . . It might be true that the *presence of thought* implies and guarantees nothing besides the *Ego*; it is certain, however, that the *presence of sensation* implies the existence of the non-Ego" (PK2, 67). This suggests that there exists a deep connection of Riehl's epistemology with his anti-Cartesian mind-body theory.

This brings us to the second ingredient of Riehl's criticism, his "critical monism." It holds that the basis of physical and psychical phenomena is neither bodily nor mental, thus connecting the Kantian doctrine of the thing in itself with the solution of the mind-body problem as proposed by psychophysical parallelism (see Heidelberger 2004, ch. 5). To accomplish this aim he relies heavily on the chapter on the paralogisms in Kant's *Critique of Pure Reason* (A 381–406; esp. 385–387) and interprets it as a straightforward endorsement of the parallelism doctrine. Kant suggests there that we have no ground for asserting that the substratum of mental phenomena is different from and independent of that of physical phenomena. Psychophysical parallelism (which Riehl calls "identity theory") is the view that mental and physical phenomena are two different aspects of one and the same object.

So far, we have considered elements in Riehl's philosophy that are close to empiricism and positivism, albeit of a realist variety, and to psychophysiology. It is time to take a look at the Kantian ingredient, which is present nevertheless. This element has precisely to do with Riehl's realism. Riehl argues that in order to have an experience in the true sense of the word, appearances must not only be *given* but must be *thought of* as related to and as united in objects that are not given. Thinking is thus an a priori condition of any possible experience. In Riehl's opinion, the "logical" a priori manifesting itself in the twelve categories as identified by Kant has come together in the unifying thought of the object that stands under the principle of identity, or, in Kantian terminology, under the transcendental unity of apperception.

As a matter of fact, Riehl sees the unity of the object manifested in two ways:

> It is a fact of our thought that we assume a much wider connection between the objects of perception than seems justified by the perceptions themselves. We assume that there is something that persists through time, or, to put it in philosophical jargon: that their properties and states hang together in the unity of *substance*. We assume further that the changes of objects stand in a necessary connection to each other; we say that they have a connection through causality. . . . With both assumptions thought transcends the facts that are given in perception. A "substance" cannot be seen; matter also, as the "substance" of the physicist, is not seen but thought . . . and causality is equally imperceptible by the senses. . . . If we call the totality of facts in perception: *pure* experience, we are already here carried to the conclusion that pure experience cannot serve as foundation of science, that it is unsuited [by itself] to constitute knowledge of an object. (Riehl 1903, 66, 68; compare 94)

Here Riehl refers to John Locke as his main authority and considers his distinction between sensation and reflection: Locke has shown, so he maintains, that in every experience we have to presuppose a substance, although he could not clarify its exact nature and significance in the end. This step was taken satisfactorily only by Kant by his introducing the concept of the thing in itself.

A similar development has taken place, according to Riehl, in relation to the concept of causality. Hume's treatment of causality had to be complemented by Kant just as Locke's treatment of substance was. Kant has shown that mere regularity has to be heightened to strict lawfulness in order to account for our experiences. He was right in claiming that "a judgment has to be made before a perception can become an experience." Experience must be seen as "a perception that is judged and understood; it is the product of thought in intuition [*Anschauung*], it is the unity of intuition and concept, thus nothing simple that could already be given in mere perception" (Riehl 1903, 103).

According to Riehl, this view carries over to the development of the concepts of space and time from Locke and Hume to Kant and the latter's "doctrine of the subjective origin and the objective significance of space and time" (Riehl 1903, 106). He identifies Newton's distinction between absolute and relative space and time as a source of this particular view. Kant agreed with Newton on the necessity of presupposing absolute space and time, but he disagreed with him in conceiving space and time as given in themselves as an absolute reality. "Absolute space (and absolute time)," Riehl says, is for Kant "not the concept of a real object, but an idea which should serve as a rule to view all movement in it as relative. This means that absolute space is necessary in order to imagine relative spaces" (Riehl 1903, 107). Absolute space and absolute time are therefore not to be conceived as objects of our intuition. They are rather forms of intuition and thus necessarily valid of all perceived objects. Riehl is at great pains to show that relative space and relative time cannot be a priori and that they depend on our empirical perceptions. Only absolute space can have objective significance.

In further reconsidering Kant's transcendental doctrines, Riehl came to the conclusion that substance and causality do not achieve separate functions in relating diverse perceptions to one unique object, but that they are fundamentally connected to each other. In this process substance is seen to play the deeper role. Riehl tries to show that causality can be reduced to substance if substance is conceived as governed by a principle of identity. This principle states that in all changes in our perceptions, something to which these perceptions refer remains identical through time. We can therefore say that "cause and effect are connected to unity through the concept of substance" (Riehl 1900, 172–173).

In this context, Riehl cites as a key witness Julius Robert Mayer, the discoverer of conservation of energy, for having shown that causality is nothing but a function of the identity principle. To state a causal relation meant for Mayer to identify cause and effect in respect of their exact numerical equality.[9] Riehl believed he had found the function of reason par excellence in Mayer's conservation principle—something Kant had longed for in vain when he said, "The functions of reason can all be found, if one can represent the functions of the unity completely in the judgments" (B 94). Riehl thought that a solid basis for the alliance between philosophy and science has been reestablished only with Mayer, an alliance that rising neo-Kantianism had hoped for after the adventures of speculative idealism were over. Moreover, Riehl sees in Mayer the true spirit of "scientific philosophy," represented more in great scientists of his time like Helmholtz and Hertz, and not so much in the contributions of the academic philosophers themselves (Riehl 1903, 236; compare Riehl 1900).

Riehl's Criticism of Helmholtz

It is very revealing to see now how Riehl applies his epistemology in criticizing Helmholtz's views. Riehl cited Helmholtz more than any other contemporary author, and over long stretches Riehl's work reads like a running commentary on Helmholtz.[10] One can very well imagine that Helmholtz's ideas were the original starting point for Riehl and that he found his own outlook only by a careful scrutiny of Helmholtz's position.

Riehl first of all criticizes Johannes Müller's principle of specific nerve energies, which Helmholtz had taken over and had made the cornerstone of his theory of signs (PK2, 50–64). He relies mainly on Wundt, who had shown that the "principle of the functional indifference" of the nerves, as one could call it, on which Helmholtz himself relied is also valid for the central nerve endings. One has to suppose an interaction between stimulus and sense organ, which leads to the eventual adaptation of the sense organ to the stimuli of its environment. Riehl claims that the principle of specific nerve energies is thus incompatible with Darwin's evolutionary theory and therefore to be rejected. The doctrine of the mere subjective nature of secondary qualities has to be abandoned too, and it has to be assumed instead that all qualities have a subjective and an objective part.

The most important component of Riehl's critique concerns Helmholtz's conception and use of the principle of causality. According to Riehl, this conception misleads Helmholtz into an idealism alien to realism and the true scientific spirit. As is well known, Helmholtz distinguishes in his theory of signs between sensation and perception. Sensations are conceived as states of our nerves that come to our consciousness. They are subjective appearances caused by the influence of external causes on our sense organs. Perceptions, on the other hand, are hypotheses about the existence, form, and position of these external objects; they are the result of an inference, leading from the sensations to their causes, and they appear as ideas (*Vorstellungen*), as interpretations of the sensations. The basis for this inference therefore lies in the law of causality, which has to be presupposed as an a priori principle of all scientific experience.

Riehl now claims that Helmholtz's account of the transition from sensation to perception through causal reasoning commits two grave errors: first in presupposing the causal principle as innate and second in the way the principle is actually applied.[11] First of all, if sensations were subjective states of which we are conscious as states of our nerves, then only physiologists would have sensations, as Riehl sarcastically puts it, since only they are scientifically competent to describe the factual connection between nerve and sensation correctly. Normally sensations are taken as part of direct perception and not as arrived at by any reflection. So perceptions (in Helmholtz's sense) are primary and given without any causal inference. The identification of a subjective ingredient in perception is the result of an abstract reflection that already presupposes successful perceptions (PK2, 195–196).

Second, a causal inference from sensations to perceptions cannot lead as far as Helmholtz wants it to. Take, for example, a sensation of blue. If we assumed that our mind came to the conclusion that this sensation is the result of a cause, what could its cause be? Since the alleged transition from sensation to perception already happens in everyday life and does not need the insight of physiology or any other science, it would be wrong to say that we infer some vibration of the ether as its cause. All we can legitimately say is that our sensation of blue is caused by—something blue! This shows that the application of the principle of causality cannot lead us any further than to our sensation and never to causes outside of sensation. We have to conclude that the existence of the idea of an external reality cannot be the result of a causal inference; it is and must already be present in our sensations. The desired explanation of the functioning of our perception can therefore come only from a more detailed "analysis of sensation" (PK2, 197). Helmholtz could have easily avoided the two mistakes if he had admitted that there are no purely subjective sensations, but that from the very beginning they have an objective ingredient. The idea of an original and pure subjectivity in experience has to be rejected. To be conscious of a perceptual content as purely subjective already presupposes some objectivity.

Sensory consciousness arises out of a state of indifference between the subjective and the objective (PK3, 53–54, Riehl 1894, 55–56).

There is more, however, to be criticized, according to Riehl, in Helmholtz's conception of causal inference. Helmholtz takes this inference as providing not only the concept of the outer object but also the idea or image (*Vorstellung*) of space. Causal reasoning as a result of our experimental interaction with the world is also supposed to somehow transform our spaceless sensations into spatial perceptions. Riehl argues against this by saying that even if we could relate sensations to an external cause in our reasoning, all we would get is the notion of an object that is different from and independent of our consciousness, but we would never get the idea (*Vorstellung*) of space through this. Causality can never create spatiality since the effects an object has on us never transmit the spatial form of the object. There is no mental act through which we can find out that an object outside our consciousness is something in outer space. "It is not only completely incomprehensible but also inconsistent to assume that through whatever acts of practice and habit spatial representations should develop out of unspatial elements, like sensations and local signs" (Riehl 1925, 106; compare PK2, 136).

We have again to conclude that to imagine an object as spatial cannot be the result of an inference in which the causal principle functions as a premise, but that spatiality, that is, the representation of an object's form and situation, is already contained in the physiological and psychological nature of the sensation beforehand! And indeed, at least our visual sensations are already extended from the start. The idea of spatial extension of external reality is thus for Riehl to be conceived as the result of experiences in time that associate themselves with the extension that is already present in the visual sense *from the beginning* (PK2, 136, 198).

Riehl claims independent evidence for his thesis in the fact that the idea of space inherent in the visual sense is different from that of the sense of touch (PK2, 137–138). The changes in the visual space are lawfully related to those in the space of touch, but these relations have to be learned. The fact that we experience the spaces of different senses as intimately related makes us forget that the spatial ideas of one sense can only be signs and not pictorial images of the spatial ideas of another sense—they are not similar to each other. As evidence for this, Riehl refers to the Molyneux problem: A person born blind who later acquires her eyesight will not be able to transpose the spatial ideas of touch to her visual impressions. Riehl reaches the conclusion that each sense has its own spatial abilities: "Visual sense and sense of touch are two languages that put the same sense into different words, such that the capability to translate from one capacity into another one has to be learned in a painfully difficult way, in spite of the organic connection that exists between the two senses" (PK2, 139; compare 137, 148).

From this diagnosis, Riehl draws the important conclusion that we have to distinguish between space as the subjective form of a particular sense and space as the objective order underlying our perceptual elements (PK2, 105). Indeed, we would not be able to relate the different sensory spaces to each other, the passage continues, "if the idea of space were nothing but 'pure intuition [reine Anschauung]' in Kant's sense" (PK2, 139). Instead, we have to oppose Kant's identification of sensory space with the mathematical one (PK2, 91). Space is therefore not just a mental representation, a form of intuition, as Kant thought. "Magnitude and distance of the objects is something [real] that is given in experience, not just something that is purely intuited [etwas rein Vorgestelltes]" (PK2, 165). Since experience is not to be identified with sensation even if sensation plays a role in it, the "properties of space are partly of a sensory and empirical nature, partly of a conceptual one" (PK2, 164; see 134).

The same difference is described by Riehl in various other ways: as one between the "sensual apprehension" of spatial properties and their "real validity" (PK2, 80), between the "empirical" and the "logical" relations of space (PK2, 182), or between the "sensory-empirical qualities" and the "logico-mathematical properties" of space, relating the latter distinction to Riemann's division of "extensional relations" and "measurable relations" (PK2, 170, 173). Riehl also distinguishes the "factual foundations of our space perception lying in the mode of our sense perception" and the "formal properties which, according to the concept of space, we have to attribute to our mental activity" (PK2, 133). This clearly foreshadows the logical empiricists' distinction between pure and physical geometry.

In another passage, which deserves to be quoted in full, Riehl distinguishes between the "image" of spatial extension as given by a particular sense, such as the visual one, and its "objective correlate" (das sachliche Correlat):

> We have to attribute an independent existence to spatial distances which rests upon itself, because spatial distances act as causes. This is true even if the image [Bild] of spatial extension, as delivered by our visual sense, has to be regarded as something subjectively produced. Here as in other cases it is not necessary that the idea [Vorstellung] as delivered by the senses be congruent with reality itself or that it repeat reality [die Sache selbst gleichsam decke oder wiederhole]; it is enough that it stands in a lawful relation to reality, and I think that this can also rightly be said of the content of our

space perception. The objective correlate of our space perception *in general* lies in the *coexistence* of the objects. The determinate, measurable relations in which the objects coexist are the [objective] correlates of our *special* spatial perceptions. (PK2, 165)

From this critical assessment of Kant's conception of space it is only a short way to a fundamental critique of Helmholtz's theory of space perception: "As great as are the merits of the *physiological optics* in explaining the separate determinate spatial intuitions," Riehl writes, "as small must be judged its success in explaining the idea of space in general." It turns out that Helmholtz's theory does not deal with the real correlate of general spatial intuition at all. Instead, it already *presupposes* objective space and asks how it turns up in our sensory experience (PK2, 83–84; compare the second edition of PK2, 106).

All this does not mean that Riehl now turns his back completely on Kant's doctrine of space and time as a priori forms of intuition. Although the conceptual part of space and time is determined by measurement, it nevertheless has an "origin a priori," because in order to unify experience we cannot help but *think* space and time as perfectly homogeneous, as strictly continuous, invariable, necessarily uniform, infinite, and with zero curvature (PK2, 115–116, 132–133). Besides the original form of intuition of objects,

> there must equally be an original thinking from which the intellectual form of the experience of objects is derived. The reason for this is that experience and mere perception are different from each other. It is true that perception gives rise to concepts which would otherwise not have developed, but this does not mean that these concepts originate from perceptions. The reason is that the function of these concepts rests in the judgment of perceptions and in the determination of the object. They therefore have to lie at the basis of this determination. That there have to be concepts that originate from thought [and not from perception] can be known a priori from the concept of experience. What these concepts are can only be found out from what they have to accomplish: to be the conditions of the idea of an object. (Riehl 1903, 111)

This passage reveals Riehl at one of his most Kantian moments.

MORITZ SCHLICK BUILDING ON RIEHL

The foregoing has now prepared the way for my claim that, in his early work, Schlick could largely build on foundations already laid by Riehl. What Schlick did to many of Riehl's ideas is, on the one hand, to energetically clarify and generalize them. On the other hand, we have to ask ourselves what he did to Riehl's doctrine of the a priori origin of concepts.

As far as Riehl's distinction between the sensory and the conceptual in spatial properties is concerned, Schlick generalizes it and arrives thereby at the distinction between *Kennen* and *Erkennen,* between acquaintance (or intuition) and cognition (or knowledge). Whereas in Riehl this distinction is primarily limited to the treatment of space and time, Schlick sees it as an all-purpose key to epistemology.[12]

Schlick also clarifies Riehl's concept of the "objective correlate" of perception that lies in the "coexistence" of objects. Schlick accomplishes this clarification through his concept of "coordination" (*Zuordnung*) or "designation" (*Bezeichnung*). Whereas for Riehl the lawful relation existing between the subjective image of space and time and its objective correlate is the expression of an identity (since all laws are ultimately derived from the identity principle), it is for Schlick an irreducible fact of nature, a coincidence of events that has to be taken as it stands. And, in addition, whereas for Riehl concepts in science are unifying instruments and thus ultimately to be reduced to laws, for Schlick they are just signs for the objects. It is unnecessary and inadequate to suppose the more intimate relation between concept and object that the Kantians assume.

We can now identify the essential correction Schlick brings about in Riehl's conception. It appears that Schlick wants to eliminate the remnants of apriorism as they still exist in Riehl's thought, namely, the special role of the unifying concepts. Schlick provides for them a positivist substitute à la Avenarius and Mach that performs the same trick but is epistemologically less problematic. This is to say that Schlick wants to melt together the "valuable parts of both the critical and positivist theories" (Schlick 1916, 251), as he had done before with his solution of the mind-body problem in an article of 1916: "In order that they once more converge upon the truth, criticism must offer up as a concession to positivism its conception of appearance, while the latter must thankfully take over from the former the concept of the thing-in-itself; in this way the two merge into a world-view [*Weltansicht*] which has nothing to fear from any critical attack" (Schlick 1916, 249).

In order to correct Riehl's epistemology one has not only to give up the (neo-)Kantian conception of appearance but also any organizing principle of unifying the manifold of the appearances other than the "sign system" of the things in themselves and the objective order among them.

Conclusion

The foregoing suggests that Alois Riehl's philosophy is of great relevance to Schlick's prepositivist philosophy of science. I would even say that Riehl's importance for logical empiricism is of the same order as Ernst Mach's. This claim immediately raises the question why Riehl has fallen into oblivion while Mach has not. The most important reason is that Riehl did not manage to connect his philosophy of science to the natural sciences, especially physics, of his day after about 1895. He kept on promoting a general view of science that was popular in the 1870s and 1880s. The most modern part of science he dealt with in print is the controversy around Wilhelm Ostwald's energetics in his *Philosophy of the Present Age* (Riehl 1903, chap. 5). It is known that he treated relativity theory in his lectures and seminars (and he had supervised at least one dissertation on the subject), but he apparently did not find the strength to integrate this new theory into his account of science. In addition, his general view of science contained too many elements that seemed old-fashioned—for example, his high valuation of Julius Robert Mayer—so that it must have looked already dated when he took his professorship in Berlin in 1905 at the age of sixty.

A second reason why his philosophy went out of fashion is its insufficient connection with the philosophy of his time. There was the idealist turn of neo-Kantianism after about 1878 (Köhnke 1991, esp. chaps. 6 and 7), which eventually led to a new valuation of the cultural and social sciences, especially in the South-West German School of neo-Kantianism. There was the rise of *Lebensphilosophie,* which worked hand in hand with Nietzsche in criticizing science as a means of a general philosophical orientation. (See Hofmann 1926; Spranger 1944, 130; and Heidelberger 2002 on the scientific worldview in Germany around 1910.)

Ernst Mach's views were much more apt to give an alternative to these movements, because he could integrate, and indeed carry on, some of their concerns in his own philosophy. As a result, it developed much more programmatic force than Riehl's did. Riehl's account was attractive enough, however, to capture the interest of Schlick and Feigl. In the end it seems that, with the physicalist turn of logical empiricism around 1930, the realist element of Riehl's heritage has taken the upper hand.

Notes

Part of the research for this paper was conducted during my time as a fellow at the Center for Philosophy of Science, University of Pittsburgh, 1998–1999.

1. For Riehl's biography and philosophy, see Graykowski 1914, Rickert 1924/1925, Jaensch 1925, Hofmann 1926, Hönigswald 1926, Maier 1926, Spranger 1944, Köhnke 1991 (see the book's index), Goller 1991, Röd 2001, and Rutte 2001.

2. Gerhardt 1999, 185–190, provides an account of Riehl's call to Berlin; see also below.

3. The other two candidates were Benno Erdmann and Wilhelm Windelband. I am grateful to Reinhard Mehring for providing me with archival material on Riehl from the archive of Humboldt University, Berlin.

4. The members of the commission were Wilhelm Dilthey, Carl Stumpf, Friedrich Paulsen, Hermann Diels, Gustav Schmoller, a certain Möbius, and Adolf Erman.

5. It is interesting to note that both Riehl and Schlick signed the petition of the 107 philosophers to the ministers of culture of the German states in 1913, demanding that no further philosophical faculty positions should go to experimental psychologists but that they should receive positions of their own (Kusch 1995, 286). On the development leading up to this petition, see Schmidt 1995, chaps. 4 and 5, and Kusch 1995, chap. 7.

6. Compare Feyerabend 1966, 7, and Feigl 1974, 8–9. For Külpe's critical realism, see Henckmann 1997.

7. See Heidelberger 2004, chapter 5, for Feigl's and Schlick's identity theory of mind and body as a close continuation of Riehl's and Gustav Theodor Fechner's conception.

8. I am grateful to Hans-Joachim Dahms for informing me of Riehl's relation to Mies van der Rohe.

9. These ideas are very similar to those of Émile Meyerson (1908). They seem also to have some similarity with Phil Dowe's recent conception of causality as conserved quantity; compare Dowe 2000, chap. 5.

10. In the preface to PK2, Riehl was at pains to inform his readers that Helmholtz's *Facts in Perception* (Helmholtz 1878) came too late to have been considered by him in this work. He stated, however, that taking it into account would not have changed his concept of space as he conceived of it in his book (PK2, iv).

11. Riehl, by the way, sees Schopenhauer committing the same two mistakes.

12. Compare, however, Riehl 1907, 93, where Riehl insists that a sharp difference must be made between the *Kenntnis* of facts and the *Erkenntnis* of a law.

XI

CRITICAL REALISM, CRITICAL IDEALISM, AND CRITICAL COMMON-SENSISM: THE SCHOOL AND WORLD PHILOSOPHIES OF RIEHL, COHEN, AND PEIRCE

Alfred Nordmann

After the excesses of German idealism and of Hegelianism, in particular, a reconciliation between science and philosophy was brought about by the return to Kant and the neo-Kantian primacy of epistemology rather than metaphysics, ontology, or philosophy of nature. This is how at least some of the neo-Kantians conceived of their project,[1] and this is how the story is told by Herbert Schnädelbach (1983) and Klaus Christian Köhnke (1991). That story gives rise to rather different, though complementary, assessments. On the one hand, we can appreciate how the neo-Kantians laid the foundations for twentieth-century philosophy of science (see, for example, Heidelberger and Richardson in this volume). On the other hand, the epistemological turn of the neo-Kantians has been viewed as a retreat of sorts: With them, philosophy abdicated its role as the most general science from which all the special sciences derive their principles, and instead became institutionalized as one academic discipline among others: "The 'golden age' of neo-Kantianism . . . is a history of schools" (Köhnke 1991, 280). To the extent that this latter assessment involves a sense of regret and complaint, the charge against the neo-Kantians needs to be specified. In their return to Kant, did they somehow diminish the significance of his philosophical program? Is the failing of neo-Kantianism that it denied Schelling or Hegel a legitimate place in the history of Kantianism rather than continue the project by going beyond them in the manner, say, of Charles Sanders Peirce? These are the guiding questions behind the following juxtaposition of the neo-Kantians Alois Riehl and Hermann Cohen with the Kantianism of Charles Sanders Peirce.

The contrast is framed by Kant's distinction of a scholastic or school philosophy and a cosmopolitan or world philosophy. While Peirce explicitly endorses Kant's cosmopolitan conception, the work of Riehl and Cohen matches the description of school philosophy. In particular, I will show in a second step that the "fact of science" makes no difference for Riehl's and Cohen's seemingly antithetical yet nearly indistinguishable "critical idealism" and "critical realism"—though both defer to the fact of science, their difference is not such that it could be motivated or resolved in respect to contemporary developments in science. The third and final section of the essay explores how Peirce's approach differs from that of the neo-Kantians precisely in the place it accords to the fact(s) of science. This becomes especially apparent where the views of Cohen and Peirce appear to converge, namely, regarding the philosophical centrality of the continuity of infinitesimals.

SCHOLASTIC AND COSMOPOLITAN IDEAS OF PHILOSOPHY

Peirce claims that his mind had been "molded by his life in the laboratory" when he "learned philosophy out of Kant . . . along with nineteen out of every twenty experimentalists who have turned to philosophy" (1905a, 331–332).[2] Instead of climbing "a tower which should reach to the heavens," Peirce thus entered what Kant calls a "dwelling house, just sufficiently commodious for our business on the level of experience, and just sufficiently high to allow of our overlooking it" (A 707). Enchanted with Kant's metaphor, which opens the "Transcendental Doctrine of Method," Peirce elaborated the connection between philosophy and architecture.

> Works of sculpture and painting can be executed for a single patron and must be by a single artist. A painting always represents a fragment of a larger whole. It is broken at its edges. It is to be shut up in a room and admired by a few. In such a work individuality of thought and feeling is an element of beauty. But a great building, such as alone can call out the depths of the architect's soul, is meant for the whole people, and is erected by the exertions of an army representative of the whole people. It is the message with which an age is charged, and which it delivers to posterity. Consequently, thought characteristic of an individual—the piquant, the nice, the clever—is too little to play any but the most subordinate role in architecture. If anybody can doubt whether this be equally true of philosophy, I can but recommend to him that splendid third chapter of the Methodology, in the *Critique of Pure Reason*. (1931–1935, 1.176)

That third chapter of Kant's "Transcendental Doctrine of Method" is entitled "The Architectonic of Pure Reason." While Peirce emphasizes the contrast between the individual character of painting and the public concern of architecture, Kant associates in this chapter the public dimension of knowledge with its scientific character. "By an architectonic I understand the art of constructing systems. As systematic unity is what first raises ordinary knowledge to the rank of science, that is, makes a system out of a mere aggregate of knowledge, architectonic is the doctrine of the scientific in our knowledge" (A 832). Such systems of knowledge are produced by philosophers, mathematicians, natural philosophers and logicians. The last three groups, however, are mere "artificers in the field of rational knowledge" (*Vernunftkünstler*) since they do not advance the essential ends of reasons that consist in the legislation of human reason with regard to nature and freedom. The philosopher is therefore placed above them but, significantly, this amounts to no one being placed above them. For instead of philosophers there is only the ideal of the philosopher and this ideal is not embodied in an academic discipline, a school, let alone in an individual thinker. Instead, the philosopher is widely distributed and represents, as Peirce might say, "the whole people." In Kant's words, "the idea of his legislation is to be found in that reason with which every human being is endowed" (A 839). Kant calls "world philosophy" (Peirce suggests "public" or "secular philosophy")[3] what thus emerges as a system of knowledge that stands above the sciences but claims no special disciplinary identity above or next to the sciences. In contrast, school philosophy seeks a system of knowledge as a science in its own right, that is, by arrogating to itself the title of philosophy.

> Hitherto the concept of philosophy has been a merely scholastic concept [*Schulbegriff*]—a concept of a system of knowledge which is sought solely in its character as a science, and which has therefore in view only the systematic unity appropriate to science, and consequently no more than *logical* perfection of knowledge. But there is also a worldly concept of philosophy [*Weltbegriff*] (conceptus cosmicus), which has always formed the real basis of the term "philosophy," especially when it has been as it were personified and its archetype represented in the ideal of the *philosopher*. On this view, philosophy is the science of the relation of all knowledge to the essential ends of human reason (teleologia rationis humanae), and the lawgiver is not an artificer in the field of reason, but himself the lawgiver of human reason. In this sense of the term it would be very vainglorious to entitle oneself a philosopher, and to pretend to have equaled the archetype that exists in the idea alone. (A 838–839)

The juxtaposition of "world" and "school" thus signifies more than a difference of scope, and more than a difference, say, between public research in a republic of scholars and parochial or esoteric concerns. Instead, it aims at the disciplinary self-understanding of philosophy, namely, whether it claims for philosophy a separate domain of knowledge. Kant thus distinguishes the philosophy of Leibniz and Wolff from the critical philosophy that in a general sense entitles "anyone a philosopher who appears to exhibit self-control under the guidance of reason, however limited his knowledge may be" (A 840). Thus, for example, "[w]hen metaphysics was declared to be the science of the first principles of human knowledge, the intention was not to mark out a special kind of knowledge, but only a certain precedence in respect of generality" (A 843). However, this critique of his precursors was lost on most of Kant's successors who responded to Kant's promise of metaphysics as a science. Therefore, one can also apply to the nineteenth century Kant's contrast between a cosmopolitan world philosophy that is distributed among the necessary interests of all who are endowed with reason and a parochial school philosophy that institutes its own system of knowledge.[4] Thus the charge of "school philosophy" applies to Schelling or Hegel in that they were taken to arrogate to themselves the title of the philosopher who legislates general principles to the special sciences. In this they seemed to be following Kant, who reconstructed Newtonian mechanics as a science by taking the concept of science from philosophy and not from "the sciences." The charge of "school philosophy" also applies to the neo-Kantians, however, who criticized Schelling and Hegel for their philosophical arrogance, who left it to the sciences to determine what science is, and who devoted themselves to the reconstruction of the given scientific methods and thus to a systematic reconciliation of the sciences and Kantian philosophy. In doing so, they turned Kantianism into a school philosophy that is no longer able to relate *all* knowledge to nature *and* freedom as the essential ends of reason.

In contrast to both these school philosophies, Peirce claims for himself a cosmopolitan middle ground by subordinating the general task of philosophy to the reasoning implicit in all attempts to make sense of the world.[5]

> The object of a theory is to render something intelligible. The object of philosophy is to render everything intelligible. Philosophy thus postulates that the processes of nature are intelligible. *Postulates*, I say, not assumes. It may not be so; but only so far as it is so can philosophy accomplish its purpose; it is therefore committed to *going upon* that assumption, true or not. It is the forlorn hope. But as far as the process of nature is intelligible, so far is the process of nature identical with the process of reason; the law of being and the law of thought must be practically assumed to be one. (Peirce 1890, 133)

While it echoes Kant's definition of the "architectonic of pure reason" as well as Schelling's philosophy of nature, this passage also anticipates Peirce's

"Objective Idealism" and especially his "Critical Common-Sensism."[6] It is critical in that it concerns the limits of reason and its, in a Kantian sense, "practical" orientation toward an aesthetic purpose. It takes a cosmopolitan interest in common sense in that its postulate regarding the intelligibility of nature is already implemented in every particular attempt to understand nature. It should not be surprising, therefore, that this passage, with its world philosophical commitment, belongs to drafts for a series of papers devoted to the "Architecture of Theories" that begins by restating that commitment.

> That systems ought to be constructed architectonically has been preached since Kant, but I do not think the full import of the maxim has by any means been apprehended. What I would recommend is that every person who wishes to form an opinion concerning fundamental problems, should first of all make a complete survey of human knowledge, should take note of all the valuable ideas in each branch of science, should observe in just what respect each has been successful and where it has failed, in order that in the light of thorough acquaintance so attained of the available materials for a philosophical theory and of the nature and strength of each, he may proceed to the study of what the problem of philosophy consists in, and of the proper way of solving it. (Peirce 1891, 286)

The problem of philosophy—to render everything intelligible—thus arises from the ways in which scientific theories make certain things intelligible. In its relation to science, philosophy should therefore do more than acknowledge merely the fact of its existence in order to transcendentally reconstruct the conditions for the possibility, in principle, of scientific knowledge. Instead, philosophy should somehow partake in or grow out of scientific inquiry. Peirce's interest in an architectonic of theories thus invites a comparison between the various ways of taking seriously the fact and the facts of science: What place is accorded to contemporary science in the disagreement between the neo-Kantians Alois Riehl and Hermann Cohen, on the one hand, in Peirce's own philosophy, on the other?

SCHOOL PHILOSOPHY: RIEHL AND COHEN

Where lies the difference that makes a difference between Alois Riehl's "critical realism" and Hermann Cohen's "critical idealism"? It is not, as one might first suspect, a metaphysical difference between realism and idealism. Instead it lies, for example, in their views regarding the future and limits of science, in particular of mechanics as their paradigm science. This does not mean, however, that they were responding to more or less contemporary developments in science.

Idealism, realism, criticism   It is commonly held that Kant left his successors with a profound ambiguity concerning the thing-in-itself. Is it the cause of our perceptions or is it a fictitious limit of experience? Or yet again, does the critique of pure reason require that we leave this alternative undecided, since any attempt to decide it would take us beyond the limits of reason? Alois Riehl and Hermann Cohen have it both ways. According to the "critical" in their "Critical Realism" and "Critical Idealism," respectively, they appreciate the impossibility of deciding the question, and yet Riehl's realism takes the thing in itself one way, Cohen's idealism, the other way. In fact, they are not confused but admirably clear about this.

Riehl invokes the causal principle to argue for realism over idealism (see Heidelberger, in this volume). The causal principle demands and suggests an explanation of how the appearances come about. At the same time, Riehl realizes that "every step towards explaining these facts will necessarily lead beyond our consciousness as it is given" (PK3, 161). He reconciles this step with Kant's criticism by justifying it as a demand of reason: Just as in science "thought complements perception" and "presupposes a greater context" (*Zusammenhang*)—for instance, by positing "substances"—so philosophy ought to presuppose the reality of the thing in itself as the cause of the appearances (1903, 66–68; compare 1908–1926, 1:521, 572; also 1907, 89) or as the best explanation of the fact that our perceptions are not subjective and individual.

> This belief in the independence of objects from their being perceived and in their persistence despite the alternation of perceptions, is the belief which we express through the concept of substance. Without this belief—[there is] no experience for it would lack an object of which anything could be said at all, an object for a judgment about perception. And since without experience there would also be no science and no possibility of agreement among the judgments of all regarding one and the same object. (1903, 67)

It is therefore a belief or a judgment (1903, 103) that assigns being priority to thought. While our thought thus "penetrates" to reality as a precondition of thoughts and things, reality itself "remains transcendent." "The secret of being cannot be explored through thought; the principle of being precedes thought: first being, then thought" (1903, 165).

Cohen raises an important objection to realism, one that applies also to Riehl's critical realism. As Kant himself has warned, instead of providing a final answer, the transcendent application of the principle of causality opens up a regress: Once we take "things in themselves" to be the cause of appearances, we have to ask what caused the things in themselves. In the case at hand this is more than an academic worry: It is "the fate of positivism and of so-called realist philosophy," Cohen writes, that they unknowingly name their system with a *Rätselwort,* or enigmatic term. How can "reality" explain "appearance" if one cannot say what reality is in the first place? Before it can serve any explanatory purpose at all, "reality" requires an explanation of its own. "What is the real? What is the positive? This is the question to which philosophy should provide the answer" (1885, 597). Cohen therefore proposes that the "telling of fables regarding the sponta-neous self-generating creation of the real must be translated into scientific language regarding its methodical production" (1885, 596). According to Cohen, this scientific language comes with the method of infinitesimals (see Richardson, in this volume). It leads from nonrelational intensive magnitude to geometrically relational extensive magnitudes. The infinitely small as an intensive "magnitude of generation" provides not a cause in an infinite causal chain of relations but the ground, *Ursprung,* or origin of what is the real in mass, force, or energy: Through it the real is scientifically prepared for its treatment by physics and mechanics, in particular (1885, 597).

Cohen thus uses the method of infinitesimals or the law of continu-ity to explicate reality as a category that, along with space and time, makes experience possible (1885, 596). "Perhaps it is required," he writes, "to begin with a concept which is more elementary than that of substance and which can precisely therefore serve as a basis for the definitions of mass, force and energy" (1914, 87). The elementary concept of the infinitely small bridges realism and idealism in that, through it, pure thought produces the most general fact that "in the infinitely small nature displays everywhere and in every sense continuity [*Stetigkeit*]" (Cohen 1914, 85; Hertz 1894, 37/43–44).[7] The law of continuity thus originates in pure thought as the principle for the realization or production of the real in reality.

We therefore have on the one hand a profound difference between Riehl's "being comes before thought" and Cohen's "pure thought offers the method for the constitution of reality." On the other hand, Riehl and Cohen acknowledge that there is no fact to the matter of their dispute, there-fore also no fact of science. While Riehl suggests that we are trapped within the real and that all of our conceptualizations always already presuppose the real, Cohen suggests that we are trapped within our principles (1885, 597–598), and that any account of the real must be an account of its realization through principles. Neither of them allows for a stepping out, for a way to assess the relation between concepts and reality.[8]

In Wittgenstein's terminology one might therefore say that both agree that the picturing relationship cannot itself be pictured (1922, 4.12). Or, to use even more anachronistic expressions, Riehl and Cohen are caught up in a performative contradiction of sorts: They are neo-Kantians debating among themselves what, on their own understanding of Kant's critical philosophy, is a pseudo-problem.

The role of science   A similar argument can be made in regard to the issue of psychologism. Here, too, the neo-Kantians are engaged in a dispute that reaches beyond the limits of Kant's critical philosophy. Alan Richardson (in this volume) is not alone in arguing that Kant's concept of transcendental philosophy is fundamentally ambiguous: In its psychological-physiological guise it attends to the subject's faculties of cognition; in its logical-analytic mode it concerns a system of propositions and the conditions of its unification (Pascher 1997, 55). While the resolution of this ambiguity is obviously of great importance to the neo-Kantians, it couldn't arise as a problem for the "historical" Kant or for the so-called psychologist Fries: Their psychology of the faculties was a rational psychology, and the consideration of mental content was a form of conceptual analysis. In the case of psychologism, as perhaps in the case of realism vs. idealism, only the intervening debates and developments rendered the issue salient. Only with Helmholtz's sense physiology, only with the beginning of an empirical cognitive psychology, did the psychological investigation of faculties emerge as a distinct alternative to the logical analysis of concepts.

Since Riehl and Cohen were united against psychologism, it matters more to attend to the intervening debates and developments, if any, that rendered salient their disagreement concerning idealism and realism. Cohen and Riehl each suggest that his answer to this question has been offered by Kant, but neither quite dispels the suspicion that their disagreement is spurious.

According to Cohen (again see Richardson in this volume), he is following Kant by taking as his point of departure the "fact of science." Kant's question "how is experience possible?" presupposes the fact and truth of Newtonian mechanics and proceeds to show first and foremost how it is possible.[9] Cohen applies Kant's approach to the differential calculus by providing a transcendental argument for the conditions of its applicability.

He thereby moves beyond Kant's specifically Newtonian commitments and (as opposed to Kant) considers not primarily the axiomatic structure of Newton's theory but focuses on the mathematical tools that helped generate this theory. One might therefore conclude that the consideration of the sciences, though not of any particular contemporary science, led Cohen to articulate his critical idealism.

However, if instead of asking what it is about science that is taken seriously, one asks why is some aspect of science taken seriously at all, one finds that Cohen's approach is motivated and defended not by its ability to specify the conditions for the applicability of the law of continuity, but by the fact that it constitutes an application of pure (mathematical) thought to the physical world (1914, 87) and therefore instantiates the "realizing power of idealism" (1914, 92). To be sure, Cohen writes that the transcendental idealist "begins with scientific experience" (1885, 605; see also 580, 582–583). But this beginning is justified only in the name of philosophy. When Cohen offers two justifications for his interest in the concept of infinitesimals, both reflect the school philosophical interest in the completion of a system of philosophical knowledge:

> Firstly, the conscience of traditional *logic* will not rest until it has done what it can to describe and explain according to its norms this basic concept of a mathematized natural science. Also, however, there remains an irreparable gap in the catalogue of the foundations and principles of knowledge as long as this fundamental tool has not been recognized and delimited as a precondition of mathematical cognition and thus of the knowledge of nature. (1883, 1)

It is also subject to debate whether there was a specific scientific context that is responsible for Riehl's critical realism. While Michael Friedman (1997) assigns to the development of science a pivotal role in the transition from Helmholtz to Schlick, Michael Heidelberger (in this volume) views it as an outcome of philosophical debate within neo-Kantianism. According to Friedman, the general theory of relativity informed Schlick's arguments regarding the conceptual coordination of spatio-temporal coincidences. Since physics no longer has any one background geometry, "the most various kinds of metrical determinations are to be employed, in general, different ones at each position" (Friedman 1997, 27; see Schlick 1917, 27).

According to Michael Heidelberger, by contrast, Schlick's method of coincidences was inspired by Riehl's work, which Schlick quotes prominently in discussions of the issue.[10] Riehl had criticized Helmholtz's commitment to the "law of specific energy," which states that the senses are indifferent to the source of sensations. Since the source of sensations is unknown to the senses, only the mind can construct a conception of space from the given sensations. Riehl objects that each of the senses correctly apprehends the source of sensations and provides a real conception space in the sensations themselves. As in Schlick's account, the really given difference of haptic and visual spaces requires conceptual clarification by the mind. Note, however, that Riehl's disagreement with Helmholtz does not rely on particular sense-physiological research. Instead, it subjects Helmholtz's conclusion to a principled critique that draws eclectically on methodological, anecdotal, conceptual, scientific arguments about haptic and visual spaces.[11]

Though the advances of science as a system of objective knowledge did not play a significant role in the disagreement between critical realism and critical idealism, the philosophical differences between scientists did play some part. Riehl and Cohen esteemed certain scientists as scientific philosophers or philosophical scientists. But in contrast to Kant, who found in Newtonian mechanics a true science to investigate for its conditions of possibility, Riehl and Cohen had to sort through scientific as well as philosophical disputes concerning the nature and future of mechanics. As they selectively recruit Helmholtz and Hertz, Mayer and Ostwald, Planck and Einstein for their attempts to advance neo-Kantianism, Riehl and Cohen end up endorsing different scientific programs and envision different developments in science.

Directing science   Riehl's program for a scientific philosophy envisions the expansion of the limits of knowledge and experience. His story begins with Locke and Hume, whom he takes to be precursors of Kant. Locke's critique of substance, Hume's critique of causality, prefigured the Kantian critique: "The discovery of Kant is not that in our cognition there are a priori elements, elements of a non-empirical origin. Locke knew this . . . Hume knew this. . . . Kant's work was to have shown how these elements a priori can and must *nevertheless* have objective validity, to have shown how and to what extent we are justified to assume that they hold for the nature of the things themselves" (1903, 105–106).

However, while Locke's critique drew the limits of knowledge very narrowly, Kant's critique expanded them somewhat. Not sufficiently familiar with mathematics, Locke considered only "push" and "motion through pushing" as comprehensible to our understanding. Kant's critique was to show that "through the application of mathematics to the natural appearances the limits of the knowledge of nature can be extended much further out than it would appear possible for the merely empirical" (1903, 84).

According to Riehl, a further expansion is necessary because the paradigm science of mechanics gives us not a picture of the world but only the outlines of the picture: The nature and reality of sensations require that in addition to quantitative measurable effects the things must be exchanging qualitative effects—"*just as surely as there are specific sensations*" (Riehl 1903, 64).

The stepping stone for this further expansion of the limits of knowledge of nature is provided by energeticism.[12] What recommends energeticism for the project of Riehl's critical philosophy is its monism. In particular, Riehl identifies one crucial move by Robert Mayer:

> Work and heat cannot be compared immediately; they do not share a common measure. . . . Experiments can never show more than proportionality. . . . Mayer, however, does not infer from proportionality to mere equivalence, he infers to the sameness of the magnitudes—more precisely, to the identity of the magnitude. There are not two magnitudes there, there is only one magnitude, it only appears in two different forms and must therefore be measured on different scales. (1903, 133)

Mayer's move prefigures the move to a monistic philosophy. For this philosophy it would not be enough to expand energeticism and simply declare consciousness to be another form of energy. But it can overcome or consolidate psycho-physical parallelism by suggesting that here, too, we are dealing with an identity. "The world is there only *once*; but it is given to the objective consciousness which relates to external things as a context [*Zusammenhang*] of quantitative physical processes and things; and all the while *a part* of that world is given to a certain organic individual as *his* conscious functions and their context" (1903, 162). Thus consciousness is not energy. But from the point of view of Riehl's philosophical monism and critical realism, consciousness and energy are two measures of the same magnitude—reality—in that energy is thought by consciousness. Here we arrive at a point where Riehl's position touches up with Cohen's:[13]

> How sensations should emerge from any combination or motion of atoms, or shorter: how atoms should have sensations, cannot be comprehended at all. But it can be comprehended that there is no real problem here and the question makes no sense in this form. Not atoms are given to us but sensations and instead of finding a path, vainly of course, from atoms to sensations, our question instead has to be this: how do we arrive from sensations at the assumption of atoms? And in this form the questions is almost as quickly answered as it is posed. The concept of atoms is a product of method. (1903, 150–151)

Being thus comes before thought not in that atoms are given to us before
sensations. Instead, scientific monism precedes philosophical monism and
provides the method to lead us from the sensations back to the thing in
itself.[14] Significantly, this serves for Riehl as an argument against Du-Bois
Reymond's "ignorabimus" and a limitation of knowledge that a scientific
philosophy aims to overcome.

"Sensation is not the real," Hermann Cohen writes, "but an index of
the real." In the transition from mathematics to physics we are concerned
"with the world of things that is thought to be affecting; these things are
announced by sensation." At precisely this point, however, he parts company
from Riehl and other realists.

> Now, the announcement of things is only the claim that is made by
> the index of sensation. How can this claim be justified? This is the great
> question which separates idealism from realism, no matter how much
> they agree in their desire for real things. This justification does not lie in
> sensation anymore. . . . The presupposition of affecting things to which
> sensation responds must inevitably be defined in transcendental language.
> (1885, 594–595)

This conception, Cohen continues, is a consequence of method, but the
transcendental method and not the method of scientific monism or any particular science: "The carriers of the *a priori* in Kant's doctrine—time, space,
as well as the categories—are to be understood as methods, not as forms of
the mind" (1885, 584).[15] And while Riehl's psycho-physical application
of the method of scientific monism leads to an extension of the limits of
knowledge, the thing in itself, and the nature of consciousness, Cohen's
method restricts or solidifies the Kantian limits of knowledge by equating
the "doctrine of the thing in itself" with the "doctrine of the limits of pure
reason" (1885, 616).

> As far as it represented by the categories, the concept of thought must be
> taken more precisely and narrowly. The "forms of thought" don't pertain
> to all thinking but solely to mechanical thought, including that thought
> which is employed already in pure mathematics alongside the methods of
> intuition. A purpose can and indeed must be thought, and an organism,
> and even a lever must be thought; but these are not basic concepts, not
> forms of thought. Instead, the latter would represent those conditions
> which have to be added to the conditions of mathematical construction
> in order for objects of a mathematized natural science to arise. (1885,
> 585–586)

Accordingly, when Cohen writes that "sensation announces the energy of the given" (1885, 594), he is referring not to energeticism but to the intensities expressed through the calculus of infinitesimals, intensities that prepare the given for a mathematized scientific treatment.

Following Heinrich Hertz, Cohen acknowledges Mayer's and Ostwald's energeticism as only one of various approaches to mechanics. In his introduction to the ninth edition of Lange's *History of Materialism,* he relates his analysis of inifinitesimals not to a particular development in science but to the philosophical ideas of prominent scientists, Hertz on the one hand, Planck and Einstein on the other.[16] Of the three, he considers Hertz the most astute philosopher and accordingly presents a very close reading of Hertz's introduction to the *Principles of Mechanics* (Cohen 1914, 72–87).[17]

Cohen points out that traditional approaches to mechanics place force before motion, while Hertz's conceptual analysis places rigid connections, that is, motion, before force. Cohen shows this by discussing the passage in which Hertz justifies his appeal to hidden masses. Unlike mysteriously special entities such as forces, Hertz writes, the hidden masses are just "mass and motion all over again" (1894, 25/30–31). According to Cohen, "motion is thus the determinative character of this new concept of mass." He continues: "[T]he new kind of mass can only attach to physical motion where one immediately thinks of the motion of *differential equations* and not primarily of those motions that are perceived or undertaken" (1914, 82). Hertz thus prepares the ground for the recognition of infinitesimals, the law of continuity, and the differential calculus as the productive origin at the foundation of knowledge (see Cohen 1914, 85, and Hertz 1894, 37/43–44). Planck only confirmed Cohen's insight when, in his 1887 essay on the conservation of force, he proposed a "theory of infinitesimals" as a successor to action-at-a-distance theories.

This recognition by a prominent scientist of infinitesimals as the "principle and foundation" of theoretical physics (Cohen 1885, 91) falls short, however, when Planck refers to the speed of light in a vacuum as a "universal constant of nature" and therefore as "the really substantial" (compare Cohen 1885, 91). The concept of infinitesimals, after all, was to be more fundamental than that of substance. Cohen therefore suspects that Planck and Einstein revert to metaphysical confusion and declares confidently that physics will move beyond Planck and Einstein in this regard:

> But the history of theoretical physics in its rigorous articulation of the mechanical principles will surely penetrate to its precondition which lies in analysis. It will thereby ensure in the last instance the realizing power

of idealism. The path of inquiry leads surely and unerringly toward idealism; at the root of physical concepts materialism will be destroyed and it
is mathematics which brings about this liberation, guaranteeing its lasting
effect. (1885, 92)

According to Riehl, reality is fully implicated in our sensations. If experience is more than can be accounted for in reference to the mechanics of
material particles, one needs to "supplement the supposition which natural
science must take as the starting-point for its calculations" (1903, 64).
Instead of calling for such an extended reach of science, Cohen wants to go
beyond Kant by further specifying and delimiting Kant's ideal of mathematized scientific experience. For Cohen, the realizing power of idealism
consists in the mathematization of knowledge and the mathematical
construction of the world.[18]

In summary, then, the facts of science did not prompt Riehl's and
Cohen's views, nor do they respond from a Kantian perspective to recent
developments in science. However, their school philosophical disagreements
recruit philosophical scientists as different as Mayer and Hertz to advance
their views and to direct the future of science philosophically.[19]

Objective Idealism with Cosmopolitan Intent

Cohen's theory of infinitesimals takes one of Kant's principles[20] and uses it
to generalize the Kantian reconstruction of physics: The very conditions of
scientific experience require the continuous realization of the real in the
perception of intensive magnitudes. And instead of reconstructing the form
and content of Newtonian physics and especially its axioms or laws of
motion, Cohen tends to the differential calculus as the mathematical instrument and method of modern physics (1885, 596–597).

Riehl appeals to reality as the common or social cause (*gemein-
schaftliche Ursache*) of all perceptions (1903, 54, 67) and thus offers a "social
proof" of the existence of the world. The real object is a necessary presupposition for the possibility of experience since it provides the rule from
which subjective perceptions can be obtained with intuitive and perceptual
consistency (Riehl 1903, 54; compare 1908–1926, 1:512). The conditions
for the possibility of scientific knowledge therefore lie in the assumption of
a single objective reality to which all subjective perceptions are referred
(1907, 89).

On first sight, Charles Sanders Peirce might be said to combine the
approaches of Cohen and Riehl. The scientific method for the fixation of

belief is the only one that agrees with the "social impulse" and it consists in positing a hypothesis of reality (Peirce 1877, 120). However, it requires a continuous process of reasoning to mediate our subjective perceptions with this hypothesis and therefore to realize the real by gradually articulating the hypothesis.

> [A]s what anything really is, is what it may finally come to be known to be in the ideal state of complete information, so that reality depends on the ultimate decision of the community; so thought is what it is, only by virtue of its addressing a future thought which is in its value as thought identical with it, though more developed. In this way, the existence of thought now, depends on what is to be hereafter; so that it has only a potential existence, dependent on the future thought of the community. (1868b, 54–55)

Peirce's view does not, after all, combine the approaches of Riehl and Cohen. That the existence and meaning of a thought should depend on its future development and reality distinguishes Peirce's "realism" from any "nominalism" that is interested either in the preparation of the real for its scientific treatment (Cohen) or in reality as the cause of perception (Riehl).[21] In particular, that all thought is implicitly interested in the future of the community establishes the world philosophical ideal of "the philosopher" as it operates in "that reason with which every human being is endowed." Accordingly, Peirce arrives at his notion of continuity, for example, in rather different and decidedly more varied ways than did Cohen. He encounters it in his analysis of reasoning, he makes it the subject of experimental researches and applies it as an explanatory hypothesis, he posits it as a principle of practical philosophy that orients us in thought and action toward an aesthetic ideal. Only a cursory sampling of these various contexts can indicate how the facts of science enter his scientific philosophy.[22]

Grounds of validity   According to Peirce, human reasoning is caught up in a self-correcting motion that leads to the fixation of belief and its attendant determination of reality. "Accordingly, just as we say that a body is in motion, and not that motion is in a body we ought to say that we are in thought, and not that thoughts are in us" (1868b, 42). In his paper entitled "The Grounds of Validity of the Laws of Logic," he therefore ascertains these grounds for the various modes of reasoning. Rather than doubt the validity of deductive or inductive reasoning, he studies them as elements of the continuous motion of thought.

In the case of deductive reasoning, continuity appears in Peirce's solution of Zeno's paradox (1869, 67–68). According to the paradox, Achilles cannot run a certain distance because, in order to do so, he would have to perform an infinite number of tasks by traversing all the fractions of this distance. The paradox therefore maintains that any number is smaller than the infinity of its possible divisions. This formulation, according to Peirce, "involves an obvious confusion between the number of terms and the value of the greatest term" (1869, 68). The confusion thus consists in using the value of the length that is run by Achilles in a context where this value ought only to be mentioned, namely, in the context where we state how many operations of division can be performed on, for example, this number. Significantly, however, this solution of Zeno's paradox allows Peirce to maintain that, without any danger to the grounds of validity of deductive logic, Achilles does indeed perform an infinite number of tasks with each discrete step he takes (1868a, 26–27). Any deductive relation of a set of premises to their conclusion represents such a discrete step or a cross-section in a continuous process:

> There is reason to believe that the action of the mind is, as it were, a continuous movement. Now the doctrine embodied in syllogistic formulæ (so far as it applies to the mind at all) is, that if two successive positions, occupied by the mind in this movement, be taken, they will be found to have certain relations. It is true that no number of successions of positions can make up a continuous movement; and this, I suppose, is what is meant by saying that a syllogism is a dead formula, while thinking is a living process. But the reply is that the syllogism is not intended to represent the mind, as to its life or deadness, but only as to the relation of its different judgments concerning the same thing. (1869, 63)

The liveliness of the continuity of thought can thus become sedimented in the rigid relation between judgments. To the extent that not only human beings but all of nature is "in thought," the productive liveliness of nature (*natura naturans*) can also become sedimented in the rigid relation of lawfulness (*natura naturata*). After establishing early on that "the mind is a sign developing according to the laws of inference" (1868b, 53), Peirce went on to develop this analogy by showing the "Universe being precisely an argument" that is "working out its conclusions in living realities" (1903a, 193–194; compare 1891, 293, 297).

As opposed to deductive or syllogistic reasoning, the validity of induction depends on the continuous life of the mind. Induction provides the paradigmatic case in which "the existence of thought now, depends on what is to be hereafter." While no particular set of premises guarantees its particular conclusion, the valid ultimate conclusion of any inductive argument is the final opinion of the community of inquirers.[23] If the interest of an inquirer were limited to the conclusions that can be reached by this individual in his or her lifetime, those conclusions would lack validity and that inquirer would be acting illogically. "[D]eath makes the number of our risks, of our inferences, finite, and so makes their mean result uncertain. The very idea of probability and of reasoning rests on the assumption that this number is indefinitely great" (1878, 149). Thus, valid induction "requires a conceived identification of one's interests with those of an unlimited community" (1878, 150).

> They must not stop at our own fate, but must embrace the whole community. This community, again, must not be limited, but must extend to all races of beings with whom we can come into immediate or mediate intellectual relation. It must reach, however vaguely, beyond this geological epoch, beyond all bounds. He who would not sacrifice his own soul to save the whole world, is, as it seems to me, illogical in all his inferences, collectively. Logic is rooted in the social principle. (1878, 149; compare 1869, 81)

This social principle resonates with Kant's world-philosophical demand that the ideal of the philosopher should be present in the reasoning of every human being. For, according to Peirce, it is not necessary that "a man should himself be capable of the heroism of self-sacrifice": "It is sufficient that he should recognize the possibility of it, should perceive that only that man's inferences who has it are really logical, and should consequently regard his own being only so far valid as they would be accepted by the hero. So far as he thus refers his inferences to that standard, he becomes identified with such a mind" (1878, 149–150).

Laboratory life   The continuity of thought fixes not only opinions but determines reality.[24] It is science, therefore, that provides the model for Peirce's unlimited community of inquirers. And philosophy becomes scientific only by adhering to the social principle of self-sacrifice, by adopting the scientific method of fixing belief and therefore by merging with science instead of asserting its normative or critical autonomy toward the special sciences.[25]

The hypothesis of reality is only the most general philosophical hypothesis which assumes meaning in the course of its articulation and in the determination of reality as the end of all inquiry. The scientific method for the fixation of belief follows this general pattern of hypothesis formation and attendant realization. Accordingly, the "hypothesis of continuity" is also viewed by Peirce as subject to the scientific method.[26]

Peirce's hypothesis originated in his antifoundationalist philosophical argument against intuitions, that is, cognitions that supposedly are not determined by previous cognitions (1868a, 25). It takes on empirical significance in the sense-physiological debate regarding perceptual minima: Is there a perceptual continuum corresponding to the continuity of thought, or is it necessary, for a stimulus to be perceptible, that it exceed a certain threshold intensity? Peirce engages Fechner's question in two empirical studies (compare Kuhn 1996 and Heidelberger 2004). His experiments and theoretical reflections in "On the Theory of Errors of Observation" maintain that a series of observations prompts a first inference to a distribution from which the observed value and the probability of error are then inferred together. If the distribution of errors follows the law of least squares (that is, if the sources of error are not discontinuous or discrete), one can recognize "the possibility of any error positive or negative, however great, although the probability of indefinitely great errors will be indefinitely small" (1870, 125–126). While the law of least squares thus suggests a continuous gradation of error, the first inference to the curve of error is already an inference to "a statistical quantity; not one which belongs to a single observation, but one which belongs to an infinite series of observations" (1870, 132).

This empirical investigation finds Peirce in agreement with Fechner. However, he later co-authored with Joseph Jastrow a second investigation that addresses a point of disagreement. Fechner had provided experimental evidence to argue for a *Unterschiedsschwelle,* or "least perceptible difference" of nerve excitations. Peirce and Jastrow find fault with Fechner's experiment and view his conception "a difficult one to seize" because it appears to run counter to Fechner's own doctrine, according to which "the intensity of the sensation increases continuously with the excitation" (Peirce and Jastrow 1885, 122). This doctrine certainly does not imply Fechner's threshold of perceptible difference, nor would its denial, namely, the notion that "a continuous increase of the excitation would be accompanied by successive discrete increments of the sensation." From this, Peirce and Jastrow conclude that Fechner's claim does not concern the continuity or discontinuity of sensation at all, but the conscious perception of differences between sensations. Even on this construal, however, Peirce and Jastrow find Fechner's thesis hard to reconcile with his own views.

> [T]he errors of our judgments in comparing our sensations seem sufficiently accounted for by the slow and doubtless complicated process by

which the impression is conveyed from the periphery to the brain; for this must be liable to more or less accidental derangement at every step of its progress. Accordingly we find that the frequencies of error of different magnitudes follow the probability curve, which is the law of an effect brought about by the sum of an infinite number of infinitesimal causes. This theory, however, does not admit of an *Unterschiedsschwelle* [threshold of difference]. On the contrary, it leads to the method of least squares, according to which the multiplication of observations will indefinitely reduce the error of their mean, so that if of two excitations one were ever so little the more intense, in the long run it would be judged to be the more intense the majority of times. (Peirce and Jastrow 1885, 123)

Fechner maintains that beneath a certain threshold of difference one's guesses as to the greater of two excitations is a matter of chance, that is, they tend to be correct 50 percent of the time. Peirce and Jastrow set out to reject Fechner's chance hypothesis. They establish that their own unconscious guesses follow the curve of error, that is, that in the long run they are correct more often than not. Fechner's threshold of difference was thus to give way to the hypothesis of continuity.[27]

Several years later, finally, Peirce used this hypothesis as a doctrine that "gives room for explanations of many facts which without it are absolutely and hopelessly inexplicable" (1892, 333; compare 1891, 297). Among those facts are the intensive continuity and spatial extension of feeling, as manifested, on the one hand, in the reaction to an irritation by an amoeba or a slime mold, and on the other hand by the affections and communications of ideas (1892, 324–325).

Peirce's "law of mind" states "that ideas tend to spread continuously and to affect certain others which stand to them in a peculiar relation of affectability." These ideas, he adds, are not "mere words" but living realities concreted out of feelings (1892, 313, 330). Accordingly, the law of mind serves not only to explain, for example, how past ideas can be present. It also "carries along with it the following doctrines: 1st, a logical realism of the most pronounced type; 2nd, objective idealism; 3rd, tychism, with its consequent thorough-going evolutionism" (1892, 333). Logical realism (as opposed to nominalism) holds reasoning to the social principle and the hypothesis of reality. Objective idealism correspondingly views matter as effete mind and views reality (*natura naturata*) as the realization (*natura naturans*) of the hypothesis. Tychism, finally, reflects continuity as the even distribution of chances, that is, as discontinuous sporting that is the source of continuous evolution.[28]

> Such are the materials out of which chiefly a philosophical theory ought to be built, in order to represent the state of knowledge to which the nineteenth century has brought us. . . . Like some of the most ancient and some of the most recent speculations it would be a Cosmogonic Philosophy. It would suppose that in the beginning—infinitely remote—there was a chaos of unpersonalized feeling, which being without connection or regularity would properly be without existence. This feeling, sporting here and there in pure arbitrariness, would have started the germ of a generalizing tendency. Its other sportings would be evanescent, but this would have a growing virtue. Thus the tendency to habit would be started; and from this with the other principles of evolution all the regularities of the universe would be evolved. At any time, however, an element of pure chance survives and will remain until the world becomes an absolutely perfect, rational, and symmetrical system, in which mind is at last crystallized in the infinitely distant future. (1891, 297)

Aside from grounding the validity of the laws of logic and aside from serving as an empirical hypothesis, the concept of continuity thus contributes to the "one intelligible theory of the universe," namely, that of "objective idealism, that matter is effete mind, inveterate habits becoming physical laws" (1891, 293).

Kant's pragmatism   The notion of continuity belongs to the mind's processes of deductive and inductive reasoning, and its judgments of sense-perception. Continuity also belongs to reality and nature—for example, in the gradual articulation of the hypothesis of reality or in the intensity and spread of feeling. While Peirce echoes Schelling when he bridges mind and nature in his cosmogonic philosophy or his law of mind, he traces his objective idealism back to the common-sensism of Kant's critical philosophy.

"Kant (whom I *more* than admire) is nothing but a somewhat confused pragmatist," Peirce writes, and he (not unlike Cohen and Riehl) attributes Kant's confusion to his notion of a thing in itself (Peirce 1931–1935, 5.525; compare 1905b, 353–354). According to Peirce, whatever is is also cognizable in that its being consists in its being related to something and thus through a continuous series of mediations also to any knowing subject. However, if one interprets Kant to say trivially that something unrelated to us and thus something unreal cannot be an object of cognition, what remains is Kant the pragmatist. Once all references to the thing in itself are "thrown out as meaningless surplusage," "we see clearly that Kant regards Space, Time, and his Categories just as everybody else does, and never doubts or has doubted their objectivity. His limitation of them to possible experience is pragmatism in the general sense; and the pragmaticist, as fully as Kant, recognizes the mental ingredient in these concepts" (Peirce 1931–1935, 5.525; compare 1905b, 353–354).

By way of time, space, and the categories, the mind is brought to nature. With their help, the appearances are transformed into experience, and by the same token, those initially vague mental constructs take on determinate meaning. Like the Kantian categories, the concept of continuity schematizes the appearances or enters into hypotheses just "in so far as its possible consequences would be of a perceptual nature." As Peirce goes on to point out, those consequences constitute the meaning of conceptions like continuity, "which agrees with my original maxim of pragmatism as far as it goes" (Peirce 1903b, 225). Thus the process of reason insinuates itself into the process of nature and the determination of the real coincides with the articulation of meaning.

By claiming that Kant thus regards the categories "just as everybody else does," Peirce takes their agreement to represent a cosmopolitan common-sensism. Since it is based on a critical rejection of Kant's "thing in itself," this common-sensism is doubly entitled to continue the legacy of Kant's critical philosophy. "This kind of Common-sensism which thus criticizes the Critical Philosophy and recognizes its own affiliation to Kant has surely a certain claim to call itself Critical Common-sensism" (Peirce 1931–1935, 5.525; compare 1905b, 353–354).

World philosophy, according to Kant, addresses the interest of all human reason in nature and freedom. According to Peirce, all mental action serves as the lawgiver of nature and is therefore oriented to hope in the intelligibility of nature and to the *summum bonum* of the continual evolutionary increase of reasonableness. Freedom, to be sure, appears in Peirce's philosophy not as Kantian autonomy but as tychism (chaos, chance) on the one hand, and on the other hand as self-sacrifice and self-control in reasoning (1869, 72; 1905a, 337, 340). And yet, Peirce's philosophy proves cosmopolitan in contrast to the neo-Kantian school philosophy also in that it preserves the relation of *all* knowledge to the theoretical *and* practical ends of reason (A 839–840).

> It may seem strange that I should put forward three sentiments, namely, interest in an indefinite community, recognition of the possibility of if this interest being made supreme, and hope in the unlimited continuance of intellectual activity, as indispensable requirements of logic. . . . It interests me to notice that these three sentiments seem to be pretty much the same as that famous trio of Charity, Faith, and Hope, which in the estimation of St. Paul, are the finest and greatest of spiritual gifts. Neither Old nor

New Testament is a text-book of the logic of science, but the latter is certainly the highest existing authority in regard to the dispositions of heart which a man ought to have. (Peirce 1878, 150–151)[29]

NOTES

I would like to thank Helmut Pape and Gerhard Gamm for critical comments on a draft.

1. "The activity of the followers of Schelling and Hegel, of the Neo-Aristotelians, and of other recent schools, is only too well calculated to justify the repugnance with which scientific men usually turn away from philosophy . . . It is, however, remarkable that the very point which the systematisers and apostles of the mechanical cosmology so carelessly pass by—the question as to the limits of natural knowledge has found full appreciation amongst deep-thinking men engaged in special researches. . . . The man who is securely master of a single field, and here sees into the heart of every problem, has won a sharpened eye for all related fields of inquiry. He will everywhere easily find his way, and so, too, will quickly attain to a general view, which may be described as genuinely philosophical, while studies which are wider in their reach may easily retain that lack of thoroughness which marks every philosophical system that evades the questions belonging to the theory of knowledge" (Lange 1866, 2:303 and 307). One of the many to echo this sentiment is Fritz Schultze, who blames the philosophy of Hegel and Schelling for the mistrust and animosity between science and philosophy, "since it brought back all that fog which Kant had dispelled with his mighty hand just then." To remedy this situation, he offers not a theory of nature but "a theory of the knowledge of nature or a natural theory of knowledge" (Schultze 1881/1882, 1:7 and 12).

2. In fact, he studied the *Critique of Pure Reason* from age sixteen to nineteen for two hours every day until he "almost knew the whole book by heart." See a manuscript from 1897 (1931–1935, 1.4). In the same manuscript, Peirce writes "Thus, in brief, my philosophy may be described as the attempt of a physicist to make such conjecture as to the constitution of the universe as the methods of science may permit, with the aid of all that has been done by previous philosophers" (1931–1935, 1.7).

3. The previously cited passage from Peirce begins as follows: "The universally and justly lauded parallel which Kant draws between a philosophical doctrine and a piece of architecture has excellencies which the beginner in philosophy might easily overlook; and not the least of these is its recognition of the cosmic character of philosophy. I use the word 'cosmic' because *cosmicus* is Kant's own choice; but I must say I think *secular* or *public* would have approached nearer to the expression of his meaning" (1931–1935, 1.176).

4. In a footnote Kant offers another version of the definition of worldly or cosmopolitan philosophy: "By 'worldly concept' is here meant the concept which relates to that in which everyone necessarily has an interest; and thus I determine the intent of a science to be according to *school concepts*, if it is regarded as a suitable instrument for certain arbitrary ends" (A 840). While this version also emphasizes the unlimited interest of philosophy as "science of the relation of all knowledge to the essential ends of human reason," it places less emphasis on what I consider the second essential element of Kant's definition, namely its rejection of "a system of knowledge which is sought solely in its

character as a science" where the proponents of such a system might vaingloriously enti-tle themselves philosophers (A 838–839). The systems of Fichte, Hegel, and Schelling are "school philosophical" in light of this second part of the definition.

5. Peirce draws on Kant, Schiller, Hegel, and Schelling, but also on Helmholtz, Herbart, and Fechner, as well as Riemann, Gauss, Grassmann, and Schröder. However, he paid lit-tle attention, if any, to Otto Liebmann's call for a return to Kant, or to the works of Her-mann Cohen, Kuno Fischer, Alois Riehl, and other neo-Kantians. One might thus call him a neo-Kantian who remained blissfully unaware of the school philosophical signif-icance of that term.

6. For a more detailed elaboration of these notions, see the last main section of this essay.

7. Cohen here quotes Hertz's *Principles of Mechanics* concerning this "experience of the most general kind." He points out that this experience is not one of sensual matter but "is produced in pure thought." See also: "In the infinitesimal lies not alone the origin of magnitude but just as much that of being itself, of the real; for in the end it serves as the origin of magnitude only to grasp the latter" (Cohen 1914, 90). Also: "Principles are the foundations of the laws of nature, the 'general laws of nature.' The laws of nature become objectified in the forces, and in these the things of nature. Thus the principles are the foundations of the things" (Cohen 1885, 590).

8. The realist Riehl, for example, states that "realism would be mistaken if it assumed that the external world existed also outside or prior to perception in the manner in which it is perceived or that things and perceptions must be similar" (quoted in Ollig 1979, 25). And the idealist Cohen rejects an idealism that annihilates the material world (1885, 608), and follows Kant in viewing the noumenon as a task that inevitably belongs to our sensibility, namely, the task to determine "whether there may not be objects that are entirely disengaged" from sensible perception (1885, 615; see A 287). He therefore declares philosophy to be the "doctrine of the thing in itself" as characterized, for exam-ple, by the mechanical law (1885, 617). See also Cohen's claim that his critical idealism tries to meet "the demands of realism while at the same time radically eliminating mate-rialism" (1914, 88).

9. According to Schlick, incidentally, Riehl disagrees with this (and thus also with Schlick's own) interpretation of Kant (1925, §36).

10. Compare Schlick 1925, § 29: "Though able to orient themselves in haptic and kines-thetic space, the patients [born blind, their eyesight surgically restored] don't know any-thing at all about optical orientation in visual space. Thus results with complete rigor the conclusion which Riehl (PK2, 139) formulates as follows: 'that all basic components of spatial construction: movement, shape, size, direction are different for the two senses, that there is therefore no other connection between the conceptions derived from each than that which is founded by *experience*.'"

11. Riehl's eclecticism is confusing at times. In his popular lectures Riehl argues that the sensations (i.e., responses to nonformal "secondary" qualities) are not exclusively deter-mined by the sense organs but also by the nature of the stimuli: Light affects the eye, sound affects the ear. To defend this point he must impeach Helmholtz, who argues the

contrary on the basis of experiments according to which we can "see light" or "hear sound" also in the absence of light and sound, as when we are hit in the head. Riehl counters as follows: Contrary to the methodological norm that anomalies are to be explained away in terms of specific exceptional conditions, Helmholtz's experiments allow that "adequate stimulation is refuted by means of anomalous stimulation." If Riehl's counterargument is less than compelling (and does not testify to his scientific acumen), matters get more confusing in the next step. There he argues that, for all we know, the anomalous stimulation may just be a variant of the adequate one—but if outwardly different stimuli are all equally adequate, and if galvanic electricity can produce perceptions of sound and of light, this would prove Helmholtz's point that "hearing sound" and "seeing light" do not depend on the nature of the stimulus (Riehl 1913, 63).

12. Riehl discerns in Mayer's work a "substantial conception of causality." Mayer's scientific work thus presents Riehl's neo-Kantian project with a unified conception of what Hume and Locke had offered Kant only separately.

13. As Geert Edel points out (1987, 44*), owing to Riehl's high regard for the sciences, his antipsychologism, and his view that the critique of reason amounts to a theory of experience, his position was frequently viewed as dependent on that of Cohen. Cohen rarely refers to Riehl (e.g., Cohen 1885, 459). Ernst Cassirer moves easily between the approaches of Riehl and Cohen, between Mayer's system of equivalences in nature and Hertz's mathematical images that allow for the treatment of transitions between concrete particulars (1910, 197–201/262–267, 229–233/305–310). The differences between Cohen and Riehl become salient only upon his sustained analysis (Cassirer 1910, 310–311/411–412). This may be further testimony to the school philosophical character of the opposition between critical realism and critical idealism.

14. Being thus comes before thought as it presents itself in sensation, but only thought can exhibit this givenness of being in the sensations. This recapitulates an argument of Riehl's that was cited at the beginning of this section: The priority of being owes itself to a transcendental argument upon what is given to us, namely, the sensations.

15. To the extent that these methods apply to the sensations, Cohen insists that the a priori is discovered not in concepts alone but also in sensibility, that the unity of consciousness includes the sensations (1885, 605, 597, 598).

16. In his popular lectures Riehl does not refer to Planck or Einstein at all and to Hertz just once: "Only its success can teach whether energeticism is destined to take the place of mechanics as the foundation of physics. After pursuing this mode of reforming the principles of mechanics for a while, Heinrich Hertz abandoned it again" (1903, 146).

17. Cohen repeatedly emphasizes the philosophical character of Hertz's contribution; see especially Cohen 1914, 72.

18. See note 7 above.

19. This does not imply, of course, that the fact of science makes no difference at all. Indeed, no matter how "school philosophical" the distinction between critical realism and critical idealism may be, it is coordinated with Mayer's substantial account of causation on the one hand and with a Hertzian, mathematically imagined realization of the

real on the other hand. And yet this coordination of philosophical positions does not result from Riehl's or Cohen's attempt to philosophically reconstruct contemporary science. Not once removed from science, like Kant, they are twice removed in that they watch Mayer and Riehl elaborate the philosophical underpinnings of their science.

20. From among Kant's principles of pure understanding, Cohen singles out the "Anticipations of Perception." In the first edition of the *Critique of Pure Reason,* Kant writes: "The principle which anticipates all perceptions, as such, is as follows: In all appearances sensation, and the *real* which corresponds to it in the object (*realitas phaenomenon*), has an *intensive magnitude,* that is, a degree" (A 166). The second edition rewords this: "In all appearances, the *real* that is an object of sensation has intensive magnitude, that is, a degree" (B 207). Kant explicates this in terms of continuity, but his references to the differential calculus remain largely implicit: "Between reality in the field of appearance and negation there is therefore a continuity [*ein kontinuierlicher Zusammenhang*] of many possible intermediate sensations, the difference between any two of which is always smaller than the difference between the given experience and zero or complete negation.... The property of magnitudes by which no part of them is the smallest possible, that is, by which no part is simple, is called their continuity" (A 168–169). Time and space are such continua. Since the synthesis of productive imagination is a progression in time which introduces limitations in space, "all appearances . . . are continuous magnitudes" (A 170). Peirce reduces Kant's definitions to this: "the essential characteristic of a continuous series is that between any two members of it a third can always be found." Since a numerable and incomplete series of rational fractions would meet the criterion and since Kant's continuous series might thus include gaps and "lose all appearance of continuity," Peirce criticizes and amends the definition accordingly (1892, 320–321).

21. Peirce uses "realism" in opposition to "nominalism" (see especially 1871, 88–89). Thus, his "objective idealism" does not contradict his commitment to realism. Though Riehl refers to reality as "the foundation of all perceptions, whether actually or only potentially given" (Riehl 1903, 54), only Peirce fully appreciates the hypothetical character of this assumption. Likewise, Cohen's continuity of intensive magnitudes concerns the affection of the senses but does not continue on via perceptual and theoretical hypotheses to the realization of the real as a process of conscious and unconscious reasoning.

22. The following survey is also incomplete, neglecting, for example, Peirce's semiotics and the continuity of interpretation. Also, none of these uses is foundational of the others. This may explain Peirce's difficulty of producing a systematic exposition of his philosophical system.

23. Compare Levi 1980, Hacking 1980, Kuhn 1996. Similarly, of course, Peirce's pragmatic maxim states that the ultimate meaning of any term consists in the final conclusions it gives rise to. This becomes particularly clear in his reformulation of the maxim. Here he refers to Kant's notion "that necessary reasoning does nothing but explicate the *meaning* of its premises" (compare A 7). Peirce agrees but adopts a far wider conception of necessary reasoning (one that includes induction). The meaning of a term thus becomes its "ultimate meaning" or "final interpretant" (1903b, 218–220).

24. The conception of reality also "essentially involves the notion of a COMMUNITY, without definite limits, and capable of an indefinite increase of knowledge" (1868b, 52).

25. In his "Outline Classification of the Sciences," Peirce considers philosophy as a "Science of Discovery," that is, a "*positive science,* in the sense of discovering what is really true." He subdivides it into phenomenology, normative science, and metaphysics. As normative science, philosophy is concerned with the duality of good and bad (aesthetics, ethics, and logic) and "the classification of possible forms." This is neither an epistemological nor a transcendental or critical normativity vis-à-vis the sciences (1903c, 258–259, 272). Indeed, it is not at all clear that Peirce classifies his own work as "philosophy" in this sense.

26. This pattern is also an instance of abductive reasoning. The grounds of validity of abduction are not discussed in Peirce 1869. However, Peirce's various elaborations of the logic of abduction clearly suggest that its ground of validity can be construed as somewhat analogous to those of induction—they rest in the continuity of reasoning that is initiated by the discontinuous intervention of an abduction.

27. To be sure, Peirce and Jastrow establish only that the power to discern differences goes somewhat beyond our ability to do so consciously. This does not suffice to exclude altogether a threshold of difference. At this point they marshal an ingenious argument. If there is continuity between conscious and unconscious judgment that follows the curve of error, then it is the curve of error (rather than sense-physiology) that measures sensibility in the first place. "[I]f this law is shown to hold good for differences so slight that the observer is not conscious of being able to discriminate between the sensations at all, all reason for believing in an *Unterschiedsschwelle* is destroyed. The mathematical theory [of least squares] has the advantage of yielding conceptions of greater definiteness than that of the physiologists, and will thus tend to improve methods of observation. Moreover, it affords a ready method for determining the sensibility or fineness of perception and allows of a comparison of one observer's results with the results of others; for, knowing the number of errors in a certain number of experiments, and accepting the conclusions of this paper, the calculated ratio to the total excitation of that variation of excitation, in judging which we should err one time out of four, measures the sensibility" (Peirce and Jastrow 1885, 124–125). The ability to compare the sensibility of one observer with that of another (a continuity of sensation among inquirers) was of special significance to astronomical transit-observations which provide the practical backdrop to both of Peirce's experimental investigations of continuity (he was working at the time on questions of weight and measure for the United States Coast and Geodetic Survey, with particular emphasis on gravity research).

28. More clearly than most of his contemporaries, Peirce appreciated that Darwin's theory showed chance variation to be sufficient to account for evolution by natural selection. Yet he preferred a somewhat Lamarckian version of evolutionism that added to the workings of chance (tychism) a continuity of feeling (synechism) that tends to an idea as to a flower and "warms it into life" (evolutionary love or agape); see Peirce 1893, 354.

29. A separate investigation is required to show how Peirce's three sentiments appear in Kant's philosophy. For a beginning see Nordmann 1995.

XII

POINCARÉ'S CIRCULARITY ARGUMENTS FOR MATHEMATICAL INTUITION

Janet Folina

Jules Henri Poincaré believed that mathematics has an irreducible essence, access to which requires a priori intuition. It is in this sense that his philosophy is Kantian. Indeed, Poincaré explicitly defended a version of Kant's philosophy of mathematics against the challenges of logicism and Hilbert's program, arguing that mathematics has significant content that cannot be reduced to logic, or even to a more substantial formal system, with no appeal to intuition. His general picture of the nature of mathematics thus mirrors that of Kant: Significant mathematics is synthetic a priori because it ultimately involves a priori intuition.

Poincaré's main arguments for intuition take the form of circularity objections against the reconstructions of elementary arithmetic proposed by logicism and formalism. These arguments have been criticized by Warren Goldfarb (1988) in his paper "Poincaré against the Logicists," on the grounds that they are psychologistic. In this paper I reassess Poincaré's circularity arguments in the light of Goldfarb's central critiques, arguing that Poincaré's strongest objections are not actually psychologistic and that "intuition" is not merely a psychological term for him. A final purpose is to urge that Poincaré's position is in fact a coherent and updated version of Kantianism about arithmetic.

PRELIMINARY REMARKS ABOUT INTUITION

Arguments for mathematical intuition are epistemological in nature. Generally antireductionistic, they support the view that mathematical knowledge is genuinely mathematical. It cannot be built out of some other, more basic, kind of knowledge such as knowledge of logic. The circularity arguments in question make two fundamental claims. First, in order to derive some nontrivial portion of mathematical theory in a formal, or axiomatic, system, some mathematics must always be presupposed. Logic alone is not enough. Second, what is presupposed is not entirely arbitrary. At least some of what is presupposed is intuitive, which means in this case that it is built into the nature of the human (or the finite thinking) mind.

In particular, it is inherent in the structure of how finite thinkers experience the world.

The first claim is in one sense uncontroversial. That some mathematics must be presupposed in the axioms in order to have theorems of mathematical significance may simply be shorthand for saying that it is impossible to prove anything within a formal system that is stronger than what is put into the axioms in the first place. Now, recent work on the program implicit in Frege's *Foundations of Arithmetic* has shown that the Peano axioms are derivable from second-order logic together with Hume's principle, which identifies sameness of number in terms of one–one correspondence.[1] However this is not a case of getting arithmetic from pure logic, for Hume's principle is not a logical truth. Though it appears harmless and even sensible, it is in effect an axiom of infinity, for it entails the existence of a Dedekind infinite set. If a necessary criterion of logical truths is that they be true in all nonempty domains, Hume's principle is thus extralogical.

Furthermore, since 1931 it has been known that analogous limitations apply to metatheoretic justifications of formal systems. It is impossible to provide purely finitistic consistency proofs for infinitary formal mathematical systems. Though these limitative results eliminated major portions of Hilbert's program, on their own they provide no positive support for the thesis that intuition is a source of mathematical knowledge.

In order to justify the essential role of mathematical intuition, one must also justify the second claim that at least some of what is presupposed for mathematics is "intuitive." One problem with doing so is that the term is used in too many different ways, from the more specific, as is found in Kant, to the more vague, for example to mean a psychological feeling that something is true (as in "women's intuition").[2]

Problematically, Poincaré himself uses "intuition" in several different ways, including, as Goldfarb points out, something resembling "women's intuition." We might describe this vague use as a "feeling" for a mathematical domain, or "mathematical sagacity."[3] But this sense of mathematical intuition is perfectly explainable as a deep understanding which comes from experience and facility in a discipline. No special mental faculty needs to be postulated here. However, we should not dismiss all of Poincaré's references to intuition just because we can explain away this one category. Some of his uses of "intuition" are much closer to Kant's; and these play a more serious role in his defense of Kant's view that mathematics requires a priori intuition as well as conceptual analysis. So our second general claim—that intuition provides a source of nontrivial mathematical knowledge—links us to Kant's famous claim that intuition is necessary for mathematical knowledge.

Admittedly, Poincaré's more Kantian use of "intuition" is still rather different from Kant's. Intuition for Kant was explicitly tied to sensibility; and Kant clearly distinguished sensibility from logic. The use of intuition we are here investigating does not make such clear distinctions. For example, Poincaré uses the term "direct intuition" to refer to the "power of the mind" that yields an understanding of the natural number sequence, and thereby of the principle of mathematical induction that merely "affirms" this power. Induction "is only the affirmation of the power of the mind which knows itself capable of conceiving the indefinite repetition of the same act when once this act is possible. The mind has a direct intuition of this power" (1894a, VI, 979–980).

But he also uses it (in the same paper) to describe our acceptance of individual axioms in mathematics; and elsewhere he refers to the intuition of pure number as "that of pure logical forms" (1894a, I, 974; 1900, VI, 1020). Even the sense of "intuition," which provides the strongest link to Kant's philosophy of mathematics, is thus much more abstract than Kantian intuition (at least according to most interpretations of Kant). Its link to logic especially blurs the boundaries that Kant imposed and the logicists presupposed.

Nevertheless the concept of intuition in question is a descendant of Kant's—so much so that Poincaré took himself in these circularity arguments to be defending Kant's conception of the epistemology of mathematics. Regardless of how different Poincaré's concept of intuition is from Kant's, it shares with Kant's the property of being a nonlogical and nonconceptual source of knowledge. Poincaré's general picture of the nature of mathematical truth, as well as epistemology, is thus Kantian because of three components: the view that there are mental structures, or faculties, necessary for providing a framework for experience, the view that these mental faculties are nonlogical and nonconceptual, and the view that at least some of the essential (synthetic) content of mathematics is derived from these a priori mental faculties.

THE CIRCULARITY ARGUMENTS

The arguments in question proceed roughly as follows. Attempts to eliminate all appeals to mathematical intuition by logical, or formal, maneuvers are circular. For the formal system used in any such endeavor will necessarily presuppose the very mathematical intuition it was designed to eliminate. The arguments focus on arithmetic, claiming that a mathematical intuition of indefinite iteration, or repetition, is presupposed in any formal system that can symbolically "construct" the natural numbers. The nature of the claimed presupposition varies from explicitly presupposing a principle of induction to merely the necessity of interpreting certain procedures as indefinitely iterable. (Though related, these arguments should be distinguished from worries about impredicativity, or other vicious-circle *principle* arguments, which I shall not address in this essay.)[4]

Poincaré's circularity complaint occurs as early as 1894, where he claims that "reasoning by recurrence" is presupposed in any attempt to prove it. "The judgement on which reasoning by recurrence rests can be put under other forms. . . . But we shall always be arrested, we shall always arrive at an undemonstrable axiom which will be in reality only the proposition to be proved translated into another language" (1894a, VI, 979). Though the antireductionist view is very clear, there is no real argument to support it. Poincaré's target is also uncertain. He only mentions Frege later, and then in a quoted passage.[5] So it seems unlikely that he ever actually read Frege's work. He discusses Dedekind's work, but again only later.

By 1905, however, Poincaré explicitly targets logicism, citing Couturat, who summarizes and explains it in French, as well as Russell, Peano, Hilbert, and others. He also has an argument. In fact he has at least four distinct circularity arguments in this paper, but two of them seem weak. In the first he appears to assimilate logicism to formalism. "What Hilbert did for geometry, others have tried to do for arithmetic and analysis" (1905b, II, 1024). He calls these others "logicians," or sometimes "logicistians," by which he apparently means anyone who rejects intuition in mathematics.

Now, in one sense comparing logicism with Hilbert's program in geometry is plausible. Hilbert's seminal work, *The Foundations of Geometry,* was the first sustained treatment of elementary geometry to remove all traces of spatial intuition. There is no need to appeal to properties or intuitions of space either to grasp the fundamental notions of "point," "line," and "plane," or to grasp the proofs. The similarity between what Hilbert had done for geometry and what the logicists were trying to do for arithmetic was, therefore, to eliminate intuition from proofs in two basic areas of mathematics.

In another sense, however, the comparison is inapt, for Poincaré takes it too far. He here regards these "logicistians" as treating the Peano axioms as implicitly defining the concept of number. "The principle of complete induction, they say, is not an assumption properly so called or a synthetic judgement a priori; it is simply the definition of whole number. It is therefore a simple convention" (1905b, IV, 1025). His criticism is that such "conventions," or "implicit definitions," form a set of postulates that requires a consistency proof. In this case the proof requires induction, since the domain of numbers is infinite. But then the process is circular, for induction is one of the postulates being proved consistent by presupposing that induction is true.

This process certainly would be circular but—pace Couturat, who represented logicism at the time to Poincaré—this is not what the main proponents of logicism were up to. In the hands of Frege, for example, the aim of logicism was of course to *derive* the Peano axioms, including induction, from definitions in purely logical vocabulary of the basic nonlogical primitives "number," "zero," and "successor." The real question is about the apparatus used to do so, a question Poincaré raises in circularity objections three and four.

A second circularity argument in this paper targets the symbolic nature of the new logic. In one such criticism Poincaré makes fun of the symbolic definitions of "zero" and "one." Couturat, he says, gives a definition of "zero"

> which means: zero is the number of things satisfying a condition never satisfied. But as never means *in no case* I do not see that the progress is great. . . . One, he says in substance, is the number of elements in a class in which any two elements are identical. It is more satisfactory, I have said, in this sense that to define 1, he does not use the word *one*; in compensation, he uses the word *two*. But I fear, if asked what is two, M. Couturat would have to use the word *one*. (1905b, VII, 1029)

More probably Couturat would have to use "one," "two," and maybe "three," given the standard constructions.[6] Poincaré's criticism is that the symbolic representation of the definitions merely obscures their circular nature. Though this complaint might be interpreted as purely psychological, he also puts the same objection in terms of understanding. We require a prior grasp of the natural numbers in order to understand the logical symbolism used to define them (1905b, XII, 1034). Underpinning this objection is the view that the symbolism has an inherently infinite-mathematical character, for which he argues in the next two objections.

The third and fourth circularity arguments, intertwined in Poincaré's paper, both seem stronger than the first two arguments, and less susceptible to the psychologism counterargument than the second. His third argument is that the new "logic" required for logicism is really mathematics in disguise, since it allows an unlimited number of combinations of symbols. "We see how much richer the new logic is than the classic logic; the symbols are multiplied and allow of varied combinations *which are no longer limited in number*. Has one the right to give this extension to the meaning of the word *logic?*" (1905b, XI, 1033).

The argument is not clearly stated by Poincaré, though the standard interpretation is as follows. The definition of "well-formed formula" in formal symbolic logic is recursive. But recursive definition is a tool that Poincaré identifies as peculiarly mathematical. For example, of the definition of addition, Poincaré says "it is of a particular nature which already distinguishes it from the purely logical definition; the equality (1) $[x+a = (x+(a-1)+1)]$ contains an infinity of distinct definitions, each having a meaning only when one knows the preceding" (1894a, III, 976). Recursive definition enables reasoning by recurrence, or inductive inference, which, according to Poincaré, gives mathematics its special power—the power to make inferences about infinite domains.[7] And this is what distinguishes a mathematical from a "purely logical" definition.

Though number theory may not be explicitly presupposed in symbolic logic, this third argument claims that it is implicitly presupposed. Since Poincaré regarded recursive definition and reasoning as peculiarly mathematical, its role in defining "well-formed formula" of modern logic may have indicated to him that mathematics is being presupposed in the foundations of the formal systems of logic used to reconstruct it.[8]

A slightly different interpretation of this objection is that it targets the logical rules themselves. Poincaré's reference to the infinite number of possible combinations of *symbols* might not be about the language of logic but about the rules of inference. Any inference rule is correctly applied to any (well-formed) formulas with the right main connectives. So the rules are indefinitely applicable, and to formulas of indefinite complexity. The combinatorial nature of logic is on this reading objectionable to Poincaré because it imports mathematical processes into logic. Why he might have thought so involves the following suggestion. The revolution in logic from Aristotelian to symbolic involves a jump from the finite to the infinite. For Poincaré, that the rules of symbolic logic determine an infinite structure means that our insight into that structure, that is, the rules which determine it, requires mathematical intuition. In any case, Poincaré objects that one cannot genuinely "derive" a part of mathematics from logic if mathematics is presupposed in the processes of logic. This version of argument three is in my view Poincaré's most penetrating criticism, and I will return to it below.

The fourth circularity objection, which follows immediately after the above,[9] gets at what is still an important problem in discerning the ultimate nature of mathematical knowledge. Poincaré begins with the following observation. Since symbolic logic is a quite different—in fact, quite mathematical—way to codify correct inference from that of traditional logic, it raises the philosophical question of what makes it logic. To Poincaré, the revolution in logic involves placing new boundaries around logic; in so doing some methods formerly considered mathematical are now included in "logic."[10] But one cannot draw new boundaries arbitrarily. One must have some justification for counting the analysis of inference as logic. This is especially acute for Poincaré if the new principles can succeed at deriving the Peano axioms. In other words, if logicism succeeds, then the logic by which it does so is especially suspicious.[11]

Clearly Poincaré is correct that merely shifting the boundary between logic and mathematics is not a victory for logicism. Certainly Russell, Poincaré's main target in this criticism, would agree. But of course it is not so simple a story. The boundary between mathematics and logic was not simply moved further into what was formerly considered mathematics. A new way to codify validity was invented. The question is whether this invention presupposes any mathematics or at least a mathematical intuition. At the heart of this objection, in any case, is a demand for a philosophical justification of the new logic.

Now, Poincaré proposes one way for the logicist to dodge the philosophical question of what makes logic logic. He could consider the logical principles as disguised definitions of the new logical constants. Poincaré's objection to this potential dodge results in the fourth circularity argument, which parallels the first one, as follows. If the logical principles are regarded as disguised definitions, then they must be proved consistent. That is, it must be shown that iterated application of them can never lead from truth to falsity. But they cannot be proved consistent without the principle of induction, for the number of possible valid inferences is infinite. Such a proof would, however, be circular for the logicist who aims to derive induction from the principles (1905b, XI, 1033–1034).

So there appear to be at least four distinct circularity arguments in Poincaré 1905b. The first argues that the logicist/formalist defines "number" by appeal to the principle of induction, which needs to be presumed true to show that the definition is consistent. The second argues that the symbolism of the logicist definitions of the numbers obscures their underlying circular nature. The third argues that even expressing a formal logical system adequate for deriving arithmetic presupposes arithmetic in the form of recursive definitions of "well-formed formula." It may also argue that the new logic is inherently mathematical, since it analyzes valid inferences using combinatorial methods and function theory. The fourth argues that if the new principles are disguised definitions of the new logical constants, then a consistency proof is required; but since the proof will need to presuppose that induction is true, any derivation of induction from the logical principles is circular.

GOLDFARB ON POINCARÉ'S CIRCULARITY ARGUMENTS

Goldfarb's main criticism of both Poincaré's concept of intuition and the circularity arguments for it is that they are psychologistic. On the concept of intuition Goldfarb claims that "[f]or Poincaré, to assert that a mathematical truth is given to us by intuition amounts to nothing more than that we recognize its truth and do not need, or do not feel a need, to argue for it. Intuition, in this sense, is a psychological term; it might as well be called 'immediate conviction'" (1988, 63). Though Poincaré does at times use the term in this way, this is inadequate, and perhaps unfair, as a general characterization of "intuition" for Poincaré. In what follows I will address both Goldfarb's critique of Poincaré's concept of intuition and his objection that Poincaré's arguments for it are likewise psychologistic. The latter first.[12]

Goldfarb considers the first three circularity objections outlined above. He argues that the first objection is a complaint against Hilbert that is only misinterpreted as directed also against logicism; and the other two are ultimately psychologistic. He concludes that all of Poincaré's circularity arguments against logicism can therefore be dismissed. In what follows I will focus on Goldfarb's response to Poincaré's third circularity objection. This is Poincaré's strongest and most interesting argument, and it draws Goldfarb's most clever counterargument.[13]

Goldfarb's clever counterargument   Goldfarb construes Poincaré's third circularity objection along the lines of the standard (first) interpretation: The definition of "well-formed formula" of symbolic logic is recursive, but recursion, and through it number theory, is presupposed in the foundations of logic (1988, 68).

Goldfarb admits that Poincaré's third objection seems to be better than the second, which appears to focus on what we think when we read logical symbols defining the numbers. What we think when we read logical symbols seems clearly a psychological issue; whereas the recursive nature of the formal definition of "well-formed formula" seems to be about logic. Goldfarb argues, however, that this appearance fades when considered more carefully. For Poincaré's argument really rests on a confusion between logic and the way we manifest logical judgments. Logic for the logicist is simply the universal framework for rational discourse; as such it need not be represented metatheoretically as a formal system. That we need to define "well-formed formula" recursively in the metatheory does not, he claims, bear on the nature of the logical inferences themselves. So it does not show that arithmetic, or its processes, are presupposed in logic.

The contrast here highlighted is between justifications and the way we express and check them. Goldfarb grants that arithmetic may be presupposed in the way we formally (metatheoretically) codify the language of logic, and in order to provide metatheoretic justifications of the rules of inference. It may even be needed to check formal logic proofs. But the need to keep track of steps and parentheses by counting does not mean that arithmetic is presupposed *in* logic. Goldfarb concludes that owing to the equivocation between logic and the way we express and check formal logical proofs, this apparently better version of Poincaré's circularity argument reduces to the former, naive version (1988, 70). That is, since it appeals ultimately to how we express and check justifications, rather than to the nature of the justifications themselves, it is also fundamentally psychologistic.

Analysis of Goldfarb's argument   Though distinguishing logic from the way we express and check logical justifications is a clever "divide-and-conquer" strategy, I wish to argue that this counterargument fails. Recursive, or generally iterative, procedures in logic are not merely psychological necessities; nor are they required in logic only when we get to the metatheoretic level—in proofs about first-order logic. Recursive procedures determine whether or not a logical step is valid. Hence they determine the logical *facts*.

For Frege, a proof is rigorous if it is gap-free—if every step is an instance of a valid rule or axiom schema. Whether or not a step is valid depends on whether or not the rule of inference (or axiom) is applied to appropriate formulas; and whether a formula is appropriate for a rule of inference depends, first, on its being a formula (well-formed), and, second, on what its main connective is. Therefore "(P & Q) $\rightarrow$ R, P & Q, so R" is, and "P & (Q $\rightarrow$ R), P & Q, so R" is not, a valid instance of modus ponens, which we might express thus: A $\rightarrow$ B, A, so B. Well-formed formulas are not strings of symbols in any old order; both the meaningfulness (that it is a well-formed formula) and the meaning of a formula depend on the syntax—on the way its parts are put together. Well-formed formulas are thus articulated; consequently, so are proofs. It is the recursive nature of this articulation that gives logic both its infinitude and, in Poincaré's view its mathematical essence.

Goldfarb claims that this recursivity is a mere fact about our access to, and understanding of, logical structures rather than a fact about logic itself. Actually we rarely have to go through a recursive procedure in order to determine whether or not a string is a well-formed formula and what its main connective is. We can often simply "see" both things: that a string is

a well-formed formula and that it matches a certain pattern. So, for sufficiently simple formulas, the psychology behind understanding what pattern
they instantiate might not be so different from the psychology behind the
pattern matching presupposed by Aristotelian logic. Psychologically, then,
symbolic logic is often on a par with Aristotelian logic. But what makes a
formula available for a particular rule of inference depends on whether or
not the recursive procedure for building formulas can result in the string in
question. In other words, to fully characterize whether a string of symbols is
a well-formed formula, no less a particular well-formed formula, and no less
a proof, depends on a recursive procedure. But this is not a fact about us or
our psychology. It is a fact about what makes a string well formed, owing to
the articulated and combinatorial nature of symbolic logic and its language.

Hence, on any reasonable conception of psychologism[14]—one that
doesn't include all philosophical questions about the nature of logic, for
example—Poincaré's arguments are not psychologistic. Furthermore, from
the contemporary point of view they seem correct. His circularity objections 2 through 4 actually build on one another as follows. The new symbolic logic forms an infinite structure that can only be investigated by
mathematical (inductive) means (argument 4). This reflects the infinite combinatorial nature of both the rules for forming a well-formed formula and
(owing to this) the inference rules themselves (argument 3, both versions).
We may be able to apply the rules in simple cases by way of pattern matching. But what makes a rule correctly applied in any case depends on recursive specifications; and any full understanding of the rules such as would
be required for logicism requires at least an implicit grasp of the recursive
procedures governing them (argument 2). For these reasons I think that
Goldfarb's characterization of Poincaré's main circularity objections as psychologistic is either incorrect or based on a conception of psychologism far
too broad to be defensible.

Now, it might be objected that this response presupposes a formal,
metatheoretic account of logic—the very one that Goldfarb denies is appropriate to the universalist picture, the very one that, on the universalist
account, equivocates between logic and our psychological access to it. However, this is the only account of symbolic logic that we have. The alternative
is to make logic opaque. This results in simply refusing to engage in any of
Poincaré's challenges to the logicists to come up with a justification of logic.
There is something deeply unsatisfying about a view that dismisses all challenges to it as "psychological."

Now, this might indeed be how Frege would respond. After all, his
famous dichotomy between logic and psychology (aligned with another

dichotomy between objective and subjective) would render all questions about the general nature of logic unanswerable in a noncircular way by logic; and thus mere psychological questions. This marries well with the universalist picture Goldfarb articulates. Questions about logic cannot be objectively meaningful, for we cannot step outside logic in order to answer them.

However, to adhere strictly to this point of view would be damaging to both Frege and Russell. For one thing, it makes the very point of their programs unable to be justified in a way that would be approved by the program itself. The point of symbolic logic, as I understand it, was to make proofs rigorous, or gap-free. But what is the problem with gaps, so long as the conclusion is a valid consequence of the premises? One problem is that of error. Gaps induce error because human psychology is frail. When the jump from one step to the next is too big, we cannot reliably tell whether or not it is valid. But this is a psychological, not a logical, issue. Hence the point of formalizing logic cannot be justified in any objective way for Frege, if he insists that psychological issues are irrelevant to objective justification.

The problem can also be expressed epistemically: For justification to be objective, it ought to be intersubjective. The importance of formalism in this account is that it eliminates all reliance on the meanings of the axioms—on how they are understood or interpreted—in proofs; and this makes proofs "radically intersubjective."[15] This epistemic concern is an important one, but it is not a purely logical concern. Of course Frege would not count it as psychological, but this leads to my next point.

Second, what we have in any case is a false dichotomy. In addition to rendering all questions about the general nature of logic psychological, the logic/psychology division casts a rather bad light on Frege's philosophical arguments—for example, those in *Foundations*. Just like Frege, Poincaré was raising a legitimate question. It may not have been strictly logical, but it was also not merely psychological. He asked a philosophical, justificatory question about the nature of logic. Poincaré's proposal to answer this question with metatheoretic proofs has become the mainstream answer, presupposing the metatheoretic view of logic that Goldfarb's "universalist" logician is trying to avoid. Russell's paradox in particular pushes us in this direction, and legitimizes Poincaré's nonuniversalist view that logic does need to be justified.[16] In the end, of course, Poincaré was right: There is a certain amount of unavoidable circularity in the metatheoretic proofs; and they do presuppose induction as well as recursive definitions. None of these points of view seems psychologistic in the current context. Hence, to the objection that the above reconstruction of Poincaré's circularity arguments equivocates, or blurs the dichotomy, between logic and the psychology of logic, one could respond that neither these nor Poincaré's remarks should be criticized on the basis of a dichotomy that is both nonexhaustive and problematic even to its proponents.[17]

GOLDFARB ON POINCARÉ'S CONCEPT OF INTUITION

Whether or not Poincaré's arguments for intuition are psychologistic, there is still the charge that the concept of intuition being defended is psychological. Recall that Goldfarb reads Poincaré's concept of intuition as something like obviousness, or immediate conviction. Goldfarb supports this psychological rendering of Poincaréan intuition by arguing that it is not genuinely Kantian. The argument is: Because his conception of intuition lacks the surrounding Kantian apparatus, Poincaré is mistaken in thinking that he is defending a Kantian epistemology of mathematics; and his concept of intuition has no content. "[I]n Poincaré's hands the notion of intuition has little in common with the Kantian one. The surrounding Kantian structure is completely lacking; there is no mention, for instance, of sensibility or the categories" (1988, 63). Goldfarb concludes that owing to the absence of such an epistemological framework, Poincaré's remarks about intuition are, in general, claims about the psychology of mathematics, not its epistemology (1988, 63).[18]

It is not immediately obvious why this conclusion should be drawn. It would seem that even if Poincaré's epistemological picture does not bear all the weight of the Kantian baggage, this does not mean that his concept of intuition is psychological. That is, even if Kant's full apparatus is absent, it seems possible that Poincaré's conception of intuition justifies mathematics in a way similar to Kant's.

Goldfarb argues against this more charitable interpretation as follows. The conception of intuition to which Poincaré appeals in his circularity arguments against logicism is an intuition that underlies the "logic" needed to reconstruct arithmetic. Now, this might seem simply correct, given the existential import of Hume's principle—or indeed any other principles that can play the role of yielding the existence of the natural numbers. However, Goldfarb points out that circularity argument 3 implies that even truth-functional logic requires intuition. For "well-formed formula" of even truth-functional logic is defined recursively. But if anything has no special content, it is truth-functional logic. For Frege and Russell, no part of logic—qua logic—could "stem from any faculty with a specialized purview" (Goldfarb 1988, 63). If mathematical intuition is taken as underlying even elementary logic, it would have no specialized purview either. Thus, Poincaré's conception of intuition is "robbed . . . of all content" (1988, 63). Since such a conception of intuition is very different from Kant's, we should be wary of granting to Poincaré's conception what we might grant to Kant's—namely, that it has some justificatory substance and is not merely psychological.

Given the infinitary nature of even truth-functional logic, I think we have to be careful here. Though truth-functional logic does not have any particular content, or specialized purview, this does not automatically mean that it is intuition-free, as argued above. The move from Aristotelian to contemporary symbolic logic involves a jump from the finite to the infinite. Even truth-functional logic—though free of any particular content—determines an infinite structure. Identifying intuition with "specialized purview" may thus be misleading. Owing to its infinitary nature Kant might even regard symbolic logic, in contrast with Aristotelian logic, as dependent on intuition, along the lines of Poincaré's argument above.

Still, Goldfarb does have a point. There is no appeal to space, time, or any ordinary notion of sensibility in Poincaré's arguments for intuition. Mathematical intuition for Poincaré is a "power of the mind." As is well known, Poincaré rejects the Kantian theory of the synthetic a priori status of Euclidean geometry in favor of conventionalism about geometry. Poincaré explicitly denies that we have a priori intuitions of space and time. For these reasons mathematical intuition does not have as obvious a role in constituting the world of experience for Poincaré as spatio-temporality does for Kant. But it may still have some constitutive role. To clarify what is at stake here, let us pause to consider what is behind the worry about psychology and psychologism.

Digression on psychology and psychologism   Anti-psychologism in this context is associated with Frege, who wished to distinguish justification conditions for a proposition from background physical and psychological conditions for knowing it, as well as from psychological causes of the belief in the proposition. This distinction is crucial,[19] for example, for the possibility of a priori knowledge. Just because some experience (and blood in our brains) is required to have any knowledge does not mean that all knowledge is therefore empirical. That is, every item of knowledge will presuppose background empirical and psychological conditions. So the presence of such conditions is irrelevant to the classification of knowledge into a priori and a posteriori. This distinction is certainly threatened if we disregard the contrast between a cause and a reason, or justification.

The charge of psychologism seems to lie in the same bag of philosophical "tricks" as the charge of the naturalistic fallacy. Keeping these two company is the fact-value distinction. By calling them "tricks" I do not wish to disparage them. These highlight important distinctions that I have no desire to collapse; but I do think they should be used sparingly and cautiously. Underlying all three is a distinction between a fact, or descriptive item, and a norm. For example, answering why we do label some things "good" might involve a psychological, historical, and/or sociological story; whereas answering why we *should,* or are correct to, label those "goods" arguably requires appeal at some level to something intrinsically valuable.[20] Analogously, answering why we do believe that *p* might involve a psychocausal story (depending on how one interprets the question). Whereas answering why we *ought,* or are *warranted,* or are *correct* to believe that *p* is the question to which a justification provides an answer. Justification, in short, is normative and psychology is not.[21]

The charge of psychologism thus consists in the following. We seek a norm when we seek justification. The problem with the psychologistic answer is that it delivers a causal story. It treats the request for justification as equivalent to a question about what causes a person to believe something. But in the psychologistic cases the causal story yields a mere fact, or description, that fails to be normative of the item for which we seek justification.

Analysis of Goldfarb's interpretation of Poincaréan "intuition"   When framed in terms of normativity, Poincaré's conception of intuition may on the surface seem psychological. For it may appear to be about why we do believe the principles of arithmetic instead of why we are correct to believe them. This is where the absence of surrounding Kantian structure, as Goldfarb puts it, may seem especially problematic. In Kant's philosophical system the constitutive role of intuition yields a source of normativity for our mathematical beliefs. Arithmetic is true of the world of experience because it is a priori imposed on it—by the a priori forms of experience. Our inclination to believe the principles of arithmetic thus has the same source as their truth. This is why we also *ought* to believe them—because they are true. For Kant, that is, the semipsychological, or causal, account merges with the account of arithmetic truth and thus of justification.

Though Poincaré's epistemology lacks much of Kant's structure, the real question relevant to the charges of psychology and psychologism is whether or not Poincaréan "intuition" is merely a causal, nonnormative notion. Put this baldly, it is clear that Poincaré's concept of intuition is not merely psychological, for it has a constitutive, normative role in Poincaré's philosophy of mathematics just as intuition does for Kant. The key is that Poincaré is a constructivist about mathematical objects and domains, and intuition plays a central role in this account. So it plays a central role in his account of mathematical truth.

Now, Poincaré is no strict constructivist, nor is he a strict Kantian, but he is far from a realist. Mathematical truth has to do with the way finite thinkers are able to construct mathematical domains, proofs, and so on. The role of intuition in this account of truth is that it allows the "construction" of infinite domains. We cannot actually construct an infinite number of objects one by one, but given the right rule we can see that there is no end to the number of objects so constructible. Owing to Poincaré's constructivism, intuition is justificatory for mathematics for it determines what objects, and thus what truths, there are. Intuition is thus normative, or justificatory, for Poincaré in roughly the same way it is for Kant. It does not provide a mere causal background for our mathematical beliefs; it also gives us grounds for holding such beliefs because it determines the boundaries of mathematical truth.

This highlights an underlying interpretational difficulty. Whether or not Poincaré's concept of intuition and his arguments for it seem psychologistic depends in part on broader background views about truth and existence in mathematics. If knowledge is something like justified true belief, then what counts as an appropriate justification depends on what makes the belief true. "I see something red" may be taken as a first-person report of one's psychological state. But it may also be taken as a justification for "There is something red" if we make the background assumption that seeing is a reliable method for getting at facts about medium-sized physical objects. So one proposition can be seen as psychology or (also) justificatory, depending on other background assumptions.

This results in a divide in the philosophy of mathematics. For what counts as the border between epistemology and psychology depends in part on one's metaphysics. From a realist perspective, for example, the necessity of defining an infinite domain iteratively is a mere contingent matter: a mere psychological fact reflecting our limited access to the independently existing mathematical domains. For the antirealist, however, it is not merely a psychological necessity that we define infinite mathematical domains iteratively, for these domains have no independent existence. Truth conditions and definition conditions are linked for the antirealist, or constructivist. What might be seen as a mere psychological condition to a realist might therefore be a justification condition for an antirealist, owing to the differing conceptions of mathematical truth.

Poincaré's account of intuition and the surrounding arguments need to be viewed in this light. When so viewed, it is easier to resist Goldfarb's psychological interpretation of Poincaré's conception of the foundations of mathematics. Though he identifies intuition in terms of our ability to conceive (the indefinite repetition of an act), and though much of the surrounding Kantian structure is absent, Poincaré is not doing armchair psychology. He is trying to get at something more fundamental than how we, in fact, do conceive—something closer to what is in principle possible for finite thinkers to conceive. His view might be reconstructed as follows. Any thinking that is sufficiently complex and abstract to do mathematics must recognize certain structures as intelligible. The ability to do so cannot come from experience, since the structures in question are infinite. The said capacity must therefore be a priori, or, as Poincaré put it, intuitive.[22]

CONCLUSION

I have argued that neither Poincaré's conception of intuition nor his circularity arguments for it can be dismissed as mere psychology. There are two main problems with the psychological interpretation. First, owing to his constructivism, Poincaré's conception of intuition is normative and not (merely) psychological. Like Kant's, it plays a justificatory role, not a mere causal role, in his account of mathematical knowledge. The normativity of intuition means that it is not simply a psychological notion.

Second, Poincaré's arguments for intuition are not psychologistic. They commit no fallacy whereby causal conditions are mistaken for justification conditions. When interpreted in their best possible light, his circularity arguments can be understood as pointing out the irreducibly recursive nature of validity in symbolic logic. He argues that even if arithmetic could be reconstructed from "logical" principles alone, this does not reveal that arithmetic knowledge reduces to logical knowledge—if by "logical knowledge" we mean that which does not require mathematical intuition. Irrespective of reservations one might have about Hume's principle or any other particular part of the apparatus of logicism, Poincaré believed that the epistemology of symbolic logic is ultimately mathematical in nature. The new logic can only be understood if we already possess a cognitive apparatus that is itself of a fundamentally combinatorial nature. Once articulated, the combinatorial principles implicit in this apparatus can be seen as equivalent to a nontrivial fragment of arithmetic—perhaps primitive recursive arithmetic. Mathematical intuition just is this cognitive apparatus that must exist in order to understand symbolic logic. In this sense, arithmetic is "prior" to logic.

Finally, these circularity arguments show that the intuition being defended is one that underlies abstract mathematical structure. The intuition here highlighted is that of indefinite iteration, by which we construct the natural numbers but also can construct other ω-sequences. Intuition thus underlies symbolic construction and manipulation in general. It is the very abstract nature of this intuition that enables it to play the role in both mathematics and logic that Poincaré attributes to it.

NOTES

I would like to thank Michael Friedman, David MacCallum, Alan Richardson, and David Stump for conversation and helpful comments on parts of this essay. Thanks also go to Peter Clark for feedback on an earlier written draft. A version of this essay was presented at Carleton College, and that audience also deserves thanks for letting me try out some of these thoughts.

1. More explicitly, Hume's principle says that the number of Fs equals the number of Gs if and only if the Fs and Gs can be one–one corresponded.

2. Perhaps the person who has done the most serious and progressive work on intuition is Charles Parsons (e.g., 1980, 1994, and some of the papers in 1983a).

3. See Goldfarb 1988, 64, and Poincaré 1900, section V, for such uses by Poincaré himself. Another use of "intuition" Poincaré discusses is the appeal to images and physical analogy in mathematics. Though fruitful, he remarks that this intuition "cannot give us certainty" (1900, III, 1016). Since he believed that the intuition we are here investigating—that of "pure number"—*does* give us certainty, it is clear that we cannot treat his references to intuition as if they are of one kind.

4. The later circularity, or impredicativity, arguments are particularly relevant to the logicist's derivation of induction via the definition of the ancestral. (See Parsons 1983b and Poincaré 1906b.) They also extend the circularity considerations and arguments for intuition to analysis where the style of argument is the same: The mathematical continuum cannot be constructed via formal mathematical procedures (such as Dedekind cuts) without presupposing the prior existence of the domain (points, real numbers) being defined. This is because impredicative definitions cannot be regarded as *sui generis* constructions. If they are, then the circularity of the definitions is vicious. So, in order to avoid the vicious circularity of impredicative definitions, one must presuppose that the domain (or its order type) preexists, which for Poincaré means given intuitively (see his 1913, 44.)

5. As far as I know. See Poincaré 1906a, section XIX, where the passage quoted is by Hilbert.

6. A standard rendering of "There is exactly one F" is $\exists x(Fx \;\&\; \forall y(Fy \rightarrow y=x))$. "There are exactly two Fs" would be $\exists x \exists y[((Fx \;\&\; Fy) \;\&\; \sim (x=y)) \;\&\; \forall z(Fz \rightarrow (x=z \;v\; y=z))]$, which we might paraphrase to oblige Poincaré as follows: There are at least two distinct Fs and any third possible F is identical to one of the other two Fs.

7. Induction is "therefore mathematical reasoning par excellence, and we must examine it more closely" (1894a, IV, 978).

8. This point may also be at the bottom of one of Poincaré's criticisms of Russell's theory of types:"In this hierarchy there occur without difficulty propositions of the first, second, . . . order, etc., and in general of the nth order, n being any finite integer. . . . [T]he theory of types is incomprehensible, if we do not suppose the theory of ordinal numbers already established. How will it then be possible to base the theory of ordinal numbers on that of types?" (1913, 52).

9. Owing to its position, it reinforces the second interpretation of the above objection over the first.

10. Of some of the new principles of Russell's logic, he says:"We regard them as intuitive when we meet them more or less explicitly enunciated in mathematical treatises; have they changed character because the meaning of the word logic has been enlarged and we now find them in a book entitled *Treatise on logic? They have not changed nature; they have only changed place*" (1905b, XI, 1033).

11. This parallels a current criticism of neologicism. That Hume's principle succeeds (in second-order logic) in entailing the existence of an infinite number of objects is what shows (for most) that it has extralogical content. Of course, as with the above, a die-hard logicist could respond that this criticism belies that the central question has been begged (assuming a die-hard logicist could believe that all domains *do* contain an infinite number of *logical* objects).

12. I will here be considering the first half of Goldfarb's paper only. In the second half Goldfarb provides an illuminating analysis of Poincaré's vicious-circle-principle arguments.

13. Goldfarb also defends Hilbert against Poincaré's first circularity argument. Of course Poincaré turned out to be right. Hilbert couldn't justify infinitary mathematics using only finitistic metamathematical principles. But Hilbert's *intended* program was not circular, according to Goldfarb. Perceiving a circle here involved Poincaré mixing up finite and transfinite induction. So on this view, even Poincaré's critique of Hilbert—though correct—ultimately involved a mix-up, and was only correct because Poincaré had good hunches.

   Though Goldfarb is right that the aim of justifying transfinite induction using "finite" induction may not have involved an explicit logical circle, Poincaré was right in spirit about Hilbert, too. For even "finite" induction presupposes indefinite iteration, and hence intuition, for Poincaré. Recall that his ultimate aim was to show that intuition cannot be eliminated from mathematics. Of course, there was no real disagreement between Hilbert and Poincaré on this.

14. More on this below.

15. For this argument, see Sieg 1994 (the quoted phrase is from p. 101).

16. This is so even though it appears that the arguments here preceded his contact with Russell's paradox. His criticism here is merely that the logicist's apparatus is unable to solve the Burali-Forti paradox. In his next two papers on the subject (1906a and b) he

begins to articulate his predicativist position, pointing out that the situation is even worse: that logicism creates its own paradoxes (e.g., 1906b, II, 1053–1054). Another case of Poincaré's having incredible insight and good hunches?

17. Perhaps Poincaré was begging the question against the universalist conception of logic, but it seems to me that Goldfarb's interpretation also begs the question against Poincaré. Goldfarb points out that the debate involves different views about what counts as a foundation. Construing Poincaré's view of foundations as psychological, however, presupposes the Fregean conception, with its dichotomy between logic and psychology. It may be true that logical foundationalism presupposes such dichotomies (as Goldfarb 1988 suggests on p. 70). But this might just be evidence against logical foundationalism, rather than evidence for Frege's conception. (Thanks to Dave MacCallum for this *modus tollens*.)

18. This is admittedly a very terse summary of Goldfarb's already concise discussion of Poincaré's uses of "intuition."

19. And, indeed, appropriate. In contrast with the logic/psychology distinction, the distinction between justification and background conditions seems closer to a dichotomy (though one that would in each case depend on context).

20. Interestingly, there is an analogous circularity argument here: Any attempt to construct values from facts smuggles in values elsewhere. As with mathematical intuition, the fact-value argument maintains that values are primitive. They cannot be reduced to, or constructed from, value-free items.

21. At least it is not in general normative of the same items. Of course, psychological evidence can justify the claims of psychology, some first-person utterances, etc.

22. The argument structure is of course reminiscent of Kantian style arguments.

XIII

Poincaré—Between Physics and Philosophy

Jeremy Gray

Philosophy, as Poincaré saw it, is an exercise in critical reflection on the essential features of a situation. It is interested in analyzing the situation, and as interested in analyzing the analysis, and asking why do we think this way about this problem. In that spirit, Poincaré wrote philosophy because around 1900 there were good reasons to reflect critically on certain topics in science, and indeed in mathematics. The way he philosophized, the topics he took up, and the conclusions he came to, closely reflect his deep and original involvement in the overlapping disciplines of mathematics and physics.

Poincaré's professional situation played a crucial role in this development. He began as a mathematician and, after getting off to a relatively slow start, was soon regarded as a genius, with the wayward habit of not explaining himself clearly. As a mathematician, he was a remarkable innovator in fields as diverse as complex function theory and algebraic topology. But for most of his working life, Poincaré was professionally a theoretical physicist, not a mathematician. His most original work, it is sometimes said, lies in the field of celestial mechanics; he was famous in his day for his work on electromagnetic theory, and in that connection his elucidation of the role of the Lorentz group is justly celebrated. He was repeatedly, if unsuccessfully, nominated for the Nobel Prize. He is undoubtedly a central figure in any exploration of how the two disciplines of mathematics and physics relate, because he united them within himself. And it is by reflecting on the relations between mathematics and physics that his philosophy grew.

As a mathematician, Poincaré presents the figure of the conservative revolutionary—one whose truly remarkable original work, amounting to the virtual creation of new fields of mathematics, was seen by himself and others as traditional, while he was perhaps even hostile to the most original aspects of the mathematics of his later days. The controversies in his fifties with Louis Couturat, Ernst Zermelo, and Bertrand Russell, his defence of Euclidean geometry, albeit on novel grounds, his emphasis on intuition all speak to his attitude of keeping abstract modern mathematics at a distance.

It is tempting to see this conservatism as linked to his appreciation of the importance of physics. He worked hard to illuminate theoretical physics at a time when experimental physics was in the ascendant in France. What he knew and cared about was the sort of mathematics that, as he put it, attended to itself and was responsive to the needs of physicists. But in responding to that need, he was careful and clear that the best answers a mathematician may give to a physicist may not be what the physicist wants or expects to hear.

POINCARÉ'S ADDRESS OF 1897

Some of Poincaré's earliest public reflections on these issues can be found in his address to the first International Congress of Mathematicians in Zurich in 1897 (Poincaré 1898a). He began his address by asking, rhetorically, what is the use of mathematics? His answer located mathematics as part of the scientific enterprise, and observed that it was the mathematicians' and the physicists' aesthetic sense that provided the very means by which mathematics and physics advance inseparably together.

It is no longer possible to believe, he said, that pure reflection will yield a law of nature; all attempts in that direction have failed. This recalls the preface to Kant's *Critique of Pure Reason,* but where that sentiment in Kant had a cutting edge intended to mark out his originality, it is only a nod in Poincaré's address to contemporary scientific opinion. It is Poincaré's account of what a law of nature is that is of more interest. Laws of nature are drawn from experiment, he said, but they are expressed in a special language created by the mathematician. A law is general, but the results of this or that experiment are particular. A law is exact, or at least pretends to be, but experiment and experience are approximate. So to formulate a law one has to generalize. However, every particular truth can be generalized in infinitely many ways, which threatens to make meaningful generalization impossible. The only way forward, said Poincaré, is by means of analogy. And among his examples he cited the vast domain of potential theory, one subject of his major paper of 1890, which embraces the theories of the electrostatic potential, magnetism, and heat diffusion.

Poincaré did not imagine that the language of mathematics had been created independently of science. On the contrary, analysis owed much to the study of nature. The study of nature not only raises problems and forces us to choose between them, said Poincaré, it raises problems we would never otherwise think of. Nature, as he put it, is much more imaginative than man (Poincaré 1898a, 87, 89). But it is not just problems that the scientist encounters. Sometimes nature affirms the existence of solutions, sometimes nature suggests the reasoning by which the solutions can be found. For example, physics can make it clear that a series which represents the solution formally must indeed converge, even when a rigorous proof is lacking. Or, that a certain analytic function exists.[1]

Poincaré modestly refrained from referring to his own (more demanding) paper on the partial differential equations of mathematical physics, where he demonstrated the benefits of taking his own advice (Poincaré 1890). There he had observed that physicists usually relied on more or less intuitive proofs of the existence of solutions to these differential equations, basing their case on an appeal to Dirichlet's principle. Such "proofs" were without value for the mathematician, who knew of counterexamples to the Dirichlet principle, and Poincaré himself had written at length on when this principle could be vindicated (Poincaré 1890, 1894b).[2] However, such arguments were of the right sort to satisfy a physicist (and not foolish or misplaced) because, as he put it, they leave the mechanism of the phenomena apparent. In contrast, the more rigorous mathematical arguments for the existence of solutions, such as his own, depended on arguments about the convergence of power series that did not acknowledge the physical insights into the problem. This disparity between the physicist's explanation and the mathematician's would grow in its importance for Poincaré over the years.

Poincaré's own work illustrates that the relationship of mathematics to physics is continually being reshaped as challenges are posed between the disciplines, not that there is some tongue-and-groove harmony waiting to be picked up. It left him with a seeming paradox: Could there be two kinds of rigor, one for scientists and one, somewhat tougher, for mathematicians? Poincaré did not believe that one should be content when a rigorous proof is lacking, and in his 1890 paper he had argued that mathematicians should not duck the challenge of solving such problems. As he pointed out, how could one be sure that something less than a rigorous proof was not actually flawed? The very idea that something inadequate for mathematics was good enough for physics struck him as misguided; the line was impossible to draw. This idea that ultimately mathematics and physics cannot be distinguished was another that grew in Poincaré's mind with the passing years.

POINCARÉ AND ELECTROMAGNETIC THEORY

Much of Poincaré's work in the 1890s can be seen as a profound attempt to work through the proper relation of mathematics to physics.[3] The first edition of his *Electricity and Optics* came out in 1890 and dealt with the ideas of

James Clerk Maxwell, Hermann von Helmholtz, and Heinrich Hertz. It was followed in 1894 by a further account of Hertz's ideas (Poincaré 1894b).[4]

Poincaré undoubtedly came to be regarded as the leading French theoretical physicist, but he matured in a climate where there were several theories of electromagnetism, and he set himself the task of adjudicating between them. He found that the experts could do very well with a wide variety of ontological commitments. Add to that his sense that mathematics could simply be correct, qua mathematics, and a picture emerges of someone unpersuaded of the need for much ontology in physics and open to a plurality of ontologically incompatible theories.

Fundamental to his *Electricity and Optics* was his insistence that many different physical hypotheses can lead to equivalent mathematical conclusions, an essential insight he attributed to Maxwell. Poincaré was willing to introduce physical hypotheses only when the mathematics demanded it. In particular, he argued that there could be no mathematical way of discriminating between one- and two-fluid theories of electricity. The only atoms he would admit were those "purely geometrical points which obey only the laws of dynamics," the atoms of the mathematician, not of the physicist, which had been barred by his earlier methodological remarks. As for how a choice between conflicting explanations could be made, he contented himself with the hope that perhaps one day physicists would take up these questions, which were inaccessible to positive methods and had been abandoned to the metaphysicians.

Poincaré therefore wrote up the work of Maxwell, Helmholtz, and Hertz in a Lagrangian formulation that makes certain conservation laws fundamental. Other laws of physics, which, as Poincaré was to explain, make the subject comprehensible, included Newton's laws, the relativity of motion, the principle of least action (Poincaré 1908d). The result was an apparent contradiction between experimental facts in optics and Newton's third law. Poincaré's preference was markedly for Newton's law, but he could not resolve the contradiction.[5]

Poincaré's attitude to fictional quantities in mathematical physics reflected and influenced his later conventionalism. He ascribed a purely notional, fictional level of existence to anything prior to the quantities described by the laws determined by experiment. It is hard to see this as a physicist's position, although it was plainly intended as such, because much of the physicist's ontological commitment has been handed over to the philosopher. Faced with a choice between incompatible physical theories incapable of further analysis, Poincaré denied that there was any logical way to proceed and implied that any choice was arbitrary.

POINCARÉ'S ADDRESS OF 1908

In 1908, in his address to the International Congress of Mathematicians in Rome, Poincaré returned to the theme of the relationship between mathematics and physics (Poincaré 1908c). He again rejected the idea that mathematics should be done solely at the behest of physicists, now on the grounds that then mathematicians would stand unarmed before scientists—the territorial disputes must have been heating up. But scientists, in their turn, have not waited and do not wait upon the call of utility, but search for laws. Poincaré now sharpened his earlier appeal to analogy. Laws do group together facts that would otherwise appear to be isolated in virtue of some analogy. But doing science and doing mathematics are also analogous: the mathematician structures the millions of mathematical facts around mathematical theorems.

The important laws of physics, and the important theorems of mathematics, in each case order the facts around them; in each domain the relevant guiding analogy permits one to decide which facts are alike, which differ significantly. There are two parallel hierarchies: The most important facts belong to the very general laws of physics, and the most elegant theorems and proofs are those that permit mathematicians to view a whole field at a glance and enable them to say (what a lengthy calculation does not) *why* something is the case. Poincaré then reminded his audience that the economy of thought Ernst Mach had regarded as the role of science depends on the clarity afforded by the discovery of general laws. This emphasis on the intellectual utility of laws and theorems, and their essential role in forming human understanding, is a novel emphasis that has crept in since 1897.

Poincaré noted that laws in physics, and theorems in mathematics, are often a place where new words can be found, often as the name of an organizing principle. In physics, energy (as in the conservation of energy) is one such word. For mathematics, Poincaré proposed his favorite topic, that of the group, and in particular the idea of isomorphic groups, which, as he put it, dispensed with matter and retained only form. Then, in an interesting passage that echoes Hilbert's views in his famous address to the ICM in Paris in 1900 (Hilbert 1901), Poincaré indicated a difference between mathematics and physics. Mathematics, according to Poincaré (and Hilbert), must always move in two opposing directions: toward critical self-reflection and toward the study of nature. The first direction rightly takes the mathematician to such matters as the study of postulates, unusual geometries, and functions with strange properties (all these Poincaré knew well). Critical self-reflection, it seems, leads to novel objects in mathematics. But the study of nature does not lead to novel objects but to novel types of answers. Here Poincaré gave vent to what became a famous aphorism: "There are not solved and unsolved problems, there are only problems which are more or less solved," meaning that first one settles for a qualitative answer, after which comes a quantitative one, perhaps to some level of approximation (Poincaré 1908c, 34).

POINCARÉ AND GROUP THEORY

The mathematical concept of a group was one of the earliest Poincaré had embraced. In 1880, in the then-unpublished *Suppléments* to his essay on linear differential equations (now Poincaré 1997), Poincaré simply said that a geometry is a group of operations formed by the displacements of a body that do not deform it. In this clear statement we can see various influences at work. The motion of rigid bodies is an idea that had been vividly presented by both Helmholtz and Eugenio Beltrami. Jules Houël wrote his book on Euclidean geometry in similar terms (Houël 1867). It is more metrical than Felix Klein's Erlangen Program, and narrower (see Klein 1872).

The sources available to Poincaré included not only this work by Houël (a friend of Jean-Gaston Darboux) on Euclidean geometry, but his translations of Beltrami's *Proof* (*Saggio*) of 1869, and Nikolai Ivanovich Lobachevskii's 1840 *Geometrical Investigations*. It is not certain that Poincaré knew the work of Helmholtz in 1880, but neither is it clear that it would have added anything to what was readily available. With or without Helmholtz's papers, Poincaré could have known from his teachers that geometry is the study of figures in a space that can be moved around rigidly, so that exact superposition is possible and there is a notion of congruence. This idea, which is easier to think through in the metrical than the projective case, works for both Euclidean and non-Euclidean geometry. To anyone aware that thinking group-theoretically is advantageous, it was then natural to observe that the rigid-body motions form a group. This idea could have been had by Camille Jordan, Darboux, Charles Hermite, or Poincaré himself; it could even have been a commonplace among the better French mathematicians of the 1870s. There is no need to attribute it to the influence of Klein.

The group idea was one of the first Poincaré put to a profound use. After he had had his famous realization, boarding the bus at Coutances, that a problem he was dealing with concerning differential equations and complex function theory involved non-Euclidean geometry, he wrote up his findings in an essay for a prize competition of the Parisian Academy of Sciences. There he wrote, "In fact, what is a geometry? It is the study of the group of operations formed by the displacements to which one can subject a body without deforming it" (1997, 11). Once he had discovered the role of non-Euclidean geometry, he moved swiftly to discover that different discrete subgroups of the group of non-Euclidean geometry led to different Riemann surfaces, and he came close to some understanding of the continuous invariants upon which they depend (the moduli). These discoveries seem to have convinced Poincaré that the group idea has priority, an idea that remained with him for the rest of his life.

In his analysis of different geometries, Poincaré used Sophus Lie's ideas about transformation groups to classify two-dimensional geometries and to show how the canonical three stood out (Poincaré 1887; see Gray 1999, 70). In an appendix to that paper Poincaré discussed how his approach differed from Riemann's, which seemed to offer infinitely many geometries. He replied: The hypotheses at the base of any geometry are not experimental facts, because geometry is not subject to the same process of endless revision that characterizes science. They are not analytic or synthetic a priori, because there is more than one geometry and so one can change the foundations of geometry, but analytic or synthetic a priori foundations would be immutable. Rather, said Poincaré, a geometry is nothing other than the study of a group. The appropriate group was independent of any particular experience, but we would learn about it from the behavior of solid bodies, which would commend to us not the correct group, but the simplest. Thus (a famous remark) one can no more say that Euclidean geometry is true and non-Euclidean geometry false than that Cartesian coordinates are true and polar coordinates false (Poincaré 1891, 76/50).

In his "On the Foundations of Geometry" (1898b), Poincaré embarked on a lengthy discussion of how sensible space is built up by the mind constantly trying to correlate different experiences—eye movements, movements of the head and of the body, and so forth. This sensible space is not geometrical space, because it is not isotropic, homogeneous, or infinite. Geometrical space is a form of understanding. We bring to that understanding the idea of a group, taught us by the motions just described; different types of motion correspond to different types of subgroup. A rich enough group narrows the choice of geometry down to the canonical three, of which Euclidean geometry is the one in which the subgroup of all translations forms a normal subgroup. He made this clear when he wrote:

> The proposition substituted will be the existence of an *invariant* subgroup, of which all the displacements are interchangeable and which is formed of all translations.

It is this that determines our choice in favour of the geometry of Euclid, because the group that corresponds to the geometry of Lobatchevskii does not contain such an invariant sub-group. (1898b, §XII, 996)

Poincaré regarded his essay as starting with Helmholtz and ending with Lie. The only difference he could see was that he started with the abstract group, they with the three-dimensional space. Then, again, geometry was nothing other than the study of a group. The appropriate one was independent of any particular experience but would be, as it were, taught to us by the behavior of displacements. Thus (another famous remark) one can no more say that Euclidean geometry is true and non-Euclidean geometry false than that the metric system is true and the yard, foot, and inch system false (1898b, 42). The Euclidean system is simpler because it has a large commutative subgroup.

For reasons that he pushed deep into the philosophy of mathematics, Poincaré's heart lay with continuous geometries in which a space is acted upon by a group; indeed, the group comes first. That is the mathematician's contribution. It is possible—this, Poincaré suggested, is the great contribution of Lie—to classify these groups in advance. The organization of the groups spills over to an organization of the corresponding spaces. This is, surprisingly, his reason for finding Euclidean geometry simpler than non-Euclidean geometry.

CONVENTIONALISM IN POINCARÉ'S PHILOSOPHY OF SCIENCE

Poincaré was an oddly solitary figure. The Parisian intellectual milieu did not encourage mathematicians to form schools and draw pupils around them, rather the reverse. But Poincaré set himself slightly further apart from the mathematicians professionally, and his work embraced more than analysis and yet gave less attention to real analysis than tradition demanded. He also found himself some way away from the physicists, who were much more inclined to experimental work in France at that time. He put himself into the gap that was opening up between mathematics and physics, one might well suppose in a deliberate attempt to bridge it, and because such a gap is antithetical to French tradition regarding their proper relation. Yet, for someone who was such a prolific and lucid author, he seems also to have been very little interested in attracting followers. I suggest that this isolation in the middle of so much activity is consistent with his special ability to stand in a new place and see clearly.

This is first apparent in his insistence, as early as 1880, that geometry is nothing other than the action of a group. Poincaré came to feel early on that the group concept was prior to the space concept. He was clear that in this he differed from everyone else, who started with the concept of space. Poincaré argued that there can be different spaces because there can be, as a matter of mathematical fact, different groups. While very few of these can be relevant to our understanding of space, at least three can, which led Poincaré to reflect upon what is the extra ingredient that allows us to form a unique concept of space. By the late 1890s he was clear that the space concept is, at least to a certain significant degree, arbitrary.

In the 1890s he turned to physics, his own domain of potential theory and partial differential equations and the overlapping domain of electromagnetic theory. Here he encountered widespread disagreement among the experts, and theories incompatible with cherished precepts. His strong view was that there can be no unique physical theory. Of necessity there will be infinitely many, with incompatible assumptions. All they have to do, in Poincaré's opinion, is be consistent with the basic laws of physics and the central experimental results. There is no possibility of arriving at one final, true description of the physical world, unless, perhaps, there is some improvement in metaphysics.[6]

How is this arbitrariness to be understood? Poincaré insisted that it is a fact of intellectual life. There are just too many things one can say in mathematics, too many things one can do in the laboratory (such, apparently, is the experience of genius). Theorems in mathematics and laws in physics not only organize one's thoughts and suggest fruitful analogies, they are essential to our understanding because they enable us to make sense of these teeming worlds. Nature, said Poincaré, teaches the scientist. But it teaches ambiguous, polysemic lessons. It forces the scientist to contemplate varieties of answers to problems in physics, from the uses of approximations (a kind of falsehood, after all) to the ultimate need for rigour. But even a seamless passage from experimental law to mathematical theorem would not reveal a simple set of truths about the objects in the world.

The arbitrariness in mathematical physics Poincaré took very seriously. Critical self-reflection in mathematics led to all sorts of novel objects. What then is going on when some of these objects, and the theories they are embedded in, are selected as true? Poincaré came to the view that a human action of selection is involved, the result, as he saw it, of evolution and our experience as infants. True descriptions are not available, but nonetheless we make our choices based on criteria of simplicity.

This arbitrariness eluded the conventional philosophical categories. Geometry was neither an a priori matter, because then it would be beyond any element of choice, nor purely empirical, which would make it too arbitrary. It had the arbitrariness of mathematics, neither more nor less, provided that it was also subject to experience. Human understanding was thus quite tightly, but not absolutely, constrained.

The element of human understanding became central to Poincaré. On the one hand, it was rooted in deep structures in the human mind. On the other hand, it was good enough. The best you could hope for, in Poincaré's view, when confronted with a problem was to make it validly understood. He does not very often say that something is true. He allows that it is true that some theorem has been proved. He allows that certain central experiments establish truths. But he does not allow that one can say truly what lies behind the laws, only that certain arguments are valid. Poincaré was not attracted to the idea that science was the accumulation of truths.

Poincaré grounded his philosophy of science in a philosophical insight about how humans acquire knowledge. According to Poincaré, we have to be able to discover different theories explaining the same facts, because historically we have and because mathematics disposes of such theories. Our social and personal histories tell us that the mind must therefore work in this sort of way, producing multiple possible theories, and that we must on occasion fail forever to distinguish between them on logical grounds.

ENRIQUES'S AND HILBERT'S RESPONSES TO POINCARÉ'S CONVENTIONALISM

The facts seem to be that outside France, Poincaré's philosophy of conventionalism found few adherents in his lifetime.[7] To give one example, the distinguished Italian mathematician and expositor Federigo Enriques, whose career paralleled Poincaré's in several ways, dismissed Poincaré's conventionalism in his *Problems of Science* on the grounds that in the real world physicists would have reasons for deciding that temperature really was a physical property (Enriques 1906, 177; see Gray 1999, 75–78). Enriques was at least as happy to historicize and psychologize mathematics as Poincaré ever was, but he evidently believed that, given the circumstances, physicists could choose, and conventionalism fails.

Enriques's point, which he does not amplify sufficiently, is to be understood along these lines. One's experience of heat, in particular its different effect on bodies of different kinds, and the fact that it can be created locally at will, make it the kind of thing one leaves behind in distilling the concept of a rigid body. Enriques's implied argument is that as a piece of philosophy of science conventionalism fails because it is contrived. Given everything we know about heat, we are unwilling to accept a uniform background radiation that varies according to a mathematical law as we move around and becomes absolutely cold at infinity. Too much physics is leeched out before the conundrum can be posed. Poincaré's heated universe is only a conundrum in a hypothetical universe quite unlike the one we inhabit. Now, Poincaré avowedly intended it as a metaphor to open his readers' eyes to a question they may not have considered before, at least from his point of view. The terms of the debate were between two infinite metrical universes, it was a question that had been around for more than seventy years, and it was highly topical when Poincaré wrote. But it is a metaphor for something else as well, or it would no longer interest anyone. It stands for all those moments when one imagines different physical theories and says: We cannot choose between them.

In a recent paper, Ulrich Majer looked at Hilbert's critique of Poincaré's conventionalism, drawing on a lecture Hilbert gave in 1921 entitled "Fundamental Ideas of Relativity Theory" (Majer 1996). There, Hilbert dismissed Poincaré's views in these terms: "The old space-time theory is recognized to be a prejudice. It had to be abandoned no matter how great the resistance was. One could not explain the observed processes in nature without gaps by merely changing the physical laws" (Majer 1996, 355).

Hilbert's objection is that in special relativity rigid bodies do not exist (else the tip of a rotating arm could move faster than light, for example). This objection, he claimed, strikes at the very foundations of Poincaré's analysis of how we learn geometry. The point is not that we can have a clear concept of rigid-body motion but find it sadly lacking in reality (any proponent of an idealized theory can cope with minor difficulties like that); it is that we cannot form the concept in a way that is logically consistent with what we want to believe about light. Therefore there are logical grounds for abandoning Poincaré's position. This, Majer argues, is the gap Poincaré failed to close. He concludes that the idea that different geometries of constant curvature are indistinguishable cannot be sustained, while semantic conventionalism (which asserts that there is a somewhat arbitrary assignment of meanings to terms) might survive (see also Friedman 1996a).

POINCARÉ ON GEOMETRY AND ARITHMETIC

Hilbert's objection is persuasive at first sight. There is no consistent theory that allows for the observed behavior of light and incorporates Euclidean rigid bodies. But the exercise of imagining how Poincaré might have replied to this argument, had he lived, is not merely a hypothetical exercise; it can shed light on how his philosophy of geometry was intended to work, and how it differs from his philosophy of arithmetic.

The crux of Poincaré's conventionalism is his resolute denial that there can be a meaningful way of distinguishing between the straight line mathematically defined and the straight line physically instantiated. This is because, ultimately, "experiments have reference not to space but to bodies" (Poincaré 1902c, 104/84).[8] There are different possible geometries, to be sure, but no way of getting behind the bodies to space, as one might say, in itself. It is worth considering in more detail what Poincaré meant by knowledge and truth here, and indeed by bodies.

In his 1895 essay "Space and Geometry," Poincaré distinguished between geometrical space, which he defined as the space that is the object of geometry, and representative space, which he defined as the framework of our representations and sensations. He had no trouble in showing that the fundamental properties of these spaces are so different that they cannot be confused or identified. As was mentioned earlier, geometrical space is homogeneous, isotropic, and three-dimensional, the representative space of vision is neither homogeneous nor isotropic, and tactile representative space has as many dimensions are there are muscles. How then do we localize objects in geometrical space? Not, Poincaré answered, by another representation. Rather, we reason about the objects *as if* they were situated in geometrical space. We localize an object by representing to ourselves the movement that must take place to reach the object.

Here Poincaré argued that the crucial element forcing us to construct geometrical space at all is that objects both move and change. Some differences in sensation can be accounted for on relativistic grounds. The idea is that some changes in a body's appearance to us are caused by its motion relative to us, and we can compensate for them by making ourselves move. The explanation Poincaré offered is a little more complicated because he was trying to account for the very idea of a geometry, so his explanation could not invoke geometrical terms. Instead he argued that in these cases we invoke voluntary changes in our own body to represent to ourselves those changes that would restore the original impression created upon us by a body. When such voluntary changes can be found, we learn to call the original changes in our representation changes in position of the rigid body. Other changes in an object cannot be reversed by our own motion; they are not the changes of a rigid body. At a later stage we learn to mentally decompose a body into small elements that move rigidly while changing their positions relative to each other, and these are the deformations of a nonrigid body. So it is, proclaimed Poincaré, that "If, then, there were no solid bodies in nature there would be no geometry" (1895, 86/61) and our experience singles out a class of phenomena called displacements, concerning which he said, "The laws of these phenomena are the object of geometry" (1895, 87/63).

For Poincaré, then, our knowledge is of bodies and their displacements. Space is a hypothetical construct, by no means uniquely determined, and we reason as if objects are located in this space. Insofar as we know what our senses tell us, we know only about bodies, whether we are small children or erudite scientists.

In "Experience and Geometry" (1902c), Poincaré imagined an experiment in which the faces and edges of two rigid bodies are compared. It might be that they could be made to agree according to displacements possible according to the Euclidean group of rigid-body motions, but not to the non-Euclidean group of rigid-body motions. Or the reverse might be true, and the displacements would necessarily be non-Euclidean and could not be Euclidean. But it would not follow that such an experiment determined the nature of space. It would remain possible that the rigid bodies are made of different kinds of matter. This being the case, we see yet again that "experiments have reference not to space but to bodies."

It is now possible to distinguish two aspects of Poincaré's position concerning our knowledge of geometry. One is epistemological. It is genetic, nativist, implicitly Darwinian: We construct geometry from our earliest experiences, doubtless as a result of our evolution as a species. The other aspect is metaphysical: Our knowledge is of bodies and their behavior, and space is a hypothetical construct. The metaphysical claim would be unaffected if space were to give way to space-time; Poincaré could argue that the nature of space would still be inaccessible to us, and questions about it would be meaningless. His account of the process whereby knowledge is acquired, however, is fundamentally tied to Lie's classification of the relevant groups, which intervenes decisively to narrow our choices down to just a few. What would happen in a high-speed universe with intelligent children able to move at nearly the speed of light? If we allow Poincaré to ignore the dangerous levels of chaos that would surely result, he would presumably have to argue that the Lorentz group would commend itself on grounds of simplicity. The analogy with the dichotomy between Euclidean and non-Euclidean geometries that Poincaré discussed is difficult to pursue, of course, because the high-speed universe is so unfamiliar. The point, however, is not to produce arguments for the simplicity of the Lorentz group, but to see why the analogy is difficult to pursue. It is difficult because we do not know from our experience what the high-speed children would learn, but we know that we ourselves picked out Euclidean geometry among the range of alternatives. Indeed, the ubiquity of this choice is part of what Poincaré was trying to explain.

The choice of geometry is governed at least as much by historical accident as by any feature of the groups involved; indeed, it is conventional, and the criteria of simplicity are patently ad hoc. A markedly non-Euclidean universe, or a high-speed one, if it contained intelligent creatures at all, would presumably contain creatures rationalizing their choice on grounds no better than those Poincaré offered for the Euclidean group. So we must conclude that for Poincaré the philosophy of geometry was compatible with a purely contingent choice of group. Once again, we see that it is the human mind whose working must be explained, not the operation of an abstract, Kantian intelligence.

Precisely this human element helps account for Poincaré's failure to incorporate the Lorentz group into his philosophy of geometry. It would have been possible for Poincaré to put the Lorentz group at the heart of other creatures' perceptions of the high-speed world; there is nothing too difficult about the idea of replacing rigid bodies with elastic bodies rigid to a high degree of approximation. But Poincaré could not so easily argue that slow-moving humans, who were only just sorting out the theory of special relativity and the Lorentz group, were evolutionarily disposed to think this way about bodies. It would indeed have been false to say that ordinary people around 1900 could formulate the concept of a geometrical space (or space-time) in a way that is compatible with their contemporary physics. This is one reason why Poincaré was not Einstein, or, perhaps more accurately, why Poincaré was not Minkowski: His commitment to space rather than space-time was very deep.[9] Poincaré too could formulate an orthodoxy.

Matters stand differently with the integers (see Folina in this volume). In his polemics with the logicists (Couturat, Russell, Zermelo) Poincaré was happy to admit that purely analytic reasoning did not increase knowledge. He gave the example of the Leibnizian proof (actually a fallacious argument) that $2 + 2 = 4$, which he said was merely a verification of a truth already known. If mathematics was not to be merely a gigantic system of tautologies, it had somehow to contain a method of increasing knowledge (a genuinely Kantian concern). This Poincaré located in the principle of induction. Now, it might seem that Poincaré had again returned to the group concept here. Allowing that multiplication is repeated addition, it would be enough to obtain the integers as a group. There are still some obstacles to be overcome, such as the familiar chicken-and-egg question, much debated in neo-Kantian and other circles in Poincaré's time, of which comes first, ordinal or cardinal numbers. There is the inconvenient fact that the positive integers and zero together only form a semigroup; negative numbers must be constructed even if one believes the positive numbers are given in nature. But there is a more fundamental objection to the yoking together of arithmetic and geometry in Poincaré's philosophy of mathematics.

According to Poincaré, our knowledge of geometry arises from our disentangling the rigid from the nonrigid bodies by the appreciation of the reciprocal effect of moving our own bodies. Poincaré nowhere offered a similar account of how we learn arithmetic. There is no convention here. Poincaré did not offer a rival candidate to the (positive) integers and a conventional criterion for selecting one over the other. Or, to make the analogy closer, he did not posit a rival hypothesis other than the principle of induction that allows us to know about the infinitude of integers. Rather, it is the integers and the principle of induction that we know directly, and to which we have access by means that are more than logical. His aim in the essays on arithmetic is precisely to show that there is no route from the truths of logic to the facts of arithmetic. But if the facts of arithmetic cannot be so derived, nor are they grounded in our experience as children or as a species. We simply assent, as soon as we are capable, to the principle of induction. It is not a posteriori or analytic; it can only be synthetic a priori.

Back in 1897 Poincaré had mentioned that one use of mathematics was to help the philosopher make precise the notion of number. Here Poincaré was at his most avowedly Kantian. Once he had been led to give a central role to the faculty of cognition, the whole question of definitions in mathematics saw him at odds with the various kinds of formalists and logicists who emerged around 1900. He had no choice but to accept Hilbert's axiomatic formulation of geometry, although he regretted its lack of psychological grounding (Poincaré 1902a, 113). But he could not accept the reduction of mathematics to logic, because that left no room for intuition and therefore made it impossible for the store of mathematical knowledge to grow. As he put it in his essay "On the Nature of Mathematical Reasoning," the ability to deduce particular cases analytically is in stark contrast with our inability to derive the general theorem analytically: "To reach it [the general theorem] we should require an infinite number of syllogisms, and we should have to cross an abyss which the patience of the analyst, restricted to the resources of formal logic, will never succeed in crossing" (1894a, 40/10–11).

Poincaré's judgment has some force. Contemporary mathematicians had formalized what they meant by the integers in Peano's axioms (Peano 1889), which define the (nonnegative) integers this way:

(i) Every integer has a unique successor,

(ii) there is an integer that is not the successor of any integer,

(iii) two distinct integers cannot have the same successor, and

(iv) if M is a set of integers such that 0 is in M and such that if an integer n is in M then its successor is in M, then every integer is in M.

But as the mathematician and logician Paul J. Cohen observed lucidly in 1966, when one attempts to formalize something so precisely that arguments about it can be followed mechanically, Peano's axioms are not good enough:

> If we examine Peano's axioms for the integers, we find that they are not capable of being transcribed in a form acceptable to a computing machine. This is because the crucial axiom of induction speaks about "sets" of integers but the axioms do not give rules for forming sets nor other basic properties of sets. . . . When we do construct a formal system corresponding to Peano's axioms we shall find that the result cannot quite live up to all our expectations. This difficulty is associated with any attempt at formalization. (Cohen 1966, 2–3)

Cohen then showed that the vague notion of a set could be replaced in a variety of ways with a precise, if infinite, list of axioms for the integers. Indeed, one could do so in several ways that captured almost the same idea. However, by Gödel's famous incompleteness theorem these formalizations could not be shown to be consistent. Poincaré's choice of the word "abyss" was surely adequate.

Poincaré lived to see the extension of formal logic in which Peano, Couturat, and Russell figured prominently (he seems never to have mentioned Frege), but he was not too impressed. He was unconvinced by the definition of the number 1 offered by Cesare Burali-Forti, as he was by Couturat's definition of the number zero (Poincaré 1906a, 152–171). Burali-Forti's definition, he thought, was based on a vicious circle; Couturat's he regarded as playing, inelegantly, with words. On the basis of these criticisms, he found not only that the new pasigraphy was inadequate to the task of giving entirely logical definitions of numbers, but that the logical task remained unanswered.

When Poincaré wrote the famous popular essays that established his philosophical credentials, there was an upsurge in philosophizing about science and its implications. Whatever philosophy had occupied itself with for most of the nineteenth century, it returned at the end to Kantian questions, taken apart and put together differently, but sharing Kant's concern that the practice of science might, after all, have something to teach the philosopher. Poincaré construed Couturat's work as an attempt to bury Kant and to prefer the claims of Leibniz. He also considered the somewhat different attempt of David Hilbert in 1904, which endeavored not to develop mathematics from logic but to develop them both together. His conclusion was that Russell and Hilbert had achieved much, "[b]ut to say that they have definitively resolved the debate between Kant and Leibniz and destroyed the Kantian theory of mathematics is evidently inexact. I do not know," he added, "if they believe they have done so, but if they do so believe, they have deceived themselves" (Poincaré 1906b, 191). The reason Poincaré demurred was that they had argued that an integer is what is obtained, à la Peano, by an appropriate process of successive addition. However, an integer must also be something upon which one can reason by recurrence or induction. These definitions, he suggested, are equivalent only in virtue of a synthetic a priori judgment, and not by a purely logical argument.

So we see here again that Poincaré insisted on the active role of the human mind. A purely logical creature, in his view, would be an intellectually impoverished thing. On the other hand, Poincaré did not want the human mind to be free to invent at will. The compelling character of mathematical reasoning was not something he was willing to give up. He was therefore comfortable with what might be called a pre- or protopsychological interpretation of Kant, according to which some capacities of the human mind are necessary for the working of the intellect, but the precise nature of these capacities need not be explored, and certainly their variation from person to person is not of interest. Poincaré's willingness to discuss the psychological aspects of his own creativity suggests that he was even happy to go a step further, whereas logicists such as Russell (and Frege) were adamant that psychology could and should be left out altogether.

CONCLUSION

Throughout the later part of his life, Poincaré reflected on how it is that we can know things and on the kind of things that we can know. He came to hold some possibly unexpected views. His position is that much can be known. If anything, too much: too many facts, too many isolated mathematical results.

Laws and theorems organize the domains of science and mathematics, and these are plainly human creations. In science, he felt that there were some crucial experimental results that simply had to be accepted. But the interpretation of these results was arbitrary in many ways, and in particular Poincaré held out against commitment to this or that kind of object. His famous conventionalism about the nature of space is a denial that it is possible to say where geometry stops and physics begins. He puts certain basic laws of nature on a par with the crucial experiments, but even then he does not insist that we hold them. Under pressure from results in contemporary optics, he allows that these laws may contradict each other, and that something will therefore have to give. He would have said the same about experiments, too. So there are few fundamental things in science, and even they are in principle revisable. All the rest is far from uniquely specifiable, if indeed it can be known at all.

As for how we can know anything, Poincaré needed a philosophy that accommodated this arbitrariness but did not say that anything goes. In particular, mathematics had to be validated. The Kantian synthetic a priori was rejected for geometry but retained for arithmetic, because there are many geometries but only one arithmetic. Geometry was made to rest on innate ideas of groups of transformations, which replace space as a form of intuition with a short list of possible forms of intuition that experience and evolution help us to select among. The feeling that there are alternative models of space but not of arithmetic, which unites Poincaré with Karl Friedrich Gauß, need not rule out logicism, and Poincaré's arguments against that are primarily mental. He was clear that what we know of the integers concerns them as an infinite totality (not a Kantian position) and he saw the capacity to reason about the infinite as intrinsically beyond logic. His insistence, in short, was not on the nature of knowledge, but on the nature of human knowledge.

NOTES

1. Poincaré referred at this point to Klein's book (Klein 1882).

2. He returned to the implications of Dirichlet's principle for mathematics in his 1898c.

3. See Buchwald 1985, and for some remarks on Poincaré, see Gray 1999, 70–73.

4. Miller (1981) has argued that it was Hertz's discovery of electromagnetic radiation and Poincaré's *Electricity and Optics* that convinced most people that an ether-based theory of electromagnetism was fecund.

5. In 1906 a similar analysis revealed a contradiction between the supposed nature of the electron and the law of conservation of energy (Poincaré 1906c). Poincaré preserved the law by hypothesizing a new physical process: nonelectrical forces designed to hold the

electron together. These, Poincaré stresses, as they are called, enable the conservation of energy, and momentum laws are to be applied consistently in the determination of the electromagnetic mass of the electron. In ontological terms, their creation sees Poincaré most clearly behaving like a physicist. See also Poincaré 1908b.

6. It is interesting that Poincaré came late to agree that atoms really do exist, convinced by Jean Perrin's account of Brownian motion.

7. It did, of course, prove very attractive to the logical positivists; see Friedman 1996a.

8. This essay "Experience and Geometry" is not, as Vuillemin claims in his edition of *Science and Hypothesis* (Poincaré 1902b, 7), a French version of Poincaré's "On the Foundations of Geometry" (1898b) but it covers much of the same ground. Unusually, it seems to have been published for the first time in *Science and Hypothesis*.

9. In the early years of the twentieth century, Poincaré lost some of his audacity. He did not go all the way with the idea that groups are of central importance in algebraic topology, where he could have formulated the idea of homotopy and homology groups but did not. His failure to ground space-time geometry in the Lorentz group, however, derives from the deeper roots suggested above.

XIV

Images and Conventions: Kantianism, Empiricism, and Conventionalism in Hertz's and Poincaré's Philosophies of Space and Mechanics

Jesper Lützen

This essay analyzes Kantian, empiricist, and conventionalist tendencies in the *Prinzipien der Mechanik* (1894) by Heinrich Hertz (1857–1894) and compares them with those found in Poincaré's *La science et l'hypothèse* (1902). It thereby throws into relief Poincaré's reaction to Kantianism.[1]

Images and Their Philosophical Problems

The mathematician Poincaré and the experimental physicist Hertz shared an interest in theoretical physics, its philosophical implications and basis. They came to somewhat different philosophical conclusions, probably partly owing to their different research backgrounds, but they were engaged in more or less the same kind of analysis and thereby arrived at more or less the same basic questions.

They both accepted that there exists an external world that we sense in various ways. We make theories (Hertz called them images) to account for our sensations. The philosophical question arises because we can make several images that will correctly account for the sensations. For example, both Poincaré and Hertz were keenly aware that there had been two or three competing theories of electromagnetic phenomena. Hertz's earliest experiments had been designed as attempts to distinguish between them, and his celebrated production of, and experiments with, radio waves (1887–1888) was interpreted by himself, and soon by most of his contemporaries, as the final experimental verification of the Maxwellian image.

However, in mechanics, to which Hertz believed one should ultimately reduce all of physics, he found another indeterminate situation. Two images were competing at the end of the nineteenth century: the Newtonian image, with the basic notions of time, space, mass, and force, and the energetic image, with the basic notions of time, space, mass, and energy. To these images Hertz added his own alternative image that operated with only three basic notions: time, space, and mass. Instead of introducing a fourth kind of basic notion, his image operated with two kinds of mass:

ordinary mass and concealed mass. The latter cannot be sensed directly but its fast cyclic motion is transferred to the ordinary mass through rigid connections and gives rise to sensible apparent forces (or potential energy). Hertz developed this image through the last three years of his life, and it was published posthumously in his *Prinzipien der Mechanik*.

As presented in the introduction to his book, Hertz's idea of images shows several similarities with Poincaré's conventions. Like Poincaré, Hertz did not attach any ontological truth to the various elements of his image. He required only that an image be *logically permissible* (consistent) and *correct* in the sense that it correspond to our experience of the outer world. (From now on I shall use the word "correct" in this particular sense.) Among the many such permissible and correct images, one should, according to Hertz, choose the most appropriate, which is the one that (1) is most distinct, in the sense that it describes most of the essential features (excludes most of the possibilities that do not occur in the external world), and (2) is simplest. Poincaré similarly required that one choose the simplest convention. Hertz even specified the vague notion of simplicity using a concept reminiscent of Ockham's razor: The simplest image is the one containing the fewest idling wheels. He agreed with Poincaré that a positivist rejection of all idling wheels or hypotheses is not the way out of the philosophical problems.

Mechanics: The Division between the A Priori and the Empirical

Both Hertz and Poincaré distinguished between a priori and a posteriori elements of their images (theories), but they placed the dividing line between them differently. Hertz believed that all of mathematics, including geometry, was a priori, whereas Poincaré divided mathematics into an a priori arithmetic and an a posteriori geometry (see Folina, this volume). Moreover, Hertz essentially assumed that all a posteriori elements were empirical. Poincaré operated with at least four distinct categories: the a priori, the conventional, the empirical, and the hypothetical. According to him, space and mechanics are conventional, and the rest of physics is empirical or hypothetical. Such are the broad outlines of their standpoints. In the rest of this essay I shall elaborate on and qualify this rather crude picture.

Let me immediately point out that it was important to Hertz to distinguish clearly between the a priori and the empirical elements of his image of mechanics. Indeed, this was a central part of his principal endeavor, namely, to give a logically permissible image of mechanics. In his opinion the traditional Newtonian image of mechanics, and in particular the existing accounts of it, were logically muddled if not directly impermissible. In particular he argued that it would be difficult (or perhaps even impossible) to introduce the concept of force as a basic notion in a consistent way, and therefore he chose to avoid it. In a similar way his sharp distinction between the a priori and the empirical elements of his image grew out of a critique of the Newtonian image: He admitted that many physicists found it unthinkable that new experience could ever lead us to change the fundamental principles of mechanics. And yet in Hertz's opinion the laws of mechanics were derived from experience and "that which is derived from experience can again be annulled by experience" (Hertz 1894, 11/9). The unwillingness on the part of most physicists to admit this possibility was, Hertz wrote, due to "the fact that the elements of experience are to a certain extent hidden in them and blended with the unalterable elements which are necessary consequences of thought" (Hertz 1894, 11/9).

According to Hertz, the resulting lack of clarity of the presentation of mechanics might have been an advantage in the formative period of the theory. Historically, when an image is developed, it is first formed to serve a certain purpose. Its appropriateness is thus the first aim. Afterward its correctness is tested and, if a problem occurs, one tries to remedy it by making an empirical fact into a definition or vice versa. And only at the end one tries to clean away all contradictions in order to make the image logically permissible. This procedure "gives the foundations an appearance of immutability" (Hertz 1894, 11/9).

Hertz seems to have accepted this route as a sound way to generate images, but he insisted that once a ripe image has been formed, one must reverse the process: "In a perfect science such a groping, such an appearance of certainty, is inadmissible. Mature knowledge regards logical clearness as of prime importance: only logically clear images does it test as to correctness; only correct images does it compare as to appropriateness" (Hertz 1894, 11/10).

In order to obtain logical clarity of his image, Hertz distinguished dynamics from time, space, mass, and the kinematics that connect them. In his opinion kinematics, which he dealt with in the first book of his *Mechanics*, concerns a priori forms of our intuition in Kant's sense, whereas dynamics, the subject of the second book, is an empirical science. The empirical element of his dynamics was reduced to a small core, namely his fundamental law of motion: "Every free system persists in its state of rest or of uniform motion in a straightest path" (Hertz 1894, §309). I shall not explain the meaning of this law, which is expressed in terms of a differential geometric formalism of configuration space (a geometry of systems of points, in Hertz's terminology),[2] but I shall return below to a clarification of its empirical character and to the a priori character of Hertz's geometry of ordinary space.

Poincaré agreed with Hertz (and with Mach, Kirchhoff, Duhem, and others) that it was a problem that the ordinary "treatises on mechanics do not clearly distinguish between what is experiment, what is mathematical reasoning, what is convention and what is hypothesis" (Poincaré 1902b, 89). He also saw it as his aim to clear up this logical mess. His discussion of the problem differed in many ways, however, from that of Hertz. For one thing, the styles adopted by the two scientists were strikingly different:

Having discussed the three alternative images of mechanics in the introduction, Hertz consciously chose a Euclidean synthetic style, interrupted by philosophical reflections. This made his own image crystal clear, but except in the introductory discussions, he did not analyze alternative foundations or philosophical standpoints. In particular, he hardly addressed historical or psychological questions of image formation. Indeed, the remarks, mentioned above, about how images are historically developed were his only reflections on this matter. For Hertz, testing and comparison of mature images are the only questions of interest. To be sure, while working on his *Prinzipien der Mechanik* Hertz also published a collection of his electromagnetic papers prefaced by a semihistorical account of how he gradually tested the Weberian, Helmholtzian, and Maxwellian images and finally was led to reject the first two alternatives (Hertz 1892). But it is conspicuous that in his mechanics he consciously dissociated himself from the psychological investigations of his mentor Helmholtz.

Poincaré's deep interest in the psychological aspects of theory building and his writing style were much closer to Helmholtz. He used a French discursive analytic style, suggesting several possible points of view but rejecting them one after the other in order to arrive at his own rather loosely formulated point of view. Contrary to Hertz, he did not follow this analysis by a synthesis showing how one could erect the usual mechanics on the foundation he had suggested.

In his discussion of mechanics he argued that some of the principles, such as Newton's third law (action equals reaction), are in fact definitions in disguise, while other laws, such as Newton's first law of inertia, are given to us through experience. However, Poincaré sharply disagreed with Hertz about the possibility that the law can be falsified by experience. Both in a review of the *Prinzipien der Mechanik* (Poincaré 1897) and in *La science et l'hypothèse,* Poincaré quoted Hertz's claim: "that which is derived from experience can again be annulled by experience" and then argued that it is not correct. His argument went along the following lines: If at a future time we detect a seemingly isolated system whose center of mass does not move in a uniform motion in a straight line relative to the fixed stars, we have two choices. Either we can reject Newton's law of inertia or we can argue that the system is not isolated after all. The latter alternative is in fact always true since no system is isolated, except the whole universe, and it makes no sense to talk about the absolute motion of the universe. Moreover, if a close inspection reveals that the known interactions between the system in question and other parts of the universe cannot account for the observed deviation from a uniform motion, then one can hypothetically assume the existence of hitherto unobserved masses or forces. Poincaré argued that the latter possibility was preferable to the rejection of the law of inertia. He therefore concluded that the laws of mechanics were conventions—not arbitrary conventions, but conventions that experience have led us to adopt. "Thus is explained how experiment may serve as a basis for the principles of mechanics, and yet will never invalidate them" (Poincaré 1902b, 105). This is the central idea behind Poincaré's conventionalism.

THE EMPIRICAL NATURE OF HERTZ'S LAW OF MOTION

The fundamental law of motion was the empirical core of Hertz's mechanics. "The question of the correctness (the agreement with experience) of our statements is thus coincident with the question of the correctness or general validity of that single statement (the Fundamental Law)" (Hertz 1894, §296). This is a clear statement of the empirical nature of the fundamental law in flat disagreement with Poincaré's later conventionalism: If we make an experiment that does not agree with the predictions of our image, we will be forced to change the fundamental law. But what would it mean for an experiment to be in contradiction with Hertz's image? The problem is not as simple as the similar problem in the ordinary Newtonian image. In ordinary mechanics a discord between experience and theory would mean that a mechanical system does not move according to the basic laws of motion when all the known interactions are taken into account. Poincaré's position was that it would always be possible to save the laws of motion by inventing new interactions or hidden masses that would restore the accord between experience and theory. This is precisely what happened when irregularities in the motion of Uranus were attributed to a new planet, Neptune, which, by the way, was observed soon after its theoretical prediction.

However, in the case of Hertz's mechanics, hidden masses are already an integrated part of the image. To show that Hertz's image correctly describes an experimental situation, one would have to construct (in one's mind) a hidden system whose cyclic motions and connections to the visible system would induce the system to move the way experiment shows it does move. In fact Hertz never proved that even the well-known systems, which are usually described by gravitational or electromagnetic forces, could be correctly accounted for in this way. In other words, Hertz did not show that his image was correct in the light of past experience. But at least we can now see what it would mean for an experiment to prove Hertz's mechanics incorrect: It would mean that one could prove that no hidden system exists that would account for the experiment.

Thus the disagreement between Hertz and Poincaré concerning the fallibility of the fundamental law(s) of mechanics seems to boil down to this: Poincaré believed that it would be possible to save the laws of motion in the face of any future experiment by hypothesizing a suitable hidden system, whereas Hertz imagined that we might in the future encounter an experiment that cannot be explained by any hidden system whatsoever.[3]

GEOMETRY

Poincaré's view of geometry parallels his conventionalist view of mechanics. Gray (in this volume) explains it in detail, so let me just recall the main points: Experience teaches the child that space is describable as a three-dimensional manifold of constant curvature, and since translations seem to commute, we choose to describe space by Euclidean geometry. As was the case with the laws of mechanics, this convention is therefore given to us by experience, but it cannot be falsified by experience. For example, if, in the future, we should come across a triangle of light rays with an angle sum different from the sum of two right angles, it is better to hypothesize a physical reason why light does not move along straight lines (corresponding to the varying temperature in Poincaré's heated disc model) than to reject the simple Euclidean geometry.

Hertz's view of geometry is much less clear. Of course one cannot expect to find a thorough analysis of the concept of space in a book on mechanics, but the few considerations he did devote to the subject are difficult to unite into an entirely consistent image. The confusion occurs because geometry plays two different epistemological roles in Hertz's mechanics: one in the first book on kinematics and another in the second book on dynamics. "2. . . . The space of the first book is space as we conceive it. It is therefore the space of Euclid's geometry, with all the properties which this geometry ascribes to it" (Hertz 1894, §2). More generally, Hertz wrote about the considerations of the first book:

> 1. Prefatory Note. The subject matter of the first book is completely independent of experience. All the assertions made are *a priori* judgments in Kant's sense. They are based on the laws of the internal intuition of, and upon the logical forms followed by, the person who makes the assertions; with his external experience they have no other connection than these intuitions and forms may have. (Hertz 1894, §1)

However, in the introduction to the second book Hertz wrote:

> 296. Prefatory Note. In this second book we shall understand times, spaces, and masses to be symbols for objects of external experience. . . . Our statements concerning the relations between times, spaces, and masses must therefore satisfy henceforth not only the demands of thought, but must also be in accordance with possible, and in particular, future experience. (Hertz 1894, §296)

Thus geometry in the second book is the geometry of physical space, though not as it is (Hertz did not make any such ontological claims), but as we perceive it. He used Helmholtz's term "symbol" (*Zeichen*) for an inner representation of outer objects that corresponds to certain sense perceptions. In other words, Hertz distinguished between the pure geometry (of the first book) and the practical (or physical or applied) geometry of the second book.

It is helpful to compare Hertz's dual concept of geometry with a similar distinction that Einstein made in his 1921 paper on geometry and experience. Rejecting Poincaré's conventionalism, Einstein distinguished between Hilbert-style pure axiomatic geometry and practical geometry, which arises from our experience with almost rigid bodies, and he concluded with the often quoted statement that "As far as the propositions of mathematics refer to reality they are not certain; and as far as they are certain, they do not refer to reality" (Einstein 1921, 233).

Is Hertz's distinction between the two types of geometry the same as Einstein's distinction? In particular, is the pure geometry of Hertz's first book simply formalistic axiomatic geometry? In fact Hertz seems to allow for this possibility: "It is immaterial to us whether these properties are regarded as being given by the laws of our internal intuition, or as consequences of thought which necessarily follow from arbitrary definitions" (Hertz 1894, §2). The latter possibility is probably Hertz's way to describe formal axiomatic geometry starting from arbitrary axioms, yet this possibility seems to come as an afterthought. Everywhere else in the book, including in the passages quoted above, Hertz described the geometry of the first book as the geometry of our intuition and not as a formal system given by arbitrary axioms. The remark seems to serve as a disclaimer: If the reader does not admit that Euclidean geometry is a priori the geometry of our intuition but considers it just one of many possible axiomatic geometries, this does not invalidate the remaining part of the book.

Already as a student Hertz had learned of non-Euclidean geometry, and during his time in Berlin he must certainly have heard that such geometries were as consistent as Euclidean geometry, and that Helmholtz considered them a priori possible candidates for the geometry of physical space. And still Hertz declared that there is one geometry of our intuition and that is Euclid's three-dimensional geometry. How could such a Kantian view still be upheld in 1894? According to Poincaré, the existence of non-Euclidean geometry shows that Kant was simply wrong, and yet a deep thinker like Hertz thought otherwise. What then did Hertz make of non-Euclidean (including higher-dimensional) geometry? He did not answer the question directly, but what he wrote seems in accordance with the following point of view: There may be consistent mathematical formalisms describing non-Euclidean geometries, but these formalisms do not represent our intuition of space. Such a point of view corresponds quite well to his remarks about higher-dimensional geometries in the introduction: He characterized such geometries as artificial, far-fetched, unnatural, perplexing, and as "supra-sensible abstractions" (Hertz 1894, 38–39/32–33) and therefore took pains to dissociate his own high-dimensional geometry of configuration space ("geometry of systems of points") from the speculations of the mathematicians, declaring that the analogies between the two theories are merely formal (see Lützen 1998). We may invent many formal systems and we may choose to call some of them spaces, but that does not change the fact that our intuitive concept of the space we live in is Euclidean. Such a defense of Kantianism in the face of non-Euclidean geometry should not surprise us. It corresponds very well to the initial reaction of students who are presented with non-Euclidean geometry for the first time.

It is more difficult to discern how Hertz thought about practical geometry, in particular how it relates to the geometry of our intuition. For Kant, pure intuition is the form of empirical intuition and therefore Euclidean geometry is necessarily applicable to the empirical world; but Hertz does not seem to accept this argument directly. He admitted that in order to apply the pure geometry of the first book to our sensations of the outer world (the practical geometry of the second book), a coordinating rule is needed:

> 299. Rule 2. We determine space relations according to the methods of practical geometry by means of a scale. The unit of length is settled by arbitrary convention. A given point in space is specified by its relative position with regard to a system of coordinates fixed with reference to the fixed stars and determined by convention. (Hertz 1894, §299)[4]

Hertz emphasized that this rule is not a new definition (1894, §302). It is a rule that maps outer experience into the sign language of our inner image of the world (the a priori geometry of our intuition), and that allows us to express the logically necessary consequences of our image as sensible experiences.

But this translation raises a philosophical problem: Is our inner image (three-dimensional Euclidean geometry) a correct image of the outer world? There is no doubt that Hertz answered this question in the affirmative. In a passage of his prefatory note in §296 (not included in the passage above), Hertz wrote that in the geometry of the second book times, spaces, and masses are "symbols for objects of external experience": "symbols, whose properties, however, are consistent with the properties that we have previously (in the first book) assigned to these quantities either by definition or as being forms of our internal intuition" (Hertz 1894, §296).

It is somewhat unclear, however, on what grounds he believed that the symbols of the second book were consistent with the a priori geometry of the first book, that is, why Euclidean geometry is correct. In fact there are traces of three different answers: (1) a Kantian answer, (2) an empiricist answer, and (3) a conventionalist answer.

(1) The Kantian answer is the most visible in Hertz's book. Thus he continued his prefatory remark, quoted above, as follows:

> These statements (about the relations between times, spaces, and times) are based therefore, not only on the laws of our intuition and thought, but in addition on experience. The part depending on the latter, in so far as it is not already contained in the fundamental ideas, will be comprised in a single general statement which we shall take for our Fundamental Law. No further appeal is made to experience. The question of the correctness of our statements is thus coincident with the question of the correctness or general validity of that single statement. (Hertz 1894, §296)

We have already appealed to this passage to show the empirical nature of the fundamental law and to highlight Hertz's separation between a priori and empirical elements. Here we shall emphasize that the last sentence implies that it is *only* the fundamental law that is empirically testable, that is, that the correctness of our a priori intuitions of (time and) space cannot be empirically falsified. Consequently, according to this point of view, three-dimensional Euclidean geometry is not only the a priori geometry of our intuition but also the a priori and synthetic geometry of sensible space. This very Kantian reading of Hertz's view of space is in agreement with almost all his statements about this subject and is in close agreement with his sharp distinction between the a priori and synthetic elements of his image.

(2) Still, Hertz nowhere explicitly stated that Euclidean geometry was an a priori and synthetically correct description of our sensations of external space. He actually never used the Kantian term "synthetic," and indeed there are two hints at an empiricist account of practical geometry, somewhat in line with Helmholtz and Einstein: In the quoted passage from the prefatory note to the second book, Hertz stated that the empirical parts of the relations between spaces, times, and masses will be comprised in the fundamental law "in so far as it is not already contained in the fundamental ideas." This seems to open a back door for an empirical element of geometry, although such a possibility would, strictly speaking, contradict the rest of the prefatory note. Moreover, having specified in rule 2 that we determine space relations by means of a scale, he continued: "We know from experience that we are never led into contradictions when we apply all the results of Euclidean geometry to space relations determined in this manner. The rule is also determinate and unique, except for the uncertainties which we always fail to eliminate from our actual experience, both past and future" (Hertz 1894, §298).

This seems to suggest that the practical geometry of our sensations is a matter of experience and that Euclidean geometry could in principle in the future be shown to be an incorrect image of outer space. However, such a Helmholtzian empiricist reading would entirely destroy the distinction Hertz makes between the a priori and the empirical. And in fact Hertz did not really say that the geometry of the outer world is empirically testable. What he suggested in the passage quoted above is rather that our dictionary, by which we translate our intuitive geometry into sensations of outer space (namely rule 2), can be empirically tested. Thus the apparently empiricist stance can be harmonized with the prevalent Kantian tone of Hertz's book if we read him as follows: Our experiences of outer space are a priori in accordance with our Euclidean intuition. Until now experience has corroborated our ability to measure lengths by way of rigid rods. If future experience shows that such a translation between sensations of outer space and intuitive geometry makes Euclidean geometry an incorrect image of external space, then we must determine another way of translation—for example, a better way to measure lengths—that will restore the correctness of the Euclidean image.

Such an interpretation of Hertz brings his conception of space closer to that of Poincaré. Indeed, Poincaré also focused on the dictionary by which we translate things in the physical world into mathematical terms. For example, we can translate "light ray" into "straight line." However, if future experience tells us that there are triangles of light rays that have angle sums different from 180°, then we should not reject Euclidean geometry but the translation.

Of course, Hertz and Poincaré had different reasons for declaring Euclidean geometry to be the right theory of space. For Hertz it was a priori and for Poincaré it was a matter of simplicity. But they seem to have shared the idea that the translation between sensations and theory (image) was a matter of experience.

(3) In §304 Hertz even hinted at the conventional nature of rule 2 and the corresponding rules for translating times and masses from our sensations into our image:

> 304. Observation 3. There is, nevertheless, some apparent warrant for the question whether our three rules furnish true or absolute measures of time, space, and mass, and this question must in all probability be answered in the negative, inasmuch as our rules are obviously in part fortuitous and arbitrary. In truth, however, this question needs no discussion here, not affecting the correctness of our statements, even if we attached to the question a definite meaning and answered it in the negative. It is sufficient that our rules determine such measures as enable us to express without ambiguity the result of past and future experiences. Should we agree to use other measures, then the form of our statements would suffer corresponding changes, but in such a manner that the experiences, both past and future, expressed thereby, would remain the same. (Hertz 1894)

This passage seems to hint at something like Helmholtz's mirror world (Helmholtz 1870, 24–25/259–260): Instead of measuring lengths by way of rigid rods (a scale), we first reflect the world in a strangely curved mirror and then measure lengths in the mirror image by way of a rigid rod. In this way the world would not be described correctly by Euclidean geometry but by way of some non-Euclidean geometry, depending on the mirror. The "form of our statements would suffer corresponding changes." Hertz did not analyze what kind of changes would result, nor did he argue why he preferred the usual way of measuring lengths (rule 2). Thus Hertz's brief anticipation of conventionalism remained an inconsequential parenthesis in the book.

COMPARISON IN SUMMARY

Let me conclude this discussion by a schematic comparison between Hertz's and Poincaré's views of space and mechanics.

(1) Both Hertz and Poincaré believed that we have an intuition of space as a three-dimensional Euclidean space.

(2) Hertz believed that this intuition was a priori in the sense of Kant, whereas Poincaré argued that we acquire it at an early age through our experiences with visible and tactile space, in particular the commutativity of translations.

(3) Both Hertz and Poincaré believed that three-dimensional Euclidean geometry was the right description of external or physical space.

(4) Hertz apparently thought that this was a priori and synthetic, in a sense close to that of Kant, although he hinted at other possibilities without explaining why he preferred the Euclidean convention. Poincaré, by contrast, emphasized that a Euclidean theory of physical space is a convention, but since it is the simplest convention (in a group theoretical sense), it would and should never be changed, even in the face of future apparent empirical falsification.

(5) Hertz and Poincaré agreed that in order to translate external sensations into geometrical language one needs a "dictionary"—for example, "light rays" are "straight lines," and "lengths" are measured by a "scale."

(6) They also agreed that dictionaries were empirically testable. The translations given above were in agreement with past experiences (of their day), but could in principle be falsified in the future, in which case other dictionaries would have to be found, such that Euclidean geometry would remain a correct image of external space.

(7) Hertz and Poincaré further agreed that the basic laws of mechanics are given to us by experience (in fact by sophisticated scientific experiments).

(8) However, where Hertz believed that his fundamental law of motion might be annulled by experience, Poincaré rejected such a possibility. In his opinion, the simple laws of mechanics were conventions, and, like the convention of Euclidean geometry, they were the simplest conventions. Possible future apparent falsifications could and should be explained away by making suitable hypotheses about hitherto unobserved interactions or hidden mechanical systems.

SIGNIFICANCE IN LIGHT OF EARLIER AND LATER VIEWS

Hertz's theory of images and Poincaré's conventionalism played a certain role in the revolutionary developments that took place in physics a few years after their books were published. In hindsight, however, we can see that they both misjudged the road geometry and mechanics were to take. What happened was that experimental results such as the Michelson-Morley experiment and the precession of the perihelion of Mercury led Einstein to formulate his special and general theories of relativity, in which the image of time, space, and mechanics was radically changed relative to the classical image. Einstein's theories were accepted not because it was proved that an explanation of the experiments was impossible within the classical image (in fact Lorentz did exactly that), but because Einstein's theories were considered simpler than the semiclassical competing theories.

This comparison of images and the preference for the simpler one is in accordance with Hertz's theory of images. However, it was precisely the concept of time and space, the two a priori elements in Hertz's image, that suffered the greatest changes, and not just the laws of motion that according to Hertz would be the only ones that could change in the face of experimental falsification. Thereby his careful separation of the a priori and the empirical elements of mechanics was ruined.

It is also conspicuous that Hertz only speculated that his image could be overturned if it was shown to be incorrect; he did not explicitly address the possibility that it might be rejected because it turned out to be less simple than another image in the light of future experiments. Yet his general theory of images clearly opened such a possibility.

Poincaré's conventionalism also failed, but for different reasons. In a sense, his general idea of conventions was in harmony with the later historical development. Indeed we can still argue that our theories about spacetime and mechanics are conventions. Where he failed was in his insistence that Euclidean geometry and Newtonian mechanics were the simplest conventions *and would remain so* even in the light of all possible future experience.

It is of course easy to point out these shortcomings with the benefit of hindsight. I wish to claim, however, that even in the light of what was known at the time about geometry, mechanics, and their development, these shortcomings in the philosophies of Hertz and Poincaré are remarkable, and may partly be ascribed to a Kantian influence. Indeed, it is a fact that many mathematicians and physicists (such as Riemann and Helmholtz) had explicitly pointed out that non-Euclidean geometries, including Riemannian geometry, were a priori possible images of space. Hertz surely knew of these ideas and even hinted at them, and still he chose to present Euclidean geometry as an a priori intuition in Kant's sense, whose correctness as an image of external space was immune to experience.

And, although Poincaré's conventionalism was a remarkably acute clarification of ideas one can find in Helmholtz, Poincaré's rejection of the possibility that these conventions could or should ever change was certainly conservative. Already Riemann had argued that "it is entirely conceivable that in the infinitely small the spatial relations of size are not in accord with the postulates of (Euclidean) geometry, and one would indeed be forced to this assumption as soon as it would permit a simpler explanation of the phenomena" (Riemann 1876a). It is interesting that Riemann admitted not only that one had to reject Euclidean geometry if it was shown to be incorrect, but that one should reject it as soon as a simpler explanation was found. In this respect he was more farsighted than Poincaré. Similarly Clifford (1882) imagined that one might simplify all of physics by assuming that space has curvature that varies in time and place.

Why did Poincaré reject the possibility of changing our convention of space? In addition to a weak Kantian influence, I think the reason can be found in the implicitly hierarchical structure of Poincaré's philosophy of nature. Of course, in one sense it is true that he considered the epistemology of time, space, mechanics, and physics as one connected whole. It is precisely such an integrated view that led to conventionalism. Still, the order of his presentation, as well as the logical structure of his epistemological investigations, convey a clear stratification of the material according to evolutionary, psychological, or scientific elementarity, a stratification that corresponds to the sequence in which we humans encounter the subjects. First comes the a priori arithmetic (see Folina, this volume); then comes geometry, which we learn early in life in a nonscientific way; then comes mechanics, whose laws we extract later in life from daily experience combined with astronomical observations and mechanical experiments; and finally comes the rest of physics, which we learn from even more sophisticated experimentation. Poincaré's rejection of the possibility that our conventions of geometry and mechanics may ever change seems to rest on an implicit doctrine to the effect that one must never change a lower-level convention (of, e.g., geometry or mechanics) in order to obtain simplicity on a higher level (e.g., physics), if doing so will make explanations on the lower level less simple—even if the overall theory gains simplicity. If he had not implicitly adhered to such a doctrine, he might have accepted that although the Lorenz group is not as simple as the Galilei group and thus gives a more complicated description of our everyday experience with space, it gives a much simpler description of the higher parts of physics and should be preferred because it offers a simpler overall image of space-time than the classical one.

Poincaré's reluctance to accept the possibility of such changes of conventions is all the more striking when we consider that the history of science has been full of changes of this kind.[5] For example, when Copernicus changed our image of the solar system, it did not happen because it was impossible to add more epicycles to the Ptolemaic image to save the phenomena, but because the Copernican image was simpler. At least it meant a simplification on the higher level of astronomy, but it also meant a complication on the lower level of everyday experience. For example, it became more complicated to explain why a stone ejected vertically falls down (virtually) at the point from where it was ejected. Of course, in our dealings with everyday phenomena we therefore still assume that the earth stands still, and even in most scientific experiments we work as though the laboratory is an inertial system. We know, however, that according to the overall scientific image this is only an approximation. Similarly, when we build houses or send people to the moon we still accept the classical images of time, space, and mechanics. In that sense Poincaré was right. We will never change those conventions as long as we deal with everyday experience. But, in contrast to Poincaré, we know that if we want to deal with fast-moving systems and cosmology, it is convenient to change convention.

As far as the Kantian elements in Hertz's and Poincaré's theories of space are concerned, we can conclude that Hertz's conception of space was rather Kantian, whereas Poincaré explicitly rejected Kantian apriorism. As for the laws of mechanics, none of them agreed with Kant as far as the a priori nature of the fundamental law(s) is concerned, but where Poincaré agreed with Kant that experience can never falsify the laws of mechanics,

Hertz believed that this could happen. Their different views may reflect their training and areas of research: Where a physicist like Hertz might remain content with a rather Kantian image of space, a mathematician like Poincaré who had actively pursued research in non-Euclidean geometry might be more inclined to a conventionalist view of space. And where a mathematician and theoretical physicist like Poincaré might tend to stress the simplicity of the classical image of mechanics, an experimental physicist like Hertz might be more open to the possibility that experience might lead to fundamental changes in the laws of mechanics.

NOTES

1. This is how the present paper is related to Gray's chapter in this volume.

2. For an explanation, see Lützen 1998 and Lützen 1999.

3. For a more subtle and comprehensive view of Hertz's point of view, see Lützen 2005.

4. This is Hertz's only hint to the hotly debated question of the existence of absolute space, and the determination of inertial frames.

5. Poincaré at least once expressed his willingness to give up a law of mechanics (see Gray, this volume). Indeed in his *Science et méthode* (Poincaré 1908e, 225) he argued that experiments in optics seem to show that one must give up Newton's third law of action and reaction. This is the more remarkable because this law, according to Poincaré, is not a convention but a definition (Poincaré 1902b, 110). Thus we seem to have an example of an empirical falsification of a definition!

References

As a rule, texts are identified by the date of their first publication, which may or may not be identical to the publication date of the edition used. For example, "Einstein 1921" refers to his text "Geometrie und Erfahrung," which is cited as it appeared in a 1954 collection of his papers.

Works by Immanuel Kant (as cited in this volume)

(A 5), (B 321). 1781/1787. *Kritik der reinen Vernunft.* In *Kants gesammelte Schriften,* vols. 3 and 4. Berlin: De Gruyter, 1903 and 1904. Translated into English as *Critique of Pure Reason* (Cambridge: Cambridge University Press, 1997).

§65, V, 374. 1790. *Kritik der Urteilskraft.* In *Kants gesammelte Schriften,* 5:165–485. Berlin: De Gruyter, 1908. Translated into English as *Critique of the Power of Judgment* (Cambridge: Cambridge University Press, 2002).

II, 385–419. 1770. De Mundi Sensibilis Atque Intelligibilis (inaugural dissertation). In *Kants gesammelte Schriften,* 2:385–419. Berlin: De Gruyter, 1905. Translated into English as "On the form and principles of the sensible and intelligible world (inaugural dissertation, 1770)," in *Kant's Latin Writings,* ed. Lewis White Beck, 145–192 (New York: Peter Lang, 1986).

II, 273–301. 1763. Untersuchung über die Deutlichkeit der Grundsätze der natürlichen Theologie und der Moral. In *Kants gesammelte Schriften,* 2:273–301. Berlin: De Gruyter, 1905.

IV, 253–383. 1783. Prolegomena zu einer jeden künftigen Metaphysik die als Wissenschaft wird auftreten können. In *Kants gesammelte Schriften,* 4:253–383. Berlin: De Gruyter, 1903. Translated into English as "Prolegomena to any future metaphysics that will be able to come forward as a science," in *Immanuel Kant: Theoretical Philosophy after 1781,* ed. Henry Allison and Peter Heath, 29–170 (Cambridge: Cambridge University Press, 2002).

IV, 465–565. 1786. Metaphysische Anfangsgründe der Naturwissenschaften. In *Kants gesammelte Schriften,* 4:465–565. Berlin: De Gruyter, 1903. Translated into English as "Metaphysical foundations of natural science," in *Immanuel Kant: Theoretical Philosophy after 1781,* ed. Henry Allison and Peter Heath, 171–270 (Cambridge: Cambridge University Press, 2002).

V, 1–163. 1788. *Kritik der praktischen Vernunft.* In *Kants gesammelte Schriften,* 5:1–163. Berlin: De Gruyter, 1908. Translated into English as *Critique of Practical Reason* (New York: Macmillan, 1993).

VIII, 45–66. 1785. Recension von Herders Ideen zur Philosophie. In *Kants gesammelte Schriften*, 8:45–66. Berlin: De Gruyter, 1912/1923.

VIII, 157–184. 1788. Über den Gebrauch teleologischer Prinzipien in der Philosophie. In *Kants gesammelte Schriften*, 8:157–184. Berlin: De Gruyter, 1912/1923.

XVII, 227–745, and XVIII, 3–725. 1926/1928. Reflexionen zur Metaphysik. In *Kants gesammelte Schriften*, 17:227–745 and 18:3–725. Berlin: De Gruyter, 1926, 1928.

OTHER SOURCES

Abbagnano, Nicola. 1967. Psychologism. In *Encyclopedia of Philosophy*, ed. Paul Edwards, 6:520–521. New York: Macmillan.

Adler, Emil. 1968. *Der junge Herder und die Aufklärung*. Vienna: Europa.

Ameriks, Karl. 2000. The practical foundation of philosophy in Kant, Fichte, and after. In *The Reception of Kant's Critical Philosophy*, ed. Sally Sedgwick, 109–129. Cambridge: Cambridge University Press.

Amir-Arjomand, Kamran. 1990. *Entdeckung und Rechtfertigung in der Wissenschaftsphilosophie des 19. Jahrhunderts unter besonderer Berücksichtigung der Entwicklung in Deutschland von ca. 1800–1875*. Frankfurt am Main: Lang.

Ash, Mitchell. 1995. *Gestalt Psychology in German Culture, 1890–1967: Holism and the Quest for Objectivity*. Cambridge: Cambridge University Press.

Beiser, Frederick. 2002. *German Idealism*. Cambridge, Mass.: Harvard University Press.

Beltrami, Edward. 1868. Saggio di interpretazione della geometria non-Euclidea. *Giornale di Matematica* 6: 284–312.

———. 1869. Essai d'interprétation de la géométrie non-Euclidienne. *Annales Ecole Normale Superieure* 6: 251–288.

Blumenbach, Johann Friedrich. 1791. *Über den Bildungstrieb*. Göttingen: Dietrich.

Bode, Johann Elert. 1823. *Anleitung zur Kenntnia des gestirnten Himmels*. 9th ed. Berlin: Nicolaische Buchhandlung.

Boisserée, Sulpiz. 1978. *Tagebücher*, vol. 1. Darmstadt: Eduard Roether Verlag.

Bonsiepen, Wolfgang. 1997. *Die Begründung einer Naturphilosophie bei Kant, Schelling, Fries und Hegel*. Frankfurt am Main: Klostermann.

Bonwetsch, Nathanael. 1918. *G. H. Schubert in seinen Briefen. Ein Lebensbild*. Stuttgart: Belser.

Boring, Edwin G. 1942. *Sensation and Perception in the History of Experimental Psychology*. New York: D. Appleton.

Boyle, Nicholas. 1991/2000. *Goethe: The Poet and the Age*. 2 vols. Oxford: Oxford University Press.

Buchdahl, Gerd. 1969. *Metaphysics and the Philosophy of Science*. Oxford: Basil Blackwell.

Buchwald, Jed Z. 1985. *From Maxwell to Microphysics*. Chicago: Chicago University Press.

Burckhardt, Guido. 1893/1897. *Die Brüdergemeine*. 2 vols. Gnadau: Unitäts-Buchhandlung.

Cahan, David, ed. 1992. *Hermann Helmholtz: Philosopher and Scientist*. Berkeley and Los Angeles: University of California Press.

————. 1993. *Hermann von Helmholtz and the Foundations of Nineteenth-Century Science*. Berkeley and Los Angeles: University of California Press.

Caneva, Kenneth L. 1990. Teleology with regrets. *Annals of Science* 47: 291–300.

————. 1993. *Robert Mayer and the Conservation of Energy*. Princeton: Princeton University Press.

————. 1997. Physics and *Naturphilosophie*: A reconnaissance. *History of Science* 35: 35–106.

Carnap, Rudolf. 1934a. *The Logical Syntax of Language*. London: Kegan, Paul, Trench, Teubner.

————. 1934b. The task of the logic of science. In *The Unity of Science,* ed. Brian MacGuinness, 46–66. Dordrecht: Reidel.

————. 1936. Von Erkenntnistheorie zur Wissenschaftslogik. *Actes du 8eme Congrès International de Philosophie Scientifique* 1: 36–41. Paris: Hermann.

Carrier, Martin. 1990. Kants Theorie der Materie und ihre Wirkung auf die zeitgenössische Chemie. *Kant-Studien* 81: 170–210.

Case, Thomas. 1911. Metaphysics. In *Encyclopaedia Britannica,* 11th ed., 18:224–253. Cambridge: Cambridge University Press.

Cassirer, Ernst. 1907. Kant und die moderne Mathematik. *Kantstudien* 12: 1–49.

————. 1910. *Substanzbegriff und Funktionsbegriff*. Berlin: Cassirer. Translated into English as *Substance and Function and Einstein's Theory of Relativity* (Chicago: Open Court, 1923; reprint, New York: Dover, 1953). References are to English/German editions.

————. 1920. *Das Erkenntnisproblem in der Philosophie und Wissenschaft der neuren Zeit*. Vol. 3, *Die nachkantischen Systeme*. Berlin: Bruno Cassirer.

————. 1923. *Das Erkenntnisproblem in der Philosophie und Wissenschaft der neueren Zeit*. Vol. 3. 2d ed. Darmstadt: Wissenschaftliche Buchgesellschaft, 1994.

————. 1950. *The Problem of Knowledge: Philosophy, Science, and History since Hegel*. New Haven: Yale University Press.

Cat, Jordi. 2001. On understanding: Maxwell on the methods of illustration and scientific metaphor. *Studies in History and Philosophy of Modern Physics* 32: 395–441.

Charpa, Ulrich. 1988. Methodologie der Verzeitlichung—Schleiden, Whewell und das entwicklungsgeschichtliche Projekt. *Philosophia Naturalis* 25: 74–109.

———. 1999. Schleidens Kritik an Hegel und Schelling: Abriß einer reliabilistischen Wissenschaftstheorie. In *Jakob Friedrich Fries. Philosoph, Naturwissenschaftler und Mathematiker,* ed. Wolfram Hogrebe and K. Hermann, 255–281. Frankfurt am Main: Lang.

Christensen, Dan. 1995. The Oersted-Ritter partnership and the birth of romantic natural philosophy. *Annals of Science* 52: 153–185.

Clifford, Willliam Kingdon. 1882. On the space-theory of matter. In *Mathematical Papers,* ed. Robert Tucker, pp. 21–22. London: Macmillan.

Coffa, J. A. 1991. *The Semantic Tradition from Kant to Carnap: To the Vienna Station.* Cambridge: Cambridge University Press.

Cohen, Hermann. 1871. *Kants Theorie der Erfahrung.* Berlin: Dümmler.

———. 1883. *Das Princip der Infinitesimal-Methode und seine Geschichte.* Berlin: Dümmler. Reprinted in volume 5 of Hermann Cohen's *Werke* (Hildesheim: Olms, 1984); reprinted with an introduction by W. Flach (Frankfurt am Main: Suhrkamp, 1968). References are to the 1883 edition.

———. 1885. *Kants Theorie der Erfahrung.* 2d ed. Berlin: Dümmler. The first edition appeared in 1871 (Berlin: Dümmler), the third edition in 1918 (Berlin: Bruno Cassirer). The first and third editions (with page references to the second edition) have been reprinted in volumes 1.3 and 1.1 of Hermann Cohen's *Werke* (Hildesheim: Olms, 1987).

———. 1904. *Rede bei der Gedenkfeier der Universität Marburg zur hundertsten Wiederkehr des Todestages von Immanuel Kant.* Marburg: Elwert'sche.

———. 1914. *Einleitung mit kritischem Nachtrag zur neunten Auflage der* Geschichte des Materialismus *von Friedrich Albert Lange.* 3d ed. Leipzig: Brandstetter. Reprinted in volume 5 of Hermann Cohen's *Werke* (Hildesheim: Olms, 1984).

Cohen, Paul J. 1966. *Set Theory and the Continuum Hypothesis.* New York: Freeman.

Couturat, Louis. 1901. *La logique de Leibniz d'après des documents inédits.* Paris: F. Alcan.

Crowe, Michael J. 1967. *A History of Vector Analysis.* Notre Dame: University of Notre Dame Press.

Dingler, Hugo. 1934. H. Helmholtz und die Grundlagen der Geometrie. *Zeitschrift für Physik* 90: 348–354.

Dirichlet, G. Lejeune. 1849. Über die Reduction der positiven quadratischen Formen mit drei unbestimmten ganzen Zahlen. *Journal für die reine und angewandte Mathematik* 40: 209–227. Reprinted in *G. Lejeune Dirichlet's Werke,* ed. Leopold Kronecker and Leopold Fuchs, 2:29–48 (Berlin: Königliche Preuaische Akademie der Wissenschaften, 1897).

DiSalle, Robert. 1993. Helmholtz's empiricist philosophy of mathematics. In *Hermann von Helmholtz and the Foundations of Nineteenth-Century Science,* ed. David Cahan, 496–519. Berkeley and Los Angeles: University of California Press.

———. 1995. Spacetime theory as physical geometry. *Erkenntnis* 42: 317–337.

———. 2002. Conventionalism and modern physics: A re-assessment. *Noûs* 36: 169–200.

Dowe, Phil. 2000. *Physical Causation.* Cambridge: Cambridge University Press.

Drobisch, Moritz Wilhelm. 1850. *Erste Grundlehren der mathematischen Psychologie.* Leipzig: Leopold Voss.

Durner, Manfred. 1991. Die Naturphilosophie im 18. Jahrhundert und der naturwissenschaftliche Unterricht in Tübingen. *Archiv für Geschichte der Philosophie* 73: 95–99.

———. 1994. Theorie der Chemie. In the supplementary volume to *Schelling, Historisch-Kritische Ausgabe, Wissenschaftlicher Bericht zu Schellings Naturphilosophischen Schriften 1797–1800,* 44–56. Stuttgart: Fromann.

Edel, Geert. 1987. Einleitung. In Hermann Cohen, *Werke,* 1.1:8*–59*. Hildesheim: Olms.

Ehlers, J., F. Pirani, and A. Schild. 1972. The geometry of free fall and light propagation. In *General Relativity: Essays in Honour of J. L. Synge,* ed. Lochlainn O'Raifeartaigh, 63–84. Oxford: Clarendon Press.

Einstein, Albert. 1916. The foundation of the general theory of relativity. In *The Principle of Relativity,* ed. Albert Einstein, H. A. Lorentz, H. Minkowski, and H. Weyl, 109–164. New York: Dover Books, 1952.

———. 1921. Geometrie und Erfahrung. In *Preussische Akademie der Wissenschaft. Physikalisch-mathematische Klasse. Sitzungsberichte,* 123–130. Translated into English as "Geometry and experience," in *Ideas and Opinions* (New York: Bonanza Books, 1954), 232–246. References are to the English edition.

Einstein, Albert, H. A. Lorentz, H. Minkowski, and H. Weyl. 1952. *The Principle of Relativity.* New York: Dover Books.

Elsenhans, Theodor. 1906. *Fries und Kant. Ein Beitrag zur Geschichte und zur systematischen Grundlegung der Erkenntnistheorie.* 2 vols. Gießen: Töpelmann.

Ende, Helga. 1973. *Der Konstruktionsbegriff im Umkreis des Deutschen Idealismus.* Meisenheim am Glan: Hain.

Engel, Friedrich. 1911. Grassmanns Leben. In *Hermann Grassmanns gesammelte mathematische und physikalische Werke,* vol. 3. Leipzig: Teubner.

Engelhardt, Dietrich von. 1998. Natur und Geist, Evolution und Geschichte: Goethe in seiner Beziehung zur romantischen Naturforschung und metaphysischen Naturphilosophie. In *Goethe und die Verzeitlichung der Natur,* ed. Peter Matussek, 58–74. Munich: C. H. Beck.

Enriques, Federigo. 1906. *Problemi della scienza*. Bologna: Zanichelli.

Erman, Adolf, and Max Planck. 1905. Report to Minister Althoff regarding the professorship for philosophy vacated by Dilthey. Berlin: Universitätsarchiv der Humboldt-Universität, Akten der Philosophischen Fakultät, 7 July 1905, I, no. 364.

Feigl, Herbert. 1963a. Physicalism, unity of science, and the foundations of psychology. In *The Philosophy of Rudolf Carnap*, ed. Paul Arthur Schilpp, 227–267. La Salle, Ill.: Open Court.

———. 1963b. Physicalism, unity of science, and the foundations of psychology. Incomplete manuscript, not later than 1963, 4 pp., HF 01-16-03. In Herbert Feigl Nachlass, Archive for Scientific Philosophy, Hillman Library, University of Pittsburgh. (Not included in Feigl 1963a.)

———. 1966. Biographical notes. 18 pp., HF 01-03-01, in Herbert Feigl Nachlass, Archive for Scientific Philosophy, Hillman Library, University of Pittsburgh.

———. 1974. No pot of message. In *Inquiries and Provocations: Selected Writings, 1929–1974,* ed. Robert S. Cohen, 1–20. Dordrecht: Reidel.

Feigl, Herbert, and Albert E. Blumberg. 1974. Introduction. In *General Theory of Knowledge,* ed. Moritz Schlick, xvii–xxvi. Vienna: Springer.

Ferrari, Massimo. Über die Ursprünge des logischen Empirismus, den Neukantianismus und Ernst Cassirer aus der Sicht der neueren Forschung. In *Von der Philosophie zur Wissenschaft. Cassirers Dialog mit der Naturwissenschaft,* ed. Enno Rudolph, 93–131. Hamburg: Meiner.

Feyerabend, Paul K. 1966. Herbert Feigl: A biographical sketch. In *Mind, Matter, and Method: Essays in Philosophy of Science in Honor of Herbert Feigl,* ed. Paul K. Feyerabend and Grover Maxwell, 3–13. Minneapolis: University of Minnesota Press.

Fick, Adolf. 1854. Die Bewegungen des menschlichen Augapfels. *Zeitschrift für rationelle Medicin* 4: 801.

Folina, Janet. 1992. *Poincaré and the Philosophy of Mathematics*. New York: Macmillan.

Förster, Eckart. 2000. *Kant's Final Synthesis*. Cambridge, Mass.: Harvard University Press.

Franks, Paul. 2003. What should Kantians learn from Maimon's skepticism? In *Salomon Maimon: Rational Dogmatist, Empirical Skeptic,* ed. G. Freudenthal, 200–232. Dordrecht: Kluwer.

Friedman, Michael. 1983. *Foundations of Space-Time Theories*. Princeton: Princeton University Press.

———. 1992. *Kant and the Exact Sciences*. Cambridge, Mass.: Harvard University Press.

———. 1996a. Poincaré's conventionalism and the logical positivists. In *Henri Poincaré: Science et philosophie,* ed. Jean-Louis Greffe, G. Heinzmann, and K. Lorenz, 333–344. Berlin: Akademie Verlag. Reprinted in Friedman 1999b, 71–88.

————. 1996b. Overcoming metaphysics: Carnap and Heidegger. In *Origins of Logical Empiricism,* ed. Ronald Giere and Alan Richardson, 45–79. Minneapolis: University of Minnesota Press.

————. 1997. Helmholtz's *Zeichentheorie* and Schlick's *Allgemeine Erkenntnislehre*: Early logical empiricism and its nineteenth-century background. *Philosophical Topics* 25, no. 2: 19–50.

————. 1999. *Reconsidering Logical Positivism.* Cambridge: Cambridge University Press.

————. 2000a. *A Parting of the Ways: Carnap, Cassirer, Heidegger.* Chicago: Open Court.

————. 2000b. Geometry, construction, and intuition in Kant and his successors. In *Between Logic and Intuition: Essays in Honor of Charles Parsons,* ed. G. Scher and R. Tieszen, 186–218. Cambridge: Cambridge University Press.

————. 2001. *Dynamics of Reason.* Stanford: CSLI Publications.

Fries, Jakob Friedrich. 1801. Versuch einer Kritik der Richterischen Stöchyometrie. In *Sämtliche Schriften,* 17:1–136. Aalen: Scientia Verlag, 1975.

————. 1802. Untersuchungen über die Natur der Wärme und des Lichtes. In *Sämtliche Schriften,* 17:149–220. Aalen: Scientia Verlag, 1975.

————. 1803. *Reinhold, Fichte, und Schelling.* In *Sämtliche Schriften,* 24:33–476. Aalen: Scientia Verlag, 1978.

————. 1804. *System der Philosophie als evidente Wissenschaft aufgestellt.* In *Sämtliche Schriften,* 3:7–410. Aalen: Scientia Verlag, 1968.

————. 1805. *Wissen, Glaube und Ahndung.* In *Sämtliche Schriften,* 3:413–755. Aalen: Scientia Verlag, 1968. Translated into English as *Knowledge, Belief, and Aesthetic Sense* (Cologne: Dinter Verlag, 1989).

————. 1807. *Atomistik und Dynamik.* In *Sämtliche Schriften,* 17:221–235. Aalen: Scientia Verlag, 1975.

————. 1813. *Entwurf des Systems der theoretischen Physik. Zum Gebrauch bei seinen Vorlesungen.* In *Sämtliche Schriften,* 14:257–400. Aalen: Scientia Verlag, 1974.

————. 1822. *Die mathematische Naturphilosophie nach philosophischer Methode bearbeitet.* Vol. 13 of *Sämtliche Schriften.* Aalen: Scientia Verlag, 1979.

————. 1824. *System der Metaphysik. Ein Handbuch für Lehrer und zum Selbstgebrauch.* In *Sämtliche Schriften,* 8:1–536. Aalen: Scientia Verlag, 1970.

————. 1826a. *Lehrbuch der Naturlehre.* Vol. 15 of *Sämtliche Schriften.* Aalen: Scientia Verlag, 1973.

————. 1826b. *Grundriß der Logik. Ein Lehrbuch zum Gebrauch für Schulen und Universitäten.* In *Sämtliche Schriften,* 7:29–152. Aalen: Scientia Verlag, 1971.

————. 1828–1831. *Neue oder anthropologische Kritik der Vernunft.* 2d ed. Vols. 4:31–478, 5, and 6 of *Sämtliche Schriften.* Aalen: Scientia Verlag, 1967.

————. 1837. *System der Logik. Ein Handbuch für Lehrer und zum Selbstgebrauch.* In *Sämtliche Schriften,* 7:153–622. Aalen: Scientia Verlag, 1971.

————. 1837/1839. *Handbuch der psychischen Anthropologie.* 2d ed. Vols. 1 and 2 of *Sämtliche Schriften.* Aalen: Scientia Verlag, 1982.

————. 1842. *Versuch einer Kritik der Prinzipien der Wahrscheinlichkeitsrechnung.* In *Sämtliche Schriften,* 14:11–254. Aalen: Scientia Verlag, 1974.

————. 1997. *Briefe I. Konvolute A–E.* Vol. 27 of *Sämtliche Schriften.* Aalen: Scientia Verlag.

Gauß, Carl Friedrich. 1829. Über ein neues allgemeines Grundgesetz der Mechanik. In *Carl Friedrich Gauss: Werke,* 5:5–28. Göttingen: W. F. Kaestner, 1877.

————. 1831. *Untersuchungen über die Eigenschaften der positiven ternären quadratischen Formen.* In *Carl Friedrich Gauß: Werke,* 2:188–96. Göttingen: W. F. Kaestner, 1876.

George, Rolf. 1997. Erstaunliche Unkenntnis: Philosophen über die exakten Wissenschaften im 19. Jahrhundert. *Bremer Philosophica* 1: 1–9.

Gerhardt, Volker, Reinhard Mehring, and Jana Rindert. 1999. *Berliner Geist: Eine Geschichte der Berliner Universitätsphilosophie bis 1946.* Berlin: Akademie.

Giere, Ronald, and Alan Richardson, eds. 1996. *Origins of Logical Empiricism.* Minneapolis: University of Minnesota Press.

Glasmacher, Thomas. 1989. *Fries—Apelt—Schleiden. Verzeichnis der Primär-und Sekundärliteratur, 1798–1988.* Cologne: Dinter.

Gloy, Karen. 1976. *Die Kantische Theorie der Naturwissenschaften.* Berlin: De Gruyter.

Glymour, Clark. 1977. The epistemology of geometry. *Noûs* 11: 227–251.

Goethe, Johann Wolfgang von. 1790. Versuch einer allgemeinen Vergleichungslehre. In *Sämtliche Werke,* 4.2:179–184. Munich: Carl Hanser Verlag, 1986.

————. 1792. Der Versuch als Vermittler von Objekt und Subjekt. In *Sämtliche Werke,* 4.2:321–332. Munich: Carl Hanser Verlag, 1986.

————. 1796. Vorträge, über die drei ersten Kapitel des Entwurfs einer allgemeinen Einleitung in die vergleichende Anatomie, ausgehend von der Osteologie. In *Sämtliche Werke,* 12:195–225. Munich: Carl Hanser Verlag, 1989.

————. 1798. Tag- und Jahreshefte. In *Sämtliche Werke,* 14:7–322. Munich: Carl Hanser Verlag, 1986.

————. 1800. Natur und Kunst. In *Sämtliche Werke,* 6.1:780. Munich: Carl Hanser Verlag.

————. 1817. Anschauende Urteilskraft. In *Sämtliche Werke,* 12:98–99. Munich: Carl Hanser Verlag, 1989.

————. 1820. Einwirkung der neueren Philosophie. In *Sämtliche Werke*, 12:94–98. Munich: Carl Hanser Verlag, 1989.

————. 1888. *Tagebücher*. In *Goethes Werke*, part 3, vols. 2 and 3. Weimar: Böhlau.

————. 1899. *Briefe*. In *Goethes Werke*, part 4, vol. 19. Weimar: Böhlau.

————. 1986. Weltseele. In *Sämtliche Werke*, 6.1:53–54. Munich: Carl Hanser Verlag.

————. 1988. *Goethes Briefe*. 4 vols. 4th ed. Ed. Karl Robert Mandelkow (Hamburger edition). Munich: C. H. Beck.

————. 1990. *Sämtliche Werke*, 8.1:14–15. Munich: Carl Hanser Verlag.

————. 1991. *Maximen und Reflexionen*. In *Sämtliche Werke*, 17:715–954. Munich: Carl Hanser Verlag.

————. 1998. *Goethes Gespräche*. 5 vols. Munich: Deutscher Taschenbuch Verlag.

Goldfarb, Warren. 1988. Poincaré against the logicists. In *History and Philosophy of Modern Mathematics*, ed. William Aspray and Philip Kitcher, 61–81. Minneapolis: University of Minnesota Press.

Goller, Peter. 1991. Alois Riehl (1844–1924): Bausteine zur Biographie eines Südtiroler Philosophen. *Der Schlern. Monatszeitschrift für Südtiroler Landeskunde* 65, no. 10: 530–558.

Gower, Barry. 1973. Speculation in physics: The history and practice of *Naturphilosophie*. *Studies in History and Philosophy of Science* 3: 301–356.

————. 2000. Cassirer, Schlick, and "structural" realism: The philosophy of the exact sciences in the background to early logical empiricism. *British Journal for the History of Philosophy* 8, no. 1: 71–106.

Grassmann, Hermann. 1844. *Die Wissenschaft der extensiven Grössen oder die Ausdehnungslehre, eine neue mathematische Disciplin*. Leipzig: Otto Wigand. Reprinted in *Hermann Grassmann's Gesammelte mathematische and physikalische Werke*, ed. Friedrich Engel (Leipzig: Teubner, 1894), vol. 1, part 1; references are to this 1894 edition.

————. 1847. *Geometrische Analyse geknüpft an die von Leibniz erfundene geometrische Charakteristik*. Leipzig: Weidmann.

————. 1853. Zur Theorie der Farbenmischung. *Annalen der Physik and der Chemie* 89: 169–184.

Gray, Jeremy. 1999. Geometry—formalisms and intuitions. In Jeremy Gray, *The Symbolic Universe: Geometry and Physics, 1890–1930,* 58–83. Oxford: Oxford University Press.

Graykowski, Georg. 1914. Alois Riehl: Zum 70. Geburtstage des Philosophen (27 April 1914). Separate offprint from *Montagsblatt: Wissenschaftliche Wochenbeilage der Magdeburgischen Zeitung*, Nr. 17, 8 pp.

Greffe, Jean-Louis, G. Heinzmann, and K. Lorenz, eds. 1996. *Henri Poincaré: Science et philosophie*. Berlin: Akademie Verlag.

Gregory, Frederick. 1983a. Die Kritik von J. F. Fries an Schellings Naturphilosophie. *Sudhoffs Archiv* 6: 146–157.

————. 1983b. Neo-Kantian foundations of geometry in the German romantic period. *Historia Mathematica* 10: 184–201.

————. 1984. Romantic Kantianism and the end of the Newtonian dream in chemistry. *Archives Internationales d'Histoire des Sciences* 34: 108–123.

————. 1989. Kant's influence on natural scientists in the German romantic period. In *New Trends in the History of Science,* ed. Robert P. W. Visser, H. J. M. Bos, L. C. Palm, and H. A. M. Snelders, 53–66. Amsterdam: Rodopi.

————. 1990. Theology and the sciences in the German romantic period. In *Romanticism and the Sciences,* ed. A. Cunningham and N. Jardine, 73–78. Cambridge: Cambridge University Press.

————. 1997. The Newtonian vitalism of J. F. Fries. In *Vitalisms from Haller to Cell Theory,* ed. Guido Cimino and François Duchesneau, 149–151. Florence: Olschki.

Gülberg, Niels. 2003. Alois Riehl und Japan. *Humanitas (Journal of the Waseda Law Association)* no. 41: 1–32.

Guyer, Paul. 1990a. Reason and reflective judgment: Kant on the significance of systematicity. *Noûs* 24: 17–43.

————. 1990b. Kant's conception of empirical law. *Proceedings of the Aristotelian Society,* supplement to vol. 64: 221–242.

————. 2000. Absolute idealism and the rejection of Kantian dualism. In *The Cambridge Companion to German Idealism,* ed. Karl Ameriks, 37–56. Cambridge: Cambridge University Press.

Hacking, Ian. 1980. The theory of probable inference: Neyman, Peirce, and Braithwaite. In *Science, Belief, and Behavior,* ed. D. H. Mellor, 141–160. Cambridge: Cambridge University Press.

Hacohen, Malachi Haim. 2000. *Karl Popper: The Formative Years, 1902–1945.* Cambridge: Cambridge University Press.

Hamilton, J. Taylor. 1900. *A History of the Church Known as the Moravian Church during the 18th and 19th Centuries.* Bethlehem, Pa.: Times Publishing.

Hammerstein, Notker. 1985. Universitäten und gelehrte Institutionen von der Aufklärung zum Neuhumanismus und Idealismus. In *Samuel Thomas Soemmerring und die Gelehrten der Goethezeit,* ed. Gunter Mann and Franz Dumont, 309–329. New York: Fischer Verlag.

Hankins, Thomas. 1985. *Science and the Enlightenment.* Cambridge: Cambridge University Press.

Hatfield, Gary. 1990. *The Natural and the Normative: Theories of Spatial Perception from Kant to Helmholtz.* Cambridge, Mass.: MIT Press.

Heath, Archibald Edward. 1917. Hermann Grassmann—the neglect of his work: The geometric analysis and its connection with Leibniz' characteristic. *Monist* 27: 1–56.

Hegel, Georg W. F. 1801. *Differenz zwischen des Fichte'schen und Schelling'schen Systems der Philosophie.* In *Werke in zwanzig Bänden,* 2:52–115. Frankfurt am Main: Suhrkamp, 1971.

———. 1802. Glauben und Wissen. In *Werke in zwanzig Bänden,* 2:301–333. Frankfurt am Main: Suhrkamp, 1971.

Heidelberger, Michael. 2002. Weltbildveränderungen in der modernen Physik vor dem Ersten Weltkrieg. In *Wissenschaften und Wissenschaftspolitik. Bestandsaufnahmen zu Formationen, Brüchen und Kontinuitäten im Deutschland des 20. Jahrhunderts,* ed. Rüdiger vom Bruch and Brigitte Kaderas, 84–96. Stuttgart: Steiner.

———. 2003. The mind–body problem in the origin of logical empiricism: Herbert Feigl and psychophysical parallelism. In *Logical Empiricism: Historical and Contemporary Perspectives,* ed. Paolo Parrini, Wesley C. Salmon, and Merrilee H. Salmon, 233–262. Pittsburgh: University of Pittsburgh Press.

———. 2004. *Nature from Within: Gustav Theodor Fechner's Psychophysical Worldview.* Pittsburgh: University of Pittsburgh Press. Amended English translation with a new chapter 5 of *Die innere Seite der Natur. Gustav Theodor Fechners wissenschaftlich-philosophische Weltauffassung* (Frankfurt am Main: Klostermann, 1993).

Heilbron, John L. 1979. *Electricity in the 17th and 18th Centuries: A Study of Early Modern Physics.* Berkeley and Los Angeles: University of California Press.

Heine, Heinrich. 1836. *Die romantische Schule.* Vol. 3 of *Sämtliche Schriften.* Munich: Deutscher Taschenbuch Verlag, 1997.

Helmholtz, Hermann von. 1850. Über die Methoden, kleinste Zeittheile zu messen, and ihre Anwendung für physiologische Zwecke. In *Wissenschaftliche Abhandlungen,* 2:862–880. Leipzig: Barth, 1883.

———. 1852. Über die Theorie der zusammengesetzten Farben. *Annalen der Physik and der Chemie* 87: 45–66. Reprinted in *Wissenschaftliche Abhandlungen* (Leipzig: Barth, 1883), 2:1–23. References are to the 1883 edition.

———. 1853. Ueber Goethes wissenschaftliche Arbeiten. In *Vorträge und Reden von Hermann Helmholtz,* 1:23–45. Braunschweig: F. Vieweg, 1903.

———. 1855a. Über das Sehen des Menschen. In *Vorträge und Reden von Hermann Helmholtz,* 1:85–117. Braunschweig: F. Vieweg, 1903.

———. 1855b. Über die Zusammensetzung von Spectralfarben. *Annalen der Physik and der Chemie* 89: 1–28. Reprinted in *Wissenschaftliche Abhandlungen,* 2:45–70 (Leipzig: Barth, 1883). References are to the 1883 edition.

———. 1856a. Über Combinationstöne. *Annalen der Physik und Chemie* 99: 526. Reprinted in *Wissenschaftliche Abhandlungen,* 1:290 (Leipzig: Barth, 1882). References are to the 1882 edition.

———. 1856b. Über Combinationstöne. *Monatsbericht der Königlichen Akademie der Wissenschaften zu Berlin*, 279–285. Reprinted in *Wissenschaftlichen Abhandlungen*, 1:256–262 (Leipzig: Barth, 1882). References are to the 1882 edition.

———. 1857. Über die physiologischen Ursachen der musikalischen Harmonie. In *Vorträge und Reden von Hermann Helmholtz*, 1:119–155. Braunschweig: F. Vieweg, 1903. Reprinted in *Selected Writings of Hermann von Helmholtz*, ed. Russell Kahl, 144–222 (Middletown: Wesleyan University Press, 1971). References are to Kahl's edition.

———. 1858. Über die subjectiven Nachbilder im Auge. In *Wissenschaftliche Abhandlungen*, 3:13–15. Leipzig: Barth, 1895.

———. 1859. Über Farbenblindheit. In *Wissenschaftliche Abhandlungen*, 2:346–349. Leipzig: Barth, 1883.

———. 1863a. *Die Lehre von den Tonempfindungen als physiologische Grundlage für die Theorie der Musik*. Braunschweig: F. Vieweg.

———. 1863b. Über die normalen Bewegungen des menschlichen Auges. *Archiv für Ophthalmologie* 9: 153–214.

———. 1866. Über die tatsächlichen Grundlagen der Geometrie. *Verhandlungen des natur-historisch-medicinischen Vereins zu Heidelberg* 4: 197–202. Reprinted in *Wissenschaftliche Abhandlungen*, 2:610–617 (Braunschweig: F. Vieweg, 1903).

———. 1867. *Handbuch der physiologischen Optik*. Hamburg: Leopold Voss, 1896.

———. 1868a. Die neuren Fortschritte in der Theorie des Sehens. In *Vorträge und Reden von Hermann Helmholtz*, 1:265–365. Braunschweig: F. Vieweg, 1903. Reprinted in *Selected Writings of Hermann von Helmholtz*, ed. Russell Kahl, 144–222 (Middletown: Wesleyan University Press, 1971). References are to Kahl's edition.

———. 1868b. Über die Thatsachen, die der Geometrie zum Grunde liegen. *Nachrichten von der königlichen Gesellschaft der Wissenschaften zu Göttingen* 9: 193–221. Reprinted in *Wissenschaftliche Abhandlungen*, 2:618 (Braunschweig: F. Vieweg, 1903).

———. 1870. Über den Ursprung und die Bedeutung der geometrischen Axiome. In *Vorträge und Reden von Hermann Helmholtz*, 2:1–31 and 381–383. Braunschweig: F. Vieweg, 1903. Reprinted in *Selected Writings of Hermann von Helmholtz*, ed. Russell Kahl, 246–265 and 360–365. References are to both editions.

———. 1878. Die Tatsachen in der Wahrnehmung. In *Vorträge und Reden von Hermann Helmholtz*, 2:213–247 and 387–406. Braunschweig: F. Vieweg, 1903. Reprinted in *Selected Writings of Hermann von Helmholtz*, ed. Russell Kahl, 366–408 (Middletown: Wesleyan University Press, 1971).

———. 1887. Zählen und Messen, erkenntnistheoretisch betrachtet. In *Wissenschaftliche Abhandlungen*, 3:356–391.

———. 1892. *Goethes Vorahnungen kommender naturwissenschaftlicher Ideen. Rede, gehalten in der Generalversammlung der Goethe-Gesellschaft zu Weimar den 11 Juni 1892*. In *Vorträge und Reden von Hermann Helmholtz*. Braunschweig: F. Vieweg, 1903.

———. 1921. *Schriften zur Erkenntnistheorie.* Ed. Paul Hertz and Moritz Schlick. Berlin: J. Springer. For English edition see Helmholtz 1977.

———. 1977. *Epistemological Writings: The Paul Hertz/Moritz Schlick Centenary Edition of 1921 with Notes and Commentary by the Editors.* Dordrecht: Reidel.

Henckmann, Wolfhart. 1997. Külpes Konzept der Realisierung. *Brentano-Studien* 7: 197–208.

Henke, Ernst Ludwig Theodor. 1867. *Jakob Friedrich Fries. Aus seinem handschriftlichen Nachlasse dargestellt.* Leipzig: Brockhaus.

———. 1937. *Jakob Friedrich Fries. Aus seinem handschriftlichen Nachlasse dargestellt.* 2d ed. Berlin: Verlag Öffentliches Leben.

Henrich, Dieter. 1994. On the unity of subjectivity. In *The Unity of Reason,* ed. Richard Velkley, 17–54. Cambridge, Mass.: Harvard University Press.

Herbart, Johann Friedrich. 1813. *Lehrbuch zur Einleitung in die Philosophie.* Vol. 1 of *Sämmtliche Werke,* ed. Gustav Hartenstein. Leipzig: Leopold Voss, 1850.

———. 1824/1825. Psychologie als Wissenschaft. Vols. 5 and 6 of *Sämmtliche Werke,* ed. Gustav Hartenstein. Leipzig: Leopold Voss, 1850; cited as 1850.

———. 1829. Schriften zur Metaphysik. Part 2, Allgemeine Metaphysik nebst den Anfängen der philosophischen Naturlehre. Vol. 4 of *Sämmtliche Werke,* ed. Gustav Hartenstein. Leipzig: Leopold Voss, 1851.

———. 1850–1851. *Sämmtliche Werke,* 6 vols. Ed. Gustav Hartenstein. Leipzig: Leopold Voss.

Hering, Ewald. 1861. *Beiträge zur Physiologie.* Leipzig: Engelmann.

Hermann, Kay. 2000. *Mathematische Naturphilosophie in der Grundlagendiskussion. Jakob Friedrich Fries und die Wissenschaften.* Göttingen: Vandenhoeck und Ruprecht.

Hertz, Heinrich. 1892. *Untersuchungen über die Ausbreitung der elektrischen Kraft.* Leipzig: Barth. Translated into English as *Electric Waves* (London: Macmillan, 1900).

———. 1894. *Die Prinzipien der Mechanik in neuem Zusammenhange Dargestellt.* Leipzig: Barth. Translated into English as *The Principles of Mechanics Presented in a New Form* (London: Macmillan, 1900; reprint, New York: Dover, 1956). References are zto the German/English editions of the introduction or to the paragraph number.

Hilbert, David. 1901. Mathematische Probleme. In *Gesammelte Abhandlungen,* 3:290–329. Berlin: Springer, 1935. Translated into English as "Mathematical problems," *Bulletin of the American Mathematical Society* 8 (1902): 437–479.

———. 1921. *Grundgedanken der Relativitätstheorie, Vorlesungen SS 1921.* Ed. Paul Bernays. Göttingen: Universität Göttingen Mathematisches Institut.

Hofmann, Paul. 1926. Riehls Kritizismus und die Probleme der Gegenwart. *Kantstudien* 31, nos. 2–3: 330–343.

Hönigswald, Richard. 1903. *Zur Kritik der Machschen Philosophie. Eine erkenntnistheoretische Studie.* Berlin: Schwetschke.

———. 1926. Alois Riehl. (Aus Anlaß der Neuauflage seines "Kritizismus.") *Beiträge zur Geschichte des Deutschen Idealismus* 4, no. 1: 38–47.

Houël, Guillaume Jules. 1867. *Essai critique sur les principes fondamentaux de la géométrie élémentaire ou commentaire sur les XXXII premières propositions des éléments d'Euclide.* Paris: Gauthiers-Villars.

Jacobi, Carl Gustav Jacob. 1848. Über die Reduction der quadratischen Formen auf die kleinste Anzahl der Glieder. *Journal für die reine und angewandte Mathematik* 39: 290–292. Reprinted in *C. G. J. Jacobis gesammelte Werke,* 6:318–321 (Berlin: Königliche Preuaische Akademie der Wissenschaften, Georg Reimer, 1891).

Jaensch, Erich. 1925. Zum Gedächtnis von Alois Riehl. *Kanttudien* 30, nos. 1–2: iii–xxxvi.

Jahnke, Hans Niels. 1990. *Mathematik und Bildung in der Humboldtschen Reform.* Göttingen: Vandenhoeck & Ruprecht.

Jelved, Karen, Andrew Jackson, and Ole Knuden, eds. 1998. *Selected Scientific Works of Hans Christian Ørsted.* Princeton: Princeton University Press.

Jodl, Friedrich. 1895. Neuere Versuche der Systembildung in der deutschen Philosophie. In *Vom Lebenswege. Gesammelte Vorträge und Aufsätze,* 1:363–380. Stuttgart: Cotta, 1916.

Kahl, Russell. 1971. *Selected Writings of Hermann von Helmholtz.* Middletown: Wesleyan University Press.

Klein, C. Felix. 1872. Vergleichende Betrachtungen über neuere geometrische Forschungen (Erlanger Programm). In *Gesammelte Mathematische Abhandlungen,* ed. Robert Frieke and Alexander Ostowski, 1:460–497. Berlin: Springer.

———. 1882. *Über Riemanns Theorie der algebraischen Funktionen und ihrer Integrae.* In *Gesammelte Mathematische Abhandlungen.* Ed. Robert Frieke and Alexander Ostowski, 3:499–573. Berlin: Springer.

Koenigsberger, Leo. 1902–1903. *Hermann von Helmholtz.* 3 vols. Braunschweig: Friedrich Vieweg & Sohn.

Köhnke, Klaus Christian. 1991. *The Rise of Neo-Kantianism: German Academic Philosophy between Idealism and Positivism.* Cambridge: Cambridge University Press. German edition, *Entstehung und Aufstieg des Neukantianismus* (Frankfurt am Main: Suhrkamp, 1993), first published with numerous additional footnotes in 1986.

König, Gert, and Lutz Geldsetzer. 1973. Vorbemerkung der Herausgeber zum 16. Band. In *Jakob Friedrich Fries, Sämtliche Schriften,* ed. Gert König and Lutz Geldsetzer, 16:v*–lii*. Aalen: Scientia Verlag, 1973.

———. 1979. Vorbemerkung der Herausgeber zum 13. Band. In *Jakob Friedrich Fries, Sämtliche Schriften,* ed. Gert König and Lutz Geldsetzer, 13:17*–94*. Aalen: Scientia Verlag, 1973.

Kremer, Richard L. 1993. Innovation through synthesis: Helmholtz and color research. In *Hermann von Helmholtz and the Foundations of Nineteenth-Century Science,* ed. David Cahan, 205–258. Berkeley and Los Angeles: University of California Press.

Kuhn, Friedrich. 1996. *Ein anderes Bild des Pragmatismus: Wahrscheinlichkeitstheorie und Begründung der Induktion als maßgebliche Einflußgrößen in den "Illustrations of the Logic of Science" von Charles Sanders Peirce.* Frankfurt am Main: Klostermann.

Kuhn, Thomas. 1959. Energy conservation as an example of simultaneous discovery. In *Critical Problems in the History of Science,* ed. Marshall Clagett, 321–356. Madison: University of Wisconsin Press.

Kusch, Martin. 1995. *Psychologism: A Case Study in the Sociology of Philosophical Knowledge.* London: Routledge.

Lakatos, Imre. 1970. Falsification and the methodology of scientific research programmes. In *Criticism and the Growth of Knowledge,* ed. Imre Lakatos and A. Musgrave, 91–196. Cambridge: Cambridge University Press.

Lange, Friedrich Albert. 1866. *Die Geschichte des Materialismus.* Iserlohn: Baedeker. 2d ed. 1873–1875; 4th ed., with a preface by Hermann Cohen, 1882; for the ninth edition, with a third introduction and critical expansion by Hermann Cohen, see Cohen 1914. References are to the English edition, *The History of Materialism,* with an introduction by Bertrand Russell (New York: Harcourt, Brace, 1925).

Larson, James. 1979. Vital forces: Regulative principles or constitutive agents? A strategy in German physiology, 1786–1802. *Isis* 70: 235–249.

Lenoir, Timothy. 1980. Kant, Blumenbach, and vital materialism in German biology. *Isis* 71: 77–108.

———. 1981. The Göttingen school and the development of transcendental Naturphilosophie in the romantic era. *Studies in the History of Biology* 5: 111–205.

———. 1989. *The Strategy of Life: Teleology and Mechanics in Nineteenth-Century German Biology.* Chicago: University of Chicago Press.

———. 1992. The eye as mathematician: Clinical practice, instrumentation, and Helmholtz's construction of an empiricist theory of vision. In *Hermann Helmholtz: Philosopher and Scientist,* ed. David Cahan, 109–153. Berkeley and Los Angeles: University of California Press.

Levi, Isaac. 1980. Induction as self-correcting according to Peirce. In *Science, Belief, and Behavior,* ed. D. H. Mellor, 127–140. Cambridge: Cambridge University Press.

Link, Heinrich. 1806. *Über Naturphilosophie.* Leipzig: Stiller.

Lobachevskii, Nicholas I. 1840. *Geometrische Untersuchungen zur Theorie der Parallellinien.* Berlin: Mayer and Müller, 1887. Translated into French as "Études géométriques des parallèles par J. N. Lobatchewsky," trans. Jules Houël, in *Mémoires de la Société des Sciences physiques et naturelles de Bordeaux* 4 (1866): 83–128.

Loemker, Leroy. 1956. *Gottfried Wilhelm Leibniz: Philosophical Papers and Letters*. Chicago: University of Chicago Press.

Lützen, Jesper. 1998. Heinrich Hertz and the geometrisation of mechanics. In *Heinrich Hertz: Classical Physicist, Modern Philosopher,* ed. D. Baird, R. I. G. Hughes, and Alfred Nordmann, 103–121. Dordrecht: Kluwer.

———. 1999. A matter of matter or a matter of space? *Archives Internationales d'Histoire des Sciences* 49: 105–121.

———. 2005. *Mechanistic Images in Geometric Form: Heinrich Hertz's Principles of Mechanics*. Oxford: Oxford University Press.

Maier, Heinrich. 1926. Alois Riehl. Gedächtnisrede, gehalten am 24. Januar 1925. *Kantstudien* 31, no. 4: 563–579.

Maimon, Salomon. 1790. *Versuch über die Transcendentalphilosophie*. In *Gesammelte Werke,* 2:62–65, 182–183, 362–364. Hildesheim: Olms, 1965.

Majer, Ulrich. 1996. Hilbert's criticism of Poincaré's conventionalism. In *Henri Poincaré, science et philosophie,* ed. Jean-Louis Greffe, G. Heinzmann, and K. Lorenz, 355–364. Berlin: Akademie Verlag.

McKendrick, John G. 1899. *Hermann Ludwig Ferdinand von Helmholtz*. New York: Longmans.

Melloni, Macedonio. 1842. Beobachtung über die Färbung der Netzhaut und der Krystall-Linse. *Annalen der Physik und Chemie* 56: 574–587.

Meyerson, Émile. 1908. *Identité et réalité*. 5th ed. Paris: F. Alcan. Translated into English as *Identity and Reality* (London: Allen & Unwin 1930), reprinted in 1964.

Mill, John Stuart. 1843. *A System of Logic*. 2 vols. London: Parker.

Miller, Arthur I. 1981. *Albert Einstein's Special Theory of Relativity*. Reading, Mass.: Addison-Wesley.

Millington, E. C. 1941. History of the Young-Helmholtz theory of colour vision. *Annals of Science* 5: 167–176.

Mourelatos, Alexander P. D. 1967. Jakob Friedrich Fries. In *Encyclopedia of Philosophy,* ed. P. Edwards, 3:253–255. New York: Macmillan.

Natorp, Paul. 1887/1981. On the objective and subjective grounding of knowledge. *Journal of the British Society for Phenomenology* 12: 245–266.

Nelson, Leonard. 1971. *Progress and Regress in Philosophy: From Hume and Kant to Hegel and Fries*. 2 vols. Oxford: Blackwell.

Neumeyer, Fritz. 1991. *The Artless Word—Mies van der Rohe on the Building Art*. Cambridge: MIT Press.

———. 2001. Mies's first project: Revisiting the atmosphere at Klösterli. In *Mies in Berlin,* ed. Terence Riley and Barry Bergdoll, 309–317, 378. New York: Museum of Modern Art.

Newton, Isaac. 1717/1730. *Opticks.* 2d English ed. London: G. Bell and Sons, 1931.

Nietzsche, Friedrich. 1886. *Beyond Good and Evil.* New York: Random House, 1966.

Nordmann, Alfred. 1995. Community, immortality, enlightenment: Kant's scholarly republic. *Proceedings of the Eighth International Kant Congress,* 2:705–712. Milwaukee: Marquette University Press.

Novalis. 1798. Vorarbeiten zu verschiedenen Fragmentsammlungen. In *Werke, Tagebücher und Briefe von Friedrich von Hardenberg,* ed. H.-J. Mähl and R. Samuel, 2:311–424. Munich: Hanser, 1978. References are to the numbered remarks.

———. 1798/1799. Das Allgemeine Brouillon: Materialien zur Enzyklopädistik. In *Werke, Tagebücher und Briefe von Friedrich von Hardenberg,* ed. H.-J. Mähl and R. Samuel, 2:473–720. Munich: Hanser, 1978. References are to the numbered remarks.

Oersted, Hans Christian. 1920. *Naturvidenskabelige Skifter.* 3 vols. Copenhagen: Royal Danish Society of Sciences.

Ollig, Hans-Ludwig. 1979. *Der Neukantianismus.* Stuttgart: Metzler.

Parsons, Charles. 1980. Mathematical intuition. *Proceedings of the Aristotelian Society,* n.s. 80: 145–168.

———. 1983a. *Mathematics in Philosophy: Selected Essays.* Ithaca: Cornell University Press.

———. 1983b. The impredicativity of induction. In *How Many Questions?* ed. Leigh S. Caumen, Isaac Levi, Charles Parsons, and Robert Schwarz, 132–153. Indianapolis: Hackett.

———. 1994. Intuition and number. In *Mathematics and Mind,* ed. A. George, 141–157. Oxford: Oxford University Press.

Partington, James Riddick. 1951/1953. Jeremias Benjamin Richter and the law of reciprocal proportions, I and II. *Annals of Science* 7: 172–198, and 9: 289–314.

Pascher, Manfred. 1997. *Einführung in den Neukantianismus.* Munich: Fink.

Peano, Giuseppe. 1889. *Arithmetices principia, nova methodo exposita.* Turin: Bocca. Translated into English as "The principles of arithmetic," in *From Frege to Gödel: A Source Book in Modern Logic* (Cambridge, Mass.: Harvard University Press, 1967), 83–97.

Peirce, Charles Sanders. 1868a. Questions concerning certain faculties claimed for man. In *The Essential Peirce,* ed. Nathan Houser and Christian Kloesel, 1:11–27. Bloomington: Indiana University Press, 1992. First published in *Journal of Speculative Philosophy* 2: 103–114; compare *Collected Papers* 5.213–263.

———. 1868b. Some consequences of four incapacities. In *The Essential Peirce,* ed. Nathan Houser and Christian Kloesel, 1:28–55. Bloomington: Indiana University Press,

1992. First published in *Journal of Speculative Philosophy* 2: 140–157; compare *Collected Papers* 5.264–317.

———. 1869. Grounds of validity of the laws of logic: Further consequences of four incapacities. In *The Essential Peirce*, ed. Nathan Houser and Christian Kloesel, 1:56–82. Bloomington: Indiana University Press, 1992. First published in *Journal of Speculative Philosophy* 2: 193–203; compare *Collected Papers* 5.318–357.

———. 1870. On the theory of errors of observations. In *Writings of Charles Sanders Peirce,* 3:114–137. Bloomington: Indiana University Press, 1986. First published in the *Coast Survey Report* of 1870, 200–224.

———. 1871. Fraser's "The works of George Berkeley." In *The Essential Peirce,* ed. Nathan Houser and Christian Kloesel, 1:83–105. Bloomington: Indiana University Press, 1992. First published in *North American Review* 113: 449–472; compare *Collected Papers* 8.7–38.

———. 1877. The fixation of belief. In *The Essential Peirce,* ed. Nathan Houser and Christian Kloesel, 1:109–123. Bloomington: Indiana University Press, 1992. First published in *Popular Science Monthly* 12: 1–15; compare *Collected Papers* 5.358–387.

———. 1878. The doctrine of chances. In *The Essential Peirce,* ed. Nathan Houser and Christian Kloesel, 1:142–154. Bloomington: Indiana University Press, 1992. First published in *Popular Science Monthly* 12: 604–615; compare *Collected Papers* 2.645–660.

———. 1890. Notizen über Evolution und die Architektur von Theorien. In *Naturordnung und Zeichenprozess,* ed. Helmut Pape, 126–135. Frankfurt am Main: Suhrkamp, 1991. This selection from Peirce's manuscript 965 has not been published in English; the original wording of the quoted passage was provided by Helmut Pape.

———. 1891. The architecture of theories. In *The Essential Peirce,* ed. Nathan Houser and Christian Kloesel, 1:285–297. Bloomington: Indiana University Press, 1992. First published in *Monist* 1: 161–176; compare *Collected Papers* 6.7–34.

———. 1892. The law of mind. In *The Essential Peirce,* ed. Nathan Houser and Christian Kloesel, 1:312–333. Bloomington: Indiana University Press, 1992. First published in *Monist* 2: 533–559; compare *Collected Papers* 6.102–163.

———. 1893. Evolutionary love. In *The Essential Peirce,* ed. Nathan Houser and Christian Kloesel, 1:352–371. Bloomington: Indiana University Press, 1992. First published in *Monist* 3: 176–200; compare *Collected Papers* 6.287–317.

———. 1903a. The seven systems of metaphysics. In *The Essential Peirce,* ed. Nathan Houser and Christian Kloesel, 2:179–195. Bloomington: Indiana University Press, 1998. This is the fourth of Peirce's Harvard lectures, published also by Turrisi (Peirce 1997).

———. 1903b. The nature of meaning. In *The Essential Peirce,* ed. Nathan Houser and Christian Kloesel, 2:208–225. Bloomington: Indiana University Press, 1998. Compare *Collected Papers* 5.151–179; this is the sixth of Peirce's Harvard lectures, published also by Turrisi (Peirce 1997).

———. 1903c. A syllabus of certain topics of logic. In *The Essential Peirce,* ed. Nathan Houser and Christian Kloesel, 2:258–299. Bloomington: Indiana University Press, 1998. Compare *Collected Papers* 1.180–202.

———. 1905a. What pragmatism is. In *The Essential Peirce,* ed. Nathan Houser and Christian Kloesel, 2:331–345. Bloomington: Indiana University Press, 1998. First published in *Monist* 15: 161–181. Compare *Collected Papers* 5.411–437.

———. 1905b. Issues of pragmaticism. In *The Essential Peirce,* ed. Nathan Houser and Christian Kloesel, 2:346–359. Bloomington: Indiana University Press, 1998. First published in *Monist* 15: 481–499; compare *Collected Papers* 5.438–463.

———. 1931–1935. *Collected Papers.* Ed. Charles Hartshorne and Paul Weiss. 6 vols., cited by volume and paragraph numbers. Cambridge, Mass.: Harvard University Press.

———. 1997. *Pragmatism as a Principle and Method of Right Thinking.* Ed. Patricia Ann Turrisi. Albany: State University of New York Press.

Peirce, Charles Sanders, and Joseph Jastrow. 1885. On small differences of sensation. In *Writings of Charles Sanders Peirce,* 5:122–135. Bloomington: Indiana University Press, 1993. First published in *Memoirs of the National Academy of Sciences* 3:75–83.

Plaass, Peter. 1965. *Kants Theorie der Naturwissenschaft.* Göttingen: Vandenhoeck and Ruprecht.

Planck, Max. 1909. Die Einheit des physikalischen Weltbildes. Vortrag, gehalten am 9. Dezember 1908 in der naturwissenschaftlichen Fakultät des Studentenkorps der Universität Leiden. Leipzig: Hirzel. Also published in *Physikalische Zeitschrift* 10 (1909): 62–75; reprinted in *Vorträge und Erinnerungen,* 8th ed. (Darmstadt: Wissenschaftliche Buchgesellschaft 1970), 28–51. Translated into English as "The unity of the physical world-picture," in *Physical Reality: Philosophical Essays on Twentieth-Century Physics,* ed. Stephen Toulmin, 1–27 (New York: Harper and Row, 1970).

Poincaré, Henri. 1887. Sur les hypothèses fondamentales de la géométrie. In *Oeuvres de Henri Poincaré,* 11:79–91. Paris: Gauthiers-Villars, 1956.

———. 1890. Sur les équations aux dérivées partielles de la physique mathématique. In *Oeuvres de Henri Poincaré,* 9:28–113. Paris: Gauthiers-Villars, 1954.

———. 1891. Les géométries non-euclidiennes. *Revue générale des Sciences pures et appliquées* 2: 769–774. Reprinted in *La science et l'hypothese* (Paris: Flammarion, 1968), 63–76; English edition *Science and Hypothesis* (New York: Walter Scott, 1952), 35–50. References are to the French/English editions of *Science and Hypothesis*.

———. 1894a. Sur la nature du raisonnement mathématique. *Revue de métaphysique et de morale* 2:371–384. Reprinted in *La science et l'hypothèse* (Paris: Flammarion, 1968), 31–45; English edition, *Science and Hypothesis* (New York: Walter Scott, 1952), 1–16; also as "On the nature of mathematical reasoning," in *From Kant to Hilbert,* ed. William Ewald, 973–982 (Oxford: Clarendon Press, 1996).

————. 1894b. Sur les équations de la physique mathématique. In *Oeuvres de Henri Poincaré,* 9:123–196. Paris: Gauthiers-Villars, 1954.

————. 1895. L'espace et la géométrie. *Revue de métaphysique et de morale* 3: 631–646. Reprinted in *La science et l'hypothèse* (Paris: Flammarion, 1968), 77–94; English edition, *Science and Hypothesis* (New York: Walter Scott, 1952), 51–71. References are to French/English editions of *Science and Hypothesis.*

————. 1897. Les idées de Hertz sur la mécanique. In *Oeuvres de Henri Poincaré,* 7:231–250. Paris: Gauthier-Villars, 1952.

————. 1898a. Sur les rapports de l'analyse pure et de la physique mathématique. In *Verhandlungen des ersten internationaler mathematiker-Kongresses, Zurich, 1897,* ed. F. Rudio, 81–90. Leipzig: Teubner.

————. 1898b. On the foundations of geometry. *Monist* 9: 1–43. Reprinted in *From Kant to Hilbert,* ed. William Ewald, 2:982–1011 (Oxford: Clarendon Press, 1996). References are to Ewald's edition.

————. 1898c. Sur les propriétés du potential et sur les fonctions Abéliennes. In *Oeuvres de Henri Poincaré,* 4:162–243. Paris: Gauthiers-Villars, 1950.

————. 1900. Intuition and logic in mathematics. In *From Kant to Hilbert,* ed. William Ewald, 1012–1020. Oxford: Clarendon Press, 1996.

————. 1901. *Électricité et optique.* 2d ed. Paris: Gauthier-Villars. Reprint, Paris: Gabay, 1990. First edition appeared in 1890.

————. 1902a. Les fondements de la géométrie. In *Oeuvres de Henri Poincaré,* 11:92–113. Paris: Gauthiers-Villars, 1956.

————. 1902b. *La science et l'hypothèse.* Paris: Flammarion, 1968. Translated into English as *Science and Hypothesis* (New York: Walter Scott, 1952). References are to the English edition.

————. 1902c. L'expérience et la géométrie. In *La science et l'hypothèse,* 95–110. Paris: Flammarion, 1968. Translated into English in *Science and Hypothesis* (New York: Walter Scott, 1952), 72–88. References are to French/English editions.

————. 1905a. *La valeur de la science.* Paris: Flammarion. Translated into English as *The Value of Science* (New York: Science Press, 1907).

————. 1905b. Les mathématiques et la logique. *Revue de métaphysique et de morale* 13: 815–825. Reprinted as "Mathematics and logic: I," in *From Kant to Hilbert,* ed. William Ewald, 1022–1038 (Oxford: Clarendon Press, 1996).

————. 1906a. Les mathématiques et la logique. *Revue de métaphysique et de morale* 14: 17–34. Reprinted in *Science et méthode* (Flammarion, Paris, 1916), 152–171; English edition, *Science and Method* (New York: Dover Books, n.d.); also as "Mathematics and logic: II," in *From Kant to Hilbert,* ed. William Ewald, 1038–1052 (Oxford: Clarendon Press, 1996).

————. 1906b. Les mathématiques et la logique. *Revue de métaphysique et de morale* 14: 294–317. Reprinted in *Science et méthode* (Paris: Flammarion, 1916), 172–191; English edition, *Science and Method* (New York: Dover Books, n.d.); also as "Mathematics and logic: III," in *From Kant to Hilbert,* ed. William Ewald, 1052–1071 (Oxford: Clarendon Press, 1996). References are to *Science et méthode.*

————. 1906c. Sur la dynamique de l'électron. *Rendiconti circolo Matematico di Palermo* 21: 129–176. Reprinted in *Oeuvres de Henri Poincaré,* 9:494–550 (Paris: Gauthiers-Villars, 1954).

————. 1908a. La dynamique de l'électron. In *Oeuvres de Henri Poincaré,* 9:551–586. Paris: Gauthiers-Villars, 1954.

————. 1908b. La mécanique et l'optique. Chapter 11 of *Science et méthode.* Paris: Flammarion, 1916. Part of Poincaré 1908a.

————. 1908c. L'avenir des mathématiques. *Rendiconti del Circolo Matematico di Palermo* 26: 152–168 (address to the International Congress of Mathematicians, Rome). Reprinted in *Science et méthode* (Paris: Flammarion, 1916), 19–42; English edition, *Science and Method* (New York: Dover Books, n.d.).

————. 1908d. La mécanique nouvelle. In *Science et méthode,* chapter 3, 215–272. Paris: Flammarion, 1916. English edition, *Science and Method* (New York: Dover Books, n.d.).

————. 1908e. *Science et méthode.* Paris: Flammarion, 1916. Translated into English as *Science and Method* (London: Thomas Nelson, 1914).

————. 1913. *Dernières pensées.* Paris, Flammarion. Translated into English as *Mathematics and Science: Last Essays* (New York: Dover, 1963).

————. 1921. Analyse de ses travaux scientifiques, faites par H. Poincaré. *Acta Mathematica* 38: 36–135.

————. 1997. *Three Supplementary Essays on the Discovery of Fuchsian Functions.* Edited with an introductory essay by Jeremy J. Gray and S. A. Walter. Berlin: Akademie Verlag.

Pollok, Konstantin. 2001. *Kants "Metaphysische Anfangsgründe der Naturwissenschaft": Ein kritischer Kommentar.* Hamburg: Meiner.

Popper, Karl. 1968. *The Logic of Scientific Discovery.* New York: Harper and Row.

Protokoll. 1893. Protokoll der Commissionssitzung betreffs Besetzung der dritten Professur für Philosophie, vom 1. Juli 1893, Universitätsarchiv der Humboldt-Universität zu Berlin, Akten der Philosophischen Fakultät, 1462, p. 103v.

Pulte, Helmut. 1999a. "… sondern Empirismus und Speculation sich verbinden sollen." Historiographische Überlegungen zur bisherigen Rezeption des wissenschaftstheoretischen und naturphilosophischen Werkes von J. F. Fries und einige Gründe für dessen Neubewertung. In *Jakob Friedrich Fries. Philosoph, Naturwissenschaftler und Mathematiker,* ed. W. Hogrebe and K. Hermann, 57–94. Frankfurt am Main: Lang.

————. 1999b. Von der Physikotheologie zur Methodologie. Eine wissenschaftstheoriegeschichtliche Analyse zur Transformation von nomothetischer Teleologie und Systemdenken bei Kant und Fries. In *Jakob Friedrich Fries. Philosoph, Naturwissenschaftler und Mathematiker,* ed. W. Hogrebe and K. Hermann, 301–353. Frankfurt am Main: Lang.

————. 2005. *Axiomatik und Empirie. Eine wissenschaftstheoriegeschichtliche Untersuchung zur mathematischen Naturphilosophie von Newton bis Neumann.* Darmstadt: Wissenschaftliche Buchgesellschaft.

Putnam, Hilary. 1974. The refutation of conventionalism. *Noûs* 8: 25–40.

Reichenbach, Hans. 1915. Der Begriff der Wahrscheinlichkeit für die mathematische Darstellung der Wirklichkeit. In *Gesammelte Werke in neun Bänden.* Vol 5, *Philosophische Grundlagen der Quantenmechanik und Wahrscheinlichkeit,* ed. Andreas Kamlah and Marie Reichenbach, 225–307. Braunschweig: Vieweg, 1989.

————. 1957. *The Philosophy of Space and Time.* New York: Dover.

Reinhold, Ernst. 1825. *K. Reinholds Leben und literarisches Werk.* Jena: Frommann.

Reute, C. G. Theodor. 1845. Das Ophthalmotrop, dessen Bau und Gebrauch. *Göttinger Studien* 1: 128–129.

————. 1857. *Ein neues Ophthalmotrop.* Leipzig: Otto Wigand.

Richards, Joan L. 1977. The evolution of empiricism: Hermann von Helmholtz and the foundations of geometry. *British Journal for the Philosophy of Science* 28: 235–253.

Richards, Robert. 1992. *The Meaning of Evolution.* Chicago: University of Chicago Press.

————. 2002. *The Romantic Conception of Life: Science and Philosophy in the Age of Goethe.* Chicago: University of Chicago Press.

————. Forthcoming. *The Tragic Sense of Life: Ernst Haeckel and the Struggle over Evolutionary Thought.* Chicago: University of Chicago Press.

Richardson, Alan. 1996. From epistemology to the logic of science: Carnap's philosophy of empirical knowledge in the 1930s. In *Origins of Logical Empiricism,* ed. Ronald Giere and Alan Richardson, 309–332. Minneapolis: University of Minnesota Press.

————. 1997a. *Carnap's Construction of the World.* Cambridge: Cambridge University Press.

————. 1997b. Toward a history of scientific philosophy. *Perspectives on Science* 5, no. 3: 418–451.

Rickert, Heinrich. 1924/1925. Alois Riehl geb. 27. IV. 1844–gest. 21. XI. 1924. *Logos* 13: 162–185.

Riehl, Alois. 1876–1887. *Der philosophische Kriticismus und seine Bedeutung für die positive Wissenschaft.* Leipzig: Wilhelm Engelmann. Vol. 1, *Geschichte und Methode des philosophischen Kriticismus* (1876; referred to as PK1); vol. 2, part 1 *Die sinnlichen und logischen Grund-*

lagen der Erkenntnis (1879; referred to as PK2); vol. 2, part 2 *Zur Wissenschaftstheorie und Metaphysik* (1887; referred to as PK3; English translation of this part: Riehl 1894).

———. 1877. Die englische Logik der Gegenwart. *Vierteljahrsschrift für wissenschaftliche Philosophie* 1: 50–80. Reprinted in Alois Riehl, *Philosophische Studien aus vier Jahrzehnten* (Leipzig: Quelle & Meyer, 1925), 175–201.

———. 1883. Über wissenschaftliche und nichtwissenschaftliche Philosophie. Eine akademische Antrittsrede (Freiburg 1883). Reprinted in Alois Riehl, *Philosophische Studien aus vier Jahrzehnten* (Leipzig: Quelle & Meyer, 1925), 227–253.

———. 1894. *The Principles of the Critical Philosophy: Introduction to the Theory of Science and Metaphysics.* London: Kegan Paul, Trench, Trübner. Amended translation of Riehl 1876–1887, vol. 2, part 2.

———. 1897. *Friedrich Nietzsche. Der Künstler und der Denker.* 4th ed. 1901; 8th ed. 1923. Stuttgart: Frommann.

———. 1900. Robert Mayers Entdeckung und Beweis des Energieprincipes. In *Philosophische Abhandlungen: Christoph Sigwart zu seinem siebzigsten Geburtstage,* ed. Benno Erdmann, 159–184. Tübingen: Mohr.

———. 1903. *Zur Einführung in die Philosophie der Gegenwart, Acht Vorträge.* Leipzig: Teubner, 1913. 2d ed. 1904; 3d ed. 1908; 5th ed. 1919; 6th ed. 1921. Cited according to the 4th edition, 1913.

———. 1904. Hermann von Helmholtz in seinem Verhältnis zu Kant. *Kantstudien* 9: 261–285. Also issued as a separate monograph (Berlin: Reuther & Reichard).

———. 1907. Logik und Erkenntnistheorie. In *Die Kultur der Gegenwart in Selbstdarstellungen,* ed. Paul Hinneberg. Part 1, sec. 6, *Systematische Philosophie,* by Wilhelm Dilthey, Alois Riehl, et al., 73–102. Leipzig: Teubner. References to 3d ed., 1921, 68–97.

———. 1908–1926. *Der philosophische Kritizismus. Geschichte und System.* 2d ed. Leipzig: Engelmann. Vol. 1 (1908), *Geschichte des philosophischen Kritizismus* (this volume also appeared in an insignificantly revised third edition [Leipzig: Kröner, 1924]). Vol. 2, *Die sinnlichen und logischen Grundlagen der Erkenntnis,* ed. Hans Heyse and Eduard Spranger (Leipzig: Kröner, 1925). Vol. 3, *Zur Wissenschaftstheorie und Metaphysik,* ed., Hans Heyse and Eduard Spranger (Leipzig: Kröner, 1926).

———. 1910–1921. Four postcards to Schlick, 1910, 1911, 1919, 1921, Moritz Schlick Nachlass, Inv.-Nr. 114. Amsterdam and Haarlem: Wiener Kreis Stichting/Vienna Circle Foundation.

———. 1913. Der Beruf der Philosophie in der Gegenwart. Reprinted in Alois Riehl, *Philosophische Studien aus vier Jahrzehnten* (Leipzig: Quelle & Meyer, 1925), 304–312; English edition, "The vocation of philosophy at the present day," in *Lectures delivered in connection with the dedication of the Graduate College of Princeton University in October, 1913 by Émile Boutroux, Alois Riehl, A. D. Godley, Arthur Shipley* (Princeton: Princeton University Press, 1914), 45–63.

———. 1915. Die geistige Kultur und der Krieg. Reprinted in Alois Riehl, *Philosophische Studien aus vier Jahrzehnten* (Leipzig: Quelle & Meyer, 1925), 313–325.

———. 1921. Helmholtz als Erkenntnistheoretiker. *Die Naturwissenschaften* 35 (August 31): 702–708. Reprinted in Riehl 1922, 223–240.

———. 1922. *Führende Denker und Forscher.* Leipzig: Quelle und Meyer. 2d ed. 1924.

———. 1925a. *Philosophische Studien aus vier Jahrzehnten.* Leipzig: Quelle & Meyer.

Riemann, Bernhard. 1854. Ueber die Hypothesen, welche der Geometrie zu Grunde liegen. In *Abhandlungen der königlichen Gesellschaft der Wissenschaften zu Göttingen,* 13:133–152; also in *Bernhard Riemann's gesammelte mathematische Werke und wissenschaftlicher Nachlaß,* Heinrich Weber (Leipzig: Teubner, 1876), 254–269. References to the 1876 edition.

———. 1876. Versuch einer Lehre von den Grundbegriffen der Mathematik und Physik als Grundlage für die Naturerklärung. In *Bernhard Riemann's gesammelte mathematische Werke und wissenschaftlicher Nachlaß,* ed. Heinrich Weber, 489–506. Leipzig: Teubner, 1876.

Ringer, Fritz. 1983. *Die Gelehrten. Der Niedergang der deutschen Mandarine, 1890–1933.* Stuttgart: Klett. Translated into English as *The Decline of the German Mandarins: The German Academic Community, 1890–1933* (Hanover, N.H.: Wesleyan University Press, 1990).

Rocke, Alan J. 1984. *Chemical Atomism in the Nineteenth Century: From Dalton to Cannizzaro.* Columbus: Ohio State University Press.

Röd, Wolfgang. 1986. Alois Riehl und der Herbartianismus in Österreich. In *Von Bolzano zu Wittgenstein. Zur Tradition der österreichischen Philosophie,* ed. J. C. Nyíri, 132–140. Vienna: Hölder, Pichler, Tempski.

———. 2001. Alois Riehl: Kritischer Realismus zwischen Transzendentalismus und Empirismus. In *Bausteine zu einer Geschichte der Philosophie an der Universität Graz,* ed. Thomas Binder et al., 111–128. Amsterdam: Rodopi.

Roe, Shirley. 1981. *Matter, Life and Generation: Eighteenth-Century Embryology and the Haller-Wolff Debate.* Cambridge: Cambridge University Press.

Rosenthal-Schneider, Ilse. 1980. *Reality and Scientific Truth: Discussions with Einstein, von Laue, and Planck.* Ed. Thomas Braun. Detroit: Wayne State University Press.

Royce, Josiah. 1919. *Lectures on Modern Idealism.* New Haven: Yale University Press.

———. 1967. *The Spirit of Modern Philosophy.* New York: G. Braziller.

Rutte, Heiner. 2001. Ergänzende Bemerkungen zu Alois Riehl, H. Spitzer, C. Siegel und zur Grazer Schultradition. In *Bausteine zu einer Geschichte der Philosophie an der Universität Graz,* ed. Thomas Binder et al., 129–141. Amsterdam: Rodopi, 2001.

Schäfer, Lothar. 1966. *Kants Metaphysik der Natur.* Berlin: De Gruyter.

Schelling, Friedrich. 1797. *Ideen zu einer Philosophie der Natur,* 1st ed. Vol. 1 of *Schellings Werke.* Munich: C. H. Beck, 1927. Corresponds to vol. 2 in the 1856–1861 edition of the *Werke* (Stuttgart: Cotta); 2d ed., see Schelling 1803; English translation, see Schelling 1988.

———. 1798. *Von der Weltseele.* In *Schellings Werke,* 1:413–421. Munich: C. H. Beck, 1927. Corresponds to vol. 2, 345–353, in the 1856–1861 edition of the *Werke* (Stuttgart: Cotta).

———. 1799. *Erster Entwurf eines Systems der Naturphilosophie.* In *Schellings Werke,* 2:1–268. Munich: C. H. Beck, 1927.

———. 1800. *System des transcendentalen Idealismus.* In *Schriften zur Naturphilosophie (1799–1801), in Schellings Werke,* 2:327–634. Munich: C. H. Beck, 1927. Corresponds to vol. 3 in the 1856–1861 edition of the *Werke* (Stuttgart: Cotta).

———. 1802. *Philosophie der Kunst.* In *Schellings Werke,* 3:375–507. Munich: C. H. Beck, 1927. Corresponds to 5:355–487 in the 1856–1861 edition of the *Werke* (Stuttgart: Cotta).

———. 1803. *Ideen zur einer Philosophie der Natur.* 2d ed. In *Sämtliche Werke,* vol. 1. Stuttgart: Cotta, 1848. English translation, see Schelling 1988.

———. 1908. *Schelling als Persönlichkeit: Briefe, Reden, Aufsätze.* Ed. Otto Braun. Leipzig: Fritz Eckardt Verlag.

———. 1962. *Briefe und Dokumente.* 3 vols. to date, vol. 1 edited by Horst Fuhrmans. Bonn: Bouvier.

———. 1988. *Ideas for a Philosophy of Nature.* Cambridge: Cambridge University Press.

Schelling, Friedrich, and Georg Wilhelm F. Hegel. 1801. Ueber den wahren Begriff der Naturphilosophie und die richtige Art, ihre Probleme aufzulösen. In *Schellings Werke,* 2:713–737. Munich: C. H. Beck, 1927. Corresponds to 4:79–103 in the 1856–1861 edition of the *Werke* (Stuttgart: Cotta).

Schilpp, Paul Arthur. 1949. *Albert Einstein: Philosopher-Scientist.* Evanston, Ill.: Library of Living Philosophers. 3d ed., La Salle, Ill.: Open Court, 1970.

Schlegel, Friedrich. 1800. *Vorlesungen über Transcendentalphilosophie [Jena 1800–1801].* Vol. 12 of *Kritische Friedrich-Schlegel-Ausgabe.* Paderborn: Ferdinand Schöningh, 1964.

———. 1890. *Friedrich Schlegels Briefe an seinen Bruder August Wilhelm.* Berlin: Speyer & Peters.

———. 1958–2001. *Kritische Friedrich-Schlegel-Ausgabe.* 35 vols. to date. Ed. Ernst Behler et al. Paderborn: Ferdinand Schöningh.

———. 1967. *Fragmente.* Vol. 2 of *Kritische Friedrich-Schlegel-Ausgabe.* Paderborn: Ferdinand Schöningh.

Schleiden, Matthias Jacob. 1857. Jakob Friedrich Fries, der Philosoph der Naturforscher. Eine biographische Skizze. *Westermann's Jahrbuch der Illustrirten Deutschen Monatshefte* 2: 264–278.

———. 1989. *Wissenschaftsphilosophische Schriften.* Cologne: Dinter.

Schlick, Moritz. 1916. Idealität des Raumes, Introjektion und psychophysisches Problem. *Vierteljahrsschrift für wissenschaftliche Philosophie und Soziologie* 40: 230–254. Translated into English as Schlick 1979, 1:190–206.

———. 1917. *Raum und Zeit in der gegenwärtigen Physik. Zur Einführung in das Verständnis der Relativitäts-und Gravitationstheorie.* Berlin: Springer.

———. 1921. Letter to Paul Hensel, Erlangen, from 21 May 1921. Moritz Schlick Nachlass, Inv.-Nr. 103. Amsterdam and Haarlem: Wiener Kreis Stichting/Vienna Circle Foundation.

———. 1922. Helmholtz als Erkenntnistheoretiker. In *Helmholtz als Physiker, Physiologe und Philosoph. Drei Vorträge gehalten zur Feier seines 100. Geburtstages,* ed. Emil Warburg, Max Rubner, and Moritz Schlick, 29–39. Karlsruhe: C. F. Müller.

———. 1925. *Allgemeine Erkenntnislehre.* 2d ed. Berlin: Springer. Translated into English as *General Theory of Knowledge,* trans. Albert Blumberg (Wien: Springer 1974); 1st ed. appeared in 1918. References are to section numbers.

———. 1979. *Philosophical Papers.* 2 vols. Ed. H. Mulder and B. van de Velde-Schlick. Dordrecht: Reidel.

Schmidt, Nicole D. 1995. *Philosophie und Psychologie. Trennungsgeschichte, Dogmen und Perspektiven.* Reinbek: Rowohlt.

Schnädelbach, Herbert. 1983. *Philosophie in Deutschland, 1831–1933,* 4th ed. Frankfurt am Main: Suhrkamp, 1991.

Schneider, Ilse. 1921. *Das Raum-Zeit-Problem bei Kant und Einstein.* Berlin: Springer. First published as *Die Beziehungen der Einsteinschen Relativitätstheorie zur Philosophie unter besonderer Berücksichtigung der Kantischen Lehre* (Leipzig: Spamer, 1921). Dissertation, University of Berlin, Philosophical Faculty, 1920.

Schubring, Gert. 1990. Das mathematisch Unendliche bei J. F. Fries. In *Konzepte des mathematisch Unendlichen im 19. Jahrhundert,* ed. Gert König, 152–164. Göttingen: Vandenhoeck and Ruprecht.

Schultze, Fritz. 1881/1882. *Philosophie der Naturwissenschaft.* 2 vols. Leipzig: Günther.

Seebeck, August. 1844. Bemerkungen über Resonanz und über Helligkeit der Farben im Spectrum. *Annalen der Physik und Chemie* 62: 571–576.

Shanahan, Timothy. 1989. Kant, *Naturphilosophie,* and Oersted's Discovery of Electromagnetism: A Reassessment. *Studies in History and Philosophy of Science* 20: 287–305.

Sherman, Paul. 1981. *Colour Vision in the Nineteenth Century: The Young-Helmholtz-Maxwell Theory.* Bristol: Adam Hilger.

Sieg, Wilfried. 1994. Mechanical procedures and mathematical experience. In *Mathematics and Mind,* ed. A. George, 71–117. Oxford: Oxford University Press.

Siegel, Carl. 1932. *Alois Riehl. Ein Beitrag zur Geschichte des Neukantianismus.* Graz: Leuschner & Lubensky.

Söderqvist, Thomas. 1996. Existential projects and existential choice in science: Science biography as an edifying genre. In *Telling Lives in Science: Essays in Scientific Biography,* ed. Michael Shortland and Richard Yeo, 45–84. Cambridge: Cambridge University Press.

Spranger, Eduard. 1944. Alois Riehl. *Forschungen und Fortschritte* 20, nos. 16–18: 129–130.

Stauffer, Robert. 1957. Speculation and experiment in the background of Oersted's discovery of electromagnetism. *Isis* 48: 33–50.

Torretti, Roberto. 1978. *Philosophy of Geometry from Riemann to Poincaré.* Dordrecht: D. Reidel.

Turner, R. Steven. 1994. *In the Mind's Eye: Vision and the Helmholtz-Hering Controversy.* Princeton: Princeton University Press.

Tuschling, Burkhard. 1991. The system of transcendental idealism: Questions raised and left open in the *Kritik der Urteilskraft. Southern Journal of Philosophy* 30, supplement: 109–127.

Ulyenbroek, Peter Johan. 1833. *Christi. Hugenii aliorumque seculi XVII. virorum celebrium exercitationes mathematicae et philosophia.* Hagae comitum.

Vaihinger, Hans. 1893. Übersicht über die philosophischen Universitätsdocenten Deutschlands.... Letter to the Prussian Minister Friedrich Althoff. In Hermann Lotze, *Briefe und Dokumente,* arranged and introduced by Reinhardt Pester, ed. Ernst Wolfgang Orth. Würzburg: Königshausen & Neumann, 2003.

van Fraassen, Bas C. 1997. Structure and perspective: Philosophical perplexity and paradox. In *Logic and Scientific Methods,* ed. Maria Luisa Dalla Chiara et al., 511–530. Dordrecht: Kluwer.

Vogel, Stephan. 1993. Sensation of tone, perception of sound, and empiricism. In *Hermann von Helmholtz and the Foundations of Nineteenth-Century Science,* ed. David Cahan, 259–287. Berkeley and Los Angeles: University of California Press.

Vorländer, Karl. 1923. *Kant-Schiller-Goethe.* Aalen: Scientia Verlag, 1984.

Waschkies, Hans-Joachim. 1987. *Physik und Physikotheologie des jungen Kant.* Amsterdam: Grüner.

Watkins, Eric, ed. 2001. *Kant and the Sciences.* Oxford: Oxford University Press.

Weber, Heinrich. 1876. *Bernhard Riemann's gesammelte mathematische Werke und wissenschaftlicher Nachlaß.* Leipzig: Teubner.

Wetzels, Walter. 1987. Organicism and Goethe's aesthetics. In *Approaches to Organic Form,* ed. Frederick Burwick, 71–85. Dordrecht: Reidel.

Weyl, Hermann. 1921. *Raum-Zeit-Materie,* 5th ed. Berlin: Springer.

———. 1923. *Mathematische Analyse des Raumproblemes.* Berlin: Springer.

Williams, L. Pearce. 1965. *Michael Faraday: A Biography.* New York: Basic Books.

———. 1966. *The Origins of Field Theory.* New York: Random House.

———. 1973. Kant, *Naturphilosophie,* and scientific method. In *Foundations of Scientific Method in the Nineteenth Century,* ed. Ronald Giere and Richard S. Westfall, 3–22. Bloomington: Indiana University Press.

Wittgenstein, Ludwig. 1922. *Tractatus Logico-Philosophicus.* London: Routledge and Kegan Paul.

Wood, Allen W. 1998. Fries, Jakob Friedrich. In *Routledge Encyclopedia of Philosophy,* ed. Edward Craig, 3:798–799. London: Routledge.

Wundt, Wilhelm. 1862a. Über die Bewegung der Augen: Part I. *Archiv für Ophthalmologie* 8: 1–87.

———. 1862b. Beschreibung eines künstlichen Augenmuskelsystems zur Untersuchung der Bewegungsgesetze des menschlichen Auges im gesunden und kranken Zustanden: Part I. *Archiv für Ophthalmologie* 8: 88–114.

Frederick Beiser is Professor of Philosophy at Syracuse University. He has taught at Harvard, Yale, Pennsylvania, Wisconsin, Colorado, and Indiana. His major publications include *The Fate of Reason* (1987), *Enlightenment, Revolution, and Romanticism* (1992), *German Idealism* (2002), and *The Romantic Imperative* (2003). He has been the recipient of Thysen, Guggenheim, NEH, and Humboldt fellowships.

Robert DiSalle is Associate Professor of Philosophy at the University of Western Ontario. He works on the history and philosophy of physics from Newton to the present, philosophical problems of space and time, and the history of the philosophy of science from the seventeenth century to the present. His recent publications include "Newton's Philosophical Analysis of Space and Time" in the *Cambridge Companion to Newton* (2002), and *Understanding Spacetime: The Philosophical Development of Physics from Newton to Einstein* (2006).

Janet Folina is Associate Professor and Chair of the Philosophy Department at Macalester College, in St. Paul, Minnesota. Her publications are mostly in the philosophy of mathematics, including a book on Poincaré's philosophy of mathematics, as well as several other articles on Poincaré. Her current interests include the philosophy of Hermann Weyl, philosophy of mathematics during the nineteenth century, the role of pictures and diagrams in mathematics, and the concept of mathematical proof.

Michael Friedman is Frederick P. Rehmus Family Professor of Humanities and Director of the Patrick Suppes Center for the Interdisciplinary Study of Science and Technology at Stanford University. He works on the relationship between the history of philosophy and the history of science from Kant through the early twentieth century, and on the prospects for post-Kuhnian philosophy of science in light of these developments. His publications include *Foundations of Space-Time Theories: Relativistic Physics and Philosophy of Science* (1983), *Kant and the Exact Sciences* (1992), *Reconsidering Logical Positivism* (1999), *A Parting of the Ways: Carnap, Cassirer, and Heidegger* (2000), and *Dynamics of Reason* (2001).

Jeremy Gray   has taught at the Open University since 1974, where he is now Professor of the History of Mathematics and Director of the Centre for the History of the Mathematical Sciences. He is also a Visiting Fellow at the University of Warwick. He works on the history of modern mathematics, and also on issues concerned with its philosophy and cultural significance. He has recently finished a book on the history of geometry in the nineteenth century, and is now working on one on mathematical modernism and the philosophy of mathematics.

Frederick Gregory   is Professor of History of Science at the University of Florida, where he has taught for over twenty-five years. A former chair of the Department of History at Florida, he has also been a Dibner Fellow and has served as president of the History of Science Society. His research focuses on German science and society in the eighteenth and nineteenth centuries. Currently he is at work on a collaborative project with German and American colleagues entitled "Mysticism and Modernity," which is sponsored by the Volkswagen Stiftung in Hanover.

Michael Heidelberger   is Chair of Logic and Philosophy of Science at the University of Tübingen. He works on topics related to causality and probability, measurement and experiment as they are relevant to the philosophy of psychology, and the philosophy of physics. He specializes in the history of the philosophy of science, especially of the late nineteenth and early twentieth century. He is the author of *Nature from Within: Gustav Theodor Fechner and His Psychophysical Worldview* (2004), and he co-edited, with F. Steinle, *Experimental Essays Versuche zum Experiment* (1998).

Timothy Lenoir   is the Kimberly Jenkins Chair for New Technologies and Society at Duke University. He has published several books and articles on the history of biomedical science from the nineteenth century to the present, and is currently engaged in an investigation of the introduction of computers into biomedical research from the early 1960s to the present, particularly the development of computer graphics, medical visualization technology, and the development of virtual reality and its applications in surgery and other fields. For more information and links to recent work, see http://www.stanford.edu/dept/HPST/TimLenoir/.

Jesper Lützen   is Professor of History of Mathematics at the Department of Mathematics, University of Copenhagen. His research deals mostly with mathematics and its interaction with physics in the nineteenth century. He has written books on the prehistory of the theory of distributions, on Joseph Liouville and his work, and most recently on Heinrich Hertz's mechanics.

Alfred Nordmann is Professor of Philosophy and History of Science at Darmstadt Technical University. His historical interests concern the negotiation of contested fields of scientific knowledge such as theories of electricity and chemistry in the eighteenth century; mechanics, evolutionary biology, and sociology in the nineteenth century; and nursing science and nanoscale research in the twentieth century. His epistemological interests concern the trajectory that leads from Immanuel Kant via Heinrich Hertz and Ludwig Wittgenstein to contemporary analyses of models, simulations, and visualizations. He is president of the Lichtenberg Society.

Helmut Pulte is Professor for History and Philosophy of Science at the Ruhr-Universität Bochum. In 1995–1996 he was Fellow of the Alexander von Humboldt Foundation at the University of Cambridge. His publications include the book *Axiomatik und Empirie: Eine wissenschaftstheoriegeschichtliche Untersuchung zur mathematischen Naturphilosophie von Newton bis Neumann* (2001). His main research areas are history of philosophy of science, current philosophy of the exact sciences, and history of mathematics and physics.

Robert J. Richards is the Morris Fishbein Professor of the History of Science at the University of Chicago and Professor in the departments of Philosophy, History, Psychology, and Conceptual and Historical Studies of Science. He is the author of *Darwin and the Emergence of Evolutionary Theories of Mind and Behavior* (1987), *The Meaning of Evolution: The Morphological Construction and Ideological Reconstruction of Darwin's Theory* (1992), and *The Romantic Conception of Life: Science and Philosophy in the Age of Goethe* (2002). His book *The Tragic Sense of Life: Ernst Haeckel and the Battle over Evolutionary Theory in Germany* will appear shortly.

Alan Richardson is Professor of Philosophy and Distinguished University Scholar at the University of British Columbia. He is the author of *Carnap's Construction of the World* (1998), and co-editor, with Gary Hardcastle, of *Logical Empiricism in North America* (2003). His current book project is tentatively entitled *Logical Empiricism as Scientific Philosophy: Toward a Science Studies of Philosophy*.